PHYSICAL CHEMISTRY

An Advanced Treatise

Volume VIIIA / Liquid State

PHYSICAL CHEMISTRY

An Advanced Treatise

Edited by

HENRY EYRING
Departments of Chemistry and Metallurgy
University of Utah
Salt Lake City, Utah

DOUGLAS HENDERSON
IBM Research Laboratories
San Jose, California

WILHELM JOST
Institut für Physikalische Chemie der Universität Göttingen
Göttingen, Germany

PHYSICAL CHEMISTRY
An Advanced Treatise

VOLUME VIIIA / Liquid State

Edited by

DOUGLAS HENDERSON
IBM Research Laboratories
San Jose, California

1971

ACADEMIC PRESS NEW YORK / LONDON

ACADEMIC PRESS, INC.
111 Fifth Avenue, New York, New York 10003

United Kingdom Edition published by
ACADEMIC PRESS, INC. (LONDON) LTD.
24/28 Oval Road, London NW1 7DD

Library of Congress Catalog Card Number: 66 - 29951

PRINTED IN THE UNITED STATES OF AMERICA

Contents

Chapter 1 / Introduction

Robert L. Scott

Chapter 2 / Structure of Liquids

Sow-Hsin Chen

Chapter 3 / Computer Calculations for Model Systems

Francis H. Ree

Chapter 4 / Distribution Functions

R. J. Baxter

Chapter 5 / The Significant Structure Theory of Liquids

Mu Shik Jhon and Henry Eyring

Chapter 6 / Perturbation Theories

Douglas Henderson and J. A. Barker

List of Contributors

Numbers in parentheses indicate the pages on which the authors' contributions begin.

J. A. Barker, IBM Research Laboratories, San Jose, California (377)

R. J. Baxter, Research School of Physical Sciences, The Australian National University, Canberra, Australia (267)

Sow-Hsin Chen, Department of Nuclear Engineering, Massachusets Institute of Technology, Cambridge, Massachusetts (85)

Henry Eyring, Departments of Chemistry and Metallurgy, University of Utah, Salt Lake City, Utah (335)

Douglas Henderson, IBM Research Laboratories, San Jose, California (377)

Mu Shik Jhon, Korea Institute of Science and Technology, Seoul, Korea (335)

Francis H. Ree, Lawrence Livermore Laboratory, University of California, Livermore, California (157)

Robert L. Scott, Department of Chemistry, University of California, Los Angeles, California (1)

Foreword

In recent years there has been a tremendous expansion in the development of the techniques and principles of physical chemistry. As a result most physical chemists find it difficult to maintain an understanding of the entire field.

The purpose of this treatise is to present a comprehensive treatment of physical chemistry for advanced students and investigators in a reasonably small number of volumes. We have attempted to include all important topics in physical chemistry together with borderline subjects which are of particular interest and importance. The treatment is at an advanced level. However, elementary theory and facts have not been excluded but are presented in a concise form with emphasis on laws which have general importance. No attempt has been made to be encyclopedic. However, the reader should be able to find helpful references to uncommon facts or theories in the index and bibliographies.

Since no single physical chemist could write authoritatively in all the areas of physical chemistry, distinguished investigators have been invited to contribute chapters in the field of their special competence.

If these volumes are even partially successful in meeting these goals we will feel rewarded for our efforts.

We would like to thank the authors for their contributions and to thank the staff of Academic Press for their assistance.

HENRY EYRING
DOUGLAS HENDERSON
WILHELM JOST

Preface

The prediction of the properties of liquids has been one of the classic problems of physical chemistry. Until very recently, it was an unsolved problem. Even now it is widely so regarded. Fortunately, this is no longer true. The equilibrium properties of simple liquids, except in the neighborhood of the critical point, are now well understood.

This volume is restricted to simple liquids because the theory is most developed for these liquids. The term simple liquid has been interpreted broadly. Thus, a chapter is devoted to liquid helium. Nonsimple liquids, such as water, are of great practical interest. Such liquids are, in general, not considered in this volume because the theory of such liquids is not well developed. However, the techniques which have proved so useful for simple liquids will form the basis of the theory of complex liquids. It is our aim that through the study of these techniques, which are described in this volume, a student will be able to read and contribute to the current literature on both simple and complex liquids.

There are four main techniques in the theory of liquids: simulation studies, integral equation methods, lattice theories, and perturbation theories. Each of these methods is treated in this volume. The only lattice theory which has received attention during the past decade is the significant structure theory and, as a result, that is the only lattice theory considered in this volume.

As has been mentioned, the critical point and nonequilibrium properties of liquids are not so well understood. Introductions to these fields, which the student should find useful, are included.

The editor would like to thank the authors for their contributions. Thanks are also due to Drs. J. A. Barker and H. L. Frisch for many valuable suggestions relating to the organization of this volume.

Douglas Henderson

Contents of Previous and Future Volumes

VOLUME II

VOLUME III

VOLUME V

Chapter 1

Introduction

Robert L. Scott

I. Introduction

A. Definition

Although the area labeled "liquid" on a pressure–temperature phase diagram is usually small compared with that occupied by "gas" and by various crystalline modifications, for many substances that area lies near the usual convenient pressures and temperatures and is of major importance. Nonetheless, it is not always easy to define the liquid state, either experimentally or theoretically, in a way which clearly includes highly viscous fluids but excludes glasses and crystalline solids, or which distinguishes it unequivocally from the less dense fluids which we prefer to call gases.

Experimentally, the most obvious way to distinguish between two "states of aggregation" is to observe a phase transition between them. In a typical isothermal expansion of a pure substance, a graph of the variation of pressure p with volume V may look like that shown in Fig. 1. As one increases the volume isothermally (line a), one finds two regions in which

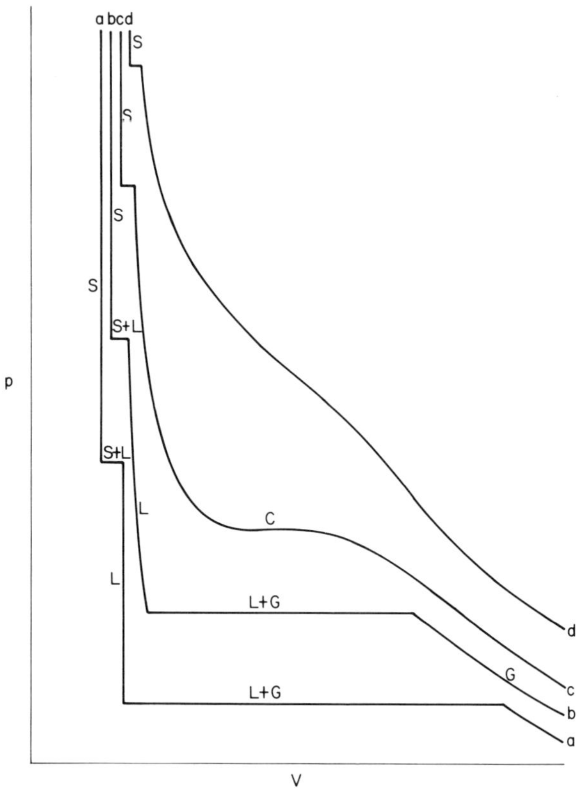

FIG. 1. Schematic pressure–volume behavior for a one-component system at a series of constant temperatures (isotherms). Lines a and b are for $T < T_c$; line c is the critical isotherm ($T = T_c$). Line d is for $T > T_c$. Here, S, L, and G denote solid, liquid, and gas phases, respectively; $S + L$ and $L + G$ denote two coexisting phases; C denotes the critical point.

the pressure remains constant; for these values of the volume, one observes visually two states of aggregation (or "phases"), obviously possessing different macroscopic properties, coexisting in the same vessel. As the volume increases, the relative proportions of the two phases change, but, until the last of the more dense phase has disappeared, the pressure remains unchanged; this pressure is the "transition pressure" for this temperature (or, in the transition from liquid to gas, the "vapor pressure"). Associated with the volume change ΔV in this constant-pressure transition are corresponding changes in other thermodynamic properties (e.g., the enthalpy and entropy changes ΔH and ΔS).

As one increases the temperature T of the isothermal path, no significant qualitative changes occur in one of these transitions, which we recognize as that between solid and liquid. However, for the other transition (liquid→gas), the volumes of the system at the beginning and end of the constant-pressure region become less different (i.e., ΔV decreases) until, above a certain temperature (the "critical temperature"), this two-phase region has disappeared entirely (e.g., line d in Fig. 1).

From these isotherms, the experimental equation of state may be built up and represented by the three-dimensional "phase diagram" of which the isotherms are constant-temperature sections. Figure 2 shows two projections (T–V and p–T) of this three-dimensional diagram. The transition (two-phase) regions are shaded in Fig. 2(a), and the isothermal paths (a, b, c, d) of Fig. 1 are shown dashed.

The point at which the gas–liquid discontinuity just disappears (labeled c in Fig. 2a, b) is the (gas–liquid) critical point, with coordinates p_c, V_c, T_c. (The critical isotherm is shown as curve c in Fig. 1). Above the critical temperature, any distinction between liquid and gas is necessarily arbitrary since the properties of the system change continuously along an isothermal path from that of a dense fluid (with properties like those of a "liquid") to those of a very-low-density fluid indistinguishable from a "gas." Indeed, since one is not restricted to isothermal paths, one may pass continuously from the "liquid" at a low temperature (e.g., that represented by isotherm a in Fig. 1) to the "gas" at the same temperature by a process of raising the temperature (at constant volume) to greater than the critical value T_c, expanding to "gas" densities at this temperature, and then cooling. Thus, "gas" and "liquid" regions are two aspects of a single continuous fluid state.

On the other hand, no such critical behavior has ever been observed for the solid→fluid transition, although it has been explored to pressures many times the critical pressure of the gas–liquid transition. No molecular

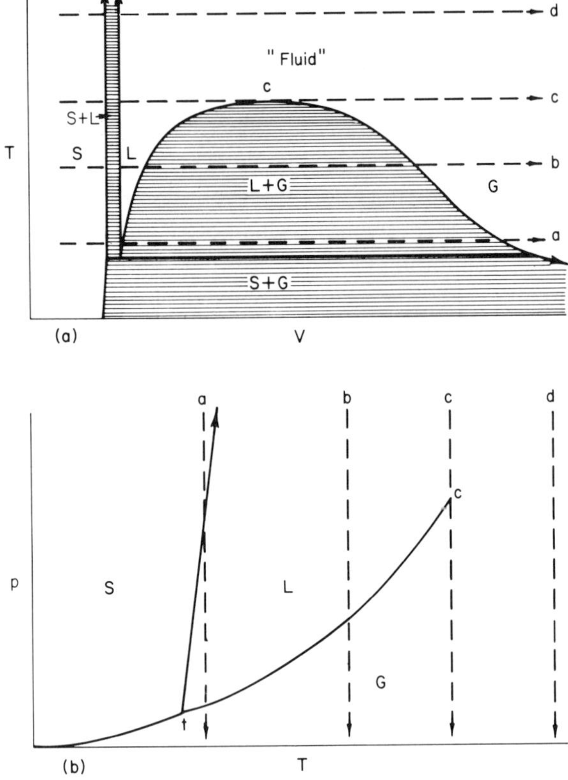

FIG. 2. Projections of the three-dimensional phase diagram: (a) T–V projection; (b) p–T projection. Labels as in Fig. 1.

theory suggests such behavior and a gradual loss of crystalline long-range order seems intuitively unlikely.

One common definition of a liquid is as a phase conforming to the shape of a vessel without filling the whole volume. This led Rowlinson (1959, 1969) to suggest that the term "liquid" should be restricted to the denser phase at the fluid→fluid transition, i.e., along the vapor pressure line. This proposal does not seem to have achieved wide acceptance, perhaps because it would make the other transition always solid→fluid, and perhaps because no one seems prepared to extend the logic and restrict the term "gas" to the less dense fluid along the same transition line.

Because of the continuity of the various regions of the fluid state, any other attempt to distinguish a liquid from a gas must draw an arbitrary line separating nearby regions which have almost identical properties.

The frequent statement that "the liquid cannot exist above the critical temperature" is (unless one accepts the very restrictive Rowlinson definition) an unhelpful one; it implies, quite erroneously, that the fluid phase in equilibrium with the crystalline solid changes character at $T = T_c$. The qualitative difference between typical "gas" and "liquid" properties is one of density; if one feels the need to separate arbitrarily the continuous fluid region into two parts, it would seem least objectionable to use some density (e.g., the critical density) as the criterion.

Experimentally, certain other states of aggregation (e.g., "liquid crystals," "glasses," and the "superfluid" phase of ^{4}He) can be distinguished from the ordinary fluid by observation of phase transitions between them. (These are sometimes second-order rather than the more familiar first-order transitions illustrated in Figs. 1 and 2.)

Ordinary fluids are also distinguished from crystalline solids by the absence of any resistance to a static shearing stress; they will flow (albeit perhaps very slowly) under the smallest stress. This property is associated with the absence of any ordered (crystalline) structure, even at distances approaching the molecular level. This brings us to a discussion of the "structure" of a fluid, which can best be specified in terms of the various molecular distribution functions describing the probabilities of finding single molecules, pairs, triplets, etc. at certain positions in space.

The probability that in a system of N molecules in volume V a molecule will be found in a small volume element $d\mathbf{r}_1$ $(= dx_1\, dy_1\, dz_1)$ at a position $\mathbf{r}_1(x_1, y_1, z_1)$ may be represented by the expression $n^{(1)}(\mathbf{r}_1)\, d\mathbf{r}_1$, where the quantity $n^{(1)}(\mathbf{r}_1)$ is called the singlet distribution function. In a fluid, there are no preferred positions or directions (unlike a crystalline solid), so $n^{(1)}$ must be independent of $\mathbf{r}_1$,

$$n^{(1)} = N/V \neq f(\mathbf{r}_1), \tag{1.1}$$

where N/V is simply the number density of molecules. Correspondingly, the probability of finding a molecule at $\mathbf{r}_1$ within $d\mathbf{r}_1$ and another at $\mathbf{r}_2$ within $d\mathbf{r}_2$ is $n^{(2)}(\mathbf{r}_1, \mathbf{r}_2)\, d\mathbf{r}_1\, d\mathbf{r}_2$, where $n^{(2)}(\mathbf{r}_1, \mathbf{r}_2)$ is the pair distribution function. In a fluid, the pair distribution function $n^{(2)}(\mathbf{r}_1, \mathbf{r}_2)$ can depend only upon the scalar distance apart $r_{12} = |\mathbf{r}_1 - \mathbf{r}_2|$, which we shall sometimes abbreviate to r,

$$n^{(2)}(\mathbf{r}_1, \mathbf{r}_2) = (N/V)^2 g(r_{12}) = (n^{(1)})^2 g(r_{12}), \tag{1.2}$$

where $g(r_{12})$ is the "radial distribution function." Similarly, the analogously defined triplet distribution function $n^{(3)}(\mathbf{r}_1, \mathbf{r}_2, \mathbf{r}_3)$ will reduce, for fluids, to $(N/V)^3 f(r_{12}, r_{13}, r_{23})$. The development of these distribution functions into detailed theories of liquids will be found in Chapter 4.[†]

The radial distribution function can be determined, to a reasonably good approximation, from the Fourier transform of experimental scattering curves (e.g., with X-rays or neutrons) (see Chapter 2). The dense fluid yields a function $g(r)$ which at small values of r is very similar to that which would be deduced from a powder X-ray photograph of a crystalline solid (i.e., for a crystal averaged over a random distribution of orientations of the crystal axes), indicating that the number and distances of separation of the nearest neighbors around a molecule in a dense fluid are like those around a molecule in the crystal of similar density.

For larger distances, however, the two radial distribution functions must diverge, for the crystal preserves its long-range order, while the fluid has none. Any completely satisfactory theory of the liquid state (i.e., of a dense fluid) must distinguish the material it describes from an ordered crystal of similar density, and must do so in a natural rather than a contrived way. Unfortunately, few theories really satisfy this condition.

B. The First "Model" and "Theory"—van der Waals

In 1873, there appeared the Ph.D. thesis of J. D. van der Waals, "On the continuity of the liquid and gaseous states." In advance of any detailed understanding of intermolecular forces and almost before the rigorous formulation of statistical mechanics, van der Waals proposed a physical model for fluids and deduced therefrom an equation of state. The pressure p, equal to $nRT/V = RT/\tilde{V}$ for a perfect gas [where n is the number of moles, R the molar gas constant, T the thermodynamic temperature, V the volume, and $\tilde{V}(=V/n)$ the molar volume], was modified to allow for the actual volume occupied by the molecules (assumed to be hard spheres, represented by a molar constant b) and for the attraction between molecules (assumed to be in a kind of a uniform potential represented by a molar constant a):

$$p = [RT/(\tilde{V} - b)] - (a/\tilde{V}^2). \tag{1.3}$$

[†] For polyatomic molecules, more than the positions of the centers of mass (i.e., the position vectors **r**) must be specified; a complete description must include the orientation of each molecule as well. For fluids, $n^{(2)}$, $n^{(3)}$, etc. will then depend upon the *relative* orientations of the molecules with respect to one another.

Subsequent discussions of the "derivation" of this equation have frequently overlooked the fact that van der Waals did not know that the interaction between molecules is a short-range one, as we now know from theory and experiment. Recently, Kac *et al.* (1963) have shown that Eq. (1.3) can be obtained quite rigorously from a one-dimensional model with an appropriate long-range interaction.

Using Eq. (1.3) to obtain the integral $\int p\, d\tilde{V}$ yields the molar Helmholtz free energy $\tilde{A}$ relative to a reference state $\tilde{A}^{\circ}_{\mathrm{id}}$ for the ideal gas at T and a reference molar volume $\tilde{V}^{\circ}$,

$$\tilde{A}(T, \tilde{V}) - \tilde{A}^{\circ}_{\mathrm{id}}(T, \tilde{V}^{\circ}) = -RT \ln[(\tilde{V} - b)/\tilde{V}^{\circ}] - (a/\tilde{V}), \quad (1.4)$$

or for the molar entropy $\tilde{S}$ and molar energy $\tilde{E}$ separately,

$$\tilde{S}(T, \tilde{V}) - \tilde{S}^{\circ}_{\mathrm{id}}(T, \tilde{V}^{\circ}) = R \ln[(\tilde{V} - b)/\tilde{V}^{\circ}], \quad (1.5)$$

$$\tilde{E}(T, \tilde{V}) - \tilde{E}^{\circ}_{\mathrm{id}}(T) = -a/\tilde{V}. \quad (1.6)$$

From Eqs. (1.3) or (1.5), one draws some interesting physical conclusions:

$$(\partial p/\partial T)_V = (\partial S/\partial V)_T = R/(\tilde{V} - b), \qquad (\partial^2 p/\partial T^2)_V = 0. \quad (1.7)$$

Equation (1.7) implies that, for a constant volume, a plot of p versus T (the "isochore") is a straight line; in fact, for liquids and gases, this is very nearly true. Thermodynamic manipulation yields the related fact that the heat capacity at constant volume C_V should be independent of volume (i.e., density or pressure) at constant temperature,

$$(\partial^2 p/\partial T^2)_V = (\partial^2 S/\partial T\, \partial V) = [\partial(C_V/T)/\partial V]_T = (1/T)(\partial C_V/\partial V)_T \overset{?}{=} 0. \quad (1.8)$$

We shall examine this prediction in a later section.

The primary accomplishment of the van der Waals theory, of course, was to account for critical behavior in fluids and for the continuous variation in properties of fluids as one goes from the liquid at low temperatures through the fluid at high temperatures (above the critical) and back at low densities to the gas at low temperatures. Moreover, the van der Waals equation of state (like any equation of state with two parameters which vary with the substance) yields a law of corresponding states, such that the properties of all fluids should fit the same curves when plotted using reduced variables proportional to RTb/a, $\tilde{V}/b$, and pb^2/a.

As more and better experimental data became available, it become obvious that the van der Waals equation could not account quantitatively for the properties of liquids and gases. Moreover, with short-range intermolecular pair energies, Eqs. (1.3)–(1.6) cannot be derived, even approximately, for the dilute gas; in particular, Eq. (1.3) leads to the absurd conclusion that the third and all higher virial coefficients are independent of temperature. As a result, the van der Waals equation went into eclipse, except as an example in textbooks, while theoreticians labored to develop better theories of liquids.

Only recently has the van der Waals model, in spirit and philosophy if not in algebraic detail, been resurrected from the dust-bin of physical chemistry. The combination of a hard-sphere entropy [akin to Eq. (1.5) but with a better algebraic form] and a "smoothed potential" energy [like Eq. (1.6)] has proved in recent years to be an extremely fruitful "zeroth" approximation for liquids, as we shall see in the final section.

II. Intermolecular Pair Potential-Energy Functions

A. Nature of the Attraction and Repulsion

Any discussion of intermolecular energies starts with the interaction between two isolated molecules, an interaction which is now known to be quite short-range in character for two molecules each with no net charge. Any realistic functional form yields an essentially zero energy at large intermolecular distances, a net attractive (negative) energy at intermediate distances, and a net repulsive (positive) energy at short distances; moreover, the curve of energy u versus distance r must be very steep at small r (see, e.g., Fig. 3). Nothing less than this will account for the "stickiness" between molecules which leads to condensation, and for the high resistance of liquids to compression beyond their low-temperature, low-pressure densities.

While this is not the place to review, even rather cursorily, modern quantum-mechanical theories of intermolecular forces† (Faraday Society, 1965; Margenau and Kestner, 1967; Hirschfelder, 1967), it is interesting

† It is a curious feature that most scientists who write about this subject call it by the general title "intermolecular forces," write an intermolecular pair potential energy function $u(r)$, and call it a "pair potential." The energy is $u(r)$, the force (which is rarely used) is $-du/dr$; neither is a potential in the correct sense. An energy is said to be "attractive" at values of r where the force $-du/dr$ is negative, and "repulsive" when $-du/dr$ is positive, but these adjectives properly apply only to the force itself.

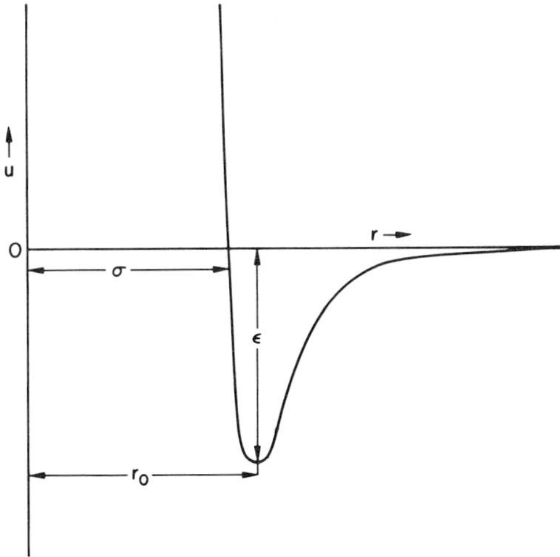

FIG. 3. Schematic representation of a pair energy function $u(r)$ as a function of the distance r between molecular centers. Here, σ is the "collision diameter" $[u(\sigma) = 0]$. The minimum energy occurs at $r = r_0$, $u = -\varepsilon$.

to consider the fundamental sources of these effects. The attraction is almost entirely electrical in origin, whether it be the electrostatic interaction of more or less fixed charges (as in the attraction of classical permanent dipoles), of fixed charges with polarizable electron clouds (as in the permanent dipole–induced dipole effect), or the net attraction due to the correlated motion of the electrons in adjacent electron clouds (as in the London dispersion energy).

The dispersion energy between two monatomic molecules (such as argon) can be shown to be of the form†

$$u_{\text{disp}}(r) = -(C_6/r^6) - (C_8/r^8) - (C_{10}/r^{10}) - \cdots, \qquad (2.1)$$

where, if the molecule is idealized as a simple three-dimensional harmonic oscillator with a single frequency ν (London, 1930),

$$C_6 = 3h\nu\alpha^2/4, \qquad (2.2)$$

† At very large distances, Eq. (2.1) ceases to be correct. Because of a quantum electrodynamic "retardation" effect (Casimir and Polder, 1938), the leading term then becomes proportional to r^{-7}, not r^{-6}. This effect becomes significant only at intermolecular separations substantially greater than 100 Å, where its contribution to the energy and other thermodynamic properties of small molecules is quite negligible.

where h is the Planck constant, and α is the molecular polarizability. In this model (Margenau, 1938), C_8 and C_{10} are more complex functions of the same parameters. For an actual system, the energy $h\nu$ may be replaced approximately by the ionization energy E_{I}. More precise quantum-mechanical evaluations (based upon spectroscopic and other molecular data) have refined the value of C_6 for simple substances, as shown by the example of argon (Table I).

TABLE I

THE ATTRACTION CONSTANT C_6 (10^{-79} J m^6) FOR Ar

London (1930, 1937)	55.4
Slater and Kirkwood (1931)	69
Dalgarno and Kingston (1961)	61.4
Barker and Leonard (1964)	62.6
Karplus and Kolker (1964)	51.7
Bell (1965)	61.0
Dalgarno *et al.* (1967)	58.2
Leonard, P. J. (reported by Barker and Pompe, 1968)	61.3
Gordon (1968)	64.7 $\pm$ 5.9

The general correctness of the estimates is confirmed by measurements (Rothe and Neynaber, 1965) of the total cross section for Ar–Ar scattering by molecular beams; this corresponds (Gordon, 1968) to $C_6 = (57 \pm 9) \times 10^{-79}$ J m^6, in good agreement with the calculated values.

Unlike these attractive energies, the repulsion at short distances, although it may have small electrical contributions, is primarily statistical in origin; it arises from the Pauli exclusion principle, which bars two electrons from being represented by the same (one-electron) wave function (i.e., effectively from occupying the same region in space). Simple analytic expressions for the repulsive energy that are theoretically sound over a substantial range of distance do not exist, but the repulsion is so steep a function of r that a useful approximation is to represent it by the infinite repulsion of two rigid spheres [i.e., $u(r) = \infty$ whenever $r < \sigma$, the "collision diameter"].

In comparing one pair energy function with another, it is convenient to express $u(r)$ in terms of energy and distance scale factors, ordinarily the depth ε of the energy well at its minimum and the (low-energy) collision diameter σ [i.e., $u(\sigma) = 0$]. An alternative scaling parameter for r is r_0, the separation at the minimum energy (see Fig. 3)

$$u(r) = \varepsilon\phi(r/\sigma) = \varepsilon\phi^*(r/r_0). \tag{2.3}$$

Equation (2.3) implies no loss of generality in $u(r)$ unless one assumes, as we shall do later in certain cases, that $\phi(r/\sigma)$ is the same function for a series of substances.

B. The Assumption of Pairwise Additivity

In statistical-mechanical theories of fluids, it has been customary for many years to assume that the potential energy U of a fluid of N molecules is simply the sum of the potential energies of the $N(N-1)/2$ molecular pairs, each calculated as if the two molecules were isolated from all the others. For spherically symmetric molecules, for which the pair energy is a function only of the distance r of separation of molecular centers, one may write

$$U = \sum_{i<j} u(r_{ij}), \tag{2.4}$$

where the summation is over all molecules i and j.

For a collection of fixed charges, it can be proved, using the equations of classical electrostatics, that the potential energy is such a sum over pairs. Similarly, to the extent that the repulsive energy resembles that of a collection of hard spheres, it too should be additive (sums of zeros and infinities). However, for the interaction of polarizable electron clouds, the situation is not so simple, and recent years have seen a determined attack upon this problem, both from the theoretical side and from the interpretation of careful experimental measurements.

The principal correction is that due to the "triplet interaction," the excess potential energy of a group of three molecules separated by distances r_{12}, r_{13}, and r_{23} over the sum of pair energies $u(r_{12}) + u(r_{13}) + u(r_{23})$. Part of this is due to the excess dispersion energy of three induced dipoles (hence the term "tripole-dipole energy") analogous to the pair energy given by Eq. (2.2) (Axilrod and Teller, 1943; Axilrod, 1951):

$$u_{123} = C_{123}(1 + 3\cos\theta_1 \cos\theta_2 \cos\theta_3)/r_{12}^3 r_{13}^3 r_{23}^3, \tag{2.5}$$

where the θ's are the interior angles of the triangle connecting the three molecules and C_{123} is a constant coefficient very nearly equal to $3\alpha C_6/4$

[where α is the polarizability and C_6 the pair coefficient given by Eq. (2.2)]. The angle-dependent factor is positive for an equilateral or a right triangle and negative for a linear array, so there is a partial cancellation in any average over configurations.

Estimates of the contribution of Eq. (2.5) to the lattice energy of a crystal (e.g., Barker and Pompe, 1968) show that the net effect is positive and that it may be as large as 10% of the total; this suggests that "quadruple-dipole," etc. terms are unlikely to contribute much more than 1%. This "triple-dipole" dispersion energy is not the only triplet interaction, but Copeland and Kestner (1968) have estimated other contributions (triple overlap, dispersion overlap, and dipole–dipole–quadrupole effects) and conclude that for argon, at distances near and beyond the minimum of $u_{12}(r)$ at $r = r_0$, all are significantly less than the triple-dipole term. At the distance $r = r_0$, however, these are not completely negligible and some (e.g., the triple overlap) are of opposite sign. Further theoretical work is needed to confirm or modify these approximate conclusions; conceivably, the net contribution of all triplet and higher-order energies to the total energy of the liquid might be substantially smaller than that estimated from the triple-dipole term alone.

Because of these uncertainties, the magnitude of the deviation from pairwise additivity must be inferred from the discrepancy between observed properties (e.g., the lattice energy or the third virial coefficient C) and those calculated assuming Eq. (2.4). Until the form of the pair energy $u(r)$ is determined with greater precision than seems presently possible, some of this uncertainty about the magnitude of the triplet contribution appears likely to continue. Thus, Rowlinson (1965b) suggests that these corrections may contribute as much as 10–15% of the total potential energy of a normal liquid, while Dymond and Adler (1968) believe they may contribute as little as 1–2%.

Specific aspects of this problem of pairwise additivity will be referred to in subsequent sections, but Eq. (2.4) will be used in much of what follows because it reduces most of the problems to more tractable forms.

C. Experimental Determination of Pair Energy Functions

Most attempts to determine the form and the parameters of an intermolecular pair potential-energy function $u(r)$ involve measurements of the thermodynamic and transport properties of moderately dilute gases, of the equilibrium properties of condensed phases, and of the collisions between molecules in molecular beams. Each of these measurements emphasizes a different range of molecular separations, and an adequate determination of $u(r)$ may well require a synthesis of the results of two or more types of measurements. Since different groups have analyzed the data in different ways, it is hardly surprising that their conclusions are sometimes somewhat discordant. Before presenting some recent attempts to extract $u(r)$ from experimental data, we summarize the principles involved in the different measurements.

1. *Determination from Thermodynamic Properties of the Dilute Gas*

It is well known that the equation of state of a gas at densities appreciably below the critical density can be expressed as a convergent series in powers of the concentration $c = n/V = 1/\tilde{V}$,

$$p/RT = c + Bc^2 + Cc^3 + Dc^4 + \cdots,$$

or, in terms of the compression factor $p\tilde{V}/RT$,

$$p\tilde{V}/RT = p/RTc = 1 + Bc + Cc^2 + Dc^3 + \cdots. \tag{2.6}$$

The coefficients 1, B, C, D, etc. are the (first), second, third, fourth, etc. "virial coefficients" in this "virial equation of state" and are functions of the temperature only. Statistical mechanics shows rigorously that B depends only upon the interactions between the two molecules in an isolated pair, C upon those between three, etc. For a classical fluid of spherically symmetric molecules [i.e., those for which the pair energy may be written simply as an angle-independent $u(r)$], more complicated general expressions for $B(T)$ reduce to

$$B(T) = 2\pi\tilde{N} \int_0^\infty (1 - e^{-u(r)/kT})r^2\, dr, \tag{2.7}$$

where $\tilde{N}$ is the number of molecules per mole (the Avogadro constant) and k is the molecular gas constant (the Boltzmann constant). [See, e.g., Mason and Spurling (1968)].

If $u(r)$ is completely specified for all distances of separation, integration of Eq. (2.7) yields B over the entire range of temperature. Unfortunately, even were $B(T)$ known accurately for all T, this would not uniquely determine $u(r)$ because u is not monotonic with r. Keller and Zumino (1959) have shown that in regions of negative potential energy (where there are two values of r, r_1 and r_2, for any given u), a complete specification of $B(T)$ yields only a kind of "width" of the energy bowl, the difference $r_2^3 - r_1^3$, as a function of u. Moreover, the experimental range of temperatures for $B(T)$ is usually rather limited and never extends much below the normal boiling point.

Consequently, except in some special cases, Eq. (2.7) is normally useful only when a form for $u(r)$—with two or more parameters to be adjusted to fit the experimental data—has been selected in advance.

The third virial coefficient C depends upon the simultaneous interaction of three molecules with distances of separation r_{12}, r_{13}, and r_{23}, and for pairwise additivity of angle-independent functions, $u(r_{ij})$ may be written in the form

$$C(T) = (8\pi^2\tilde{N}^2/3) \iiint (1 - e^{-u(r_{12})/kT})(1 - e^{-u(r_{13})/kT}) \times (1 - e^{-u(r_{23})/kT}) r_{12} r_{13} r_{23}\, dr_{12}\, dr_{13}\, dr_{23}. \tag{2.8}$$

A sensitive test of the essential correctness of a pair energy $u(r)$ selected to fit $B(T)$ or of the assumption of pairwise additivity would be the comparison of calculated with observed third virial coefficients. Unfortunately, the uncertainties about $u(r)$ from $B(T)$ and other data and about the experimental $C(T)$ are still sufficiently large to preclude an unequivocal assessment of the magnitude of deviations from pairwise additivity, although various workers (Sherwood and Prausnitz, 1964a; Rowlinson, 1965b; Barker and Pompe, 1968) have drawn the conclusion that the contribution must be substantial.

2. *Determination from Molecular Beam Experiments*

In the simplest approximate kinetic theories of dilute gases, the molecules are assumed to behave like hard spheres, and their collisions are like those of billiard balls with no moment of inertia. In actual fact, the details of the collision and the subsequent classical trajectories of the molecules depend significantly upon the details of the actual pair energy $u(r)$; moreover, the proper description of the interaction is one of wave scattering (diffraction) and quantum-mechanical corrections can be important, especially for low masses and low kinetic energies. Figure 4 is a planar representation of a classical binary collision. The quantity b is the "impact parameter," the distance of closest approach of the molecules in the absence of the energy $u(r)$. The angle of deflection θ depends upon b and upon the initial relative kinetic energy $E = \mu g^2$ of the colliding molecules [where μ is the reduced mass $m_1 m_2/(m_1 + m_2)$ of the two-molecule

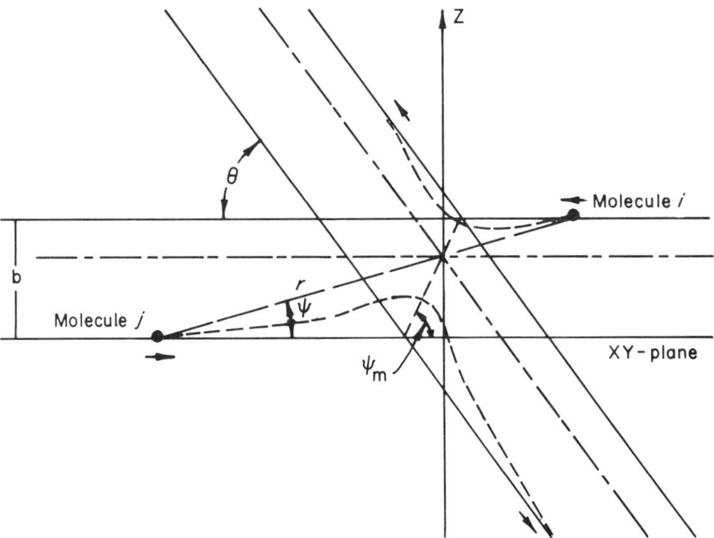

FIG. 4. Planar representation of a binary collision (according to classical mechanics). Here, r is the intermolecular distance; ψ is an angle specifying the orientation at a particular time; ψ_m is the value of ψ at the distance of closest approach; b is the impact parameter; and θ is the angle of deflection. (From Hirschfelder *et al.*, 1954).

system and g is the initial relative speed]. This classical relationship can be shown (Hirschfelder *et al.*, 1954) to depend upon an integral over a function of b, E, and $u(r)$ from the distance r_0 of the potential-energy minimum to infinity,

$$\theta(E, b) = \pi - 2b \int_{r_0}^{\infty} (1/r^2)\{1 - (b^2/r^2) - [u(r)/E]\}^{-1/2}\, dr. \tag{2.9}$$

Figure 5 shows the deflection angle θ calculated for the Lennard-Jones (12, 6) equation for $u(r)$ (Section II, D, Eq. (2.24)). For any particular negative value of θ, there are two impact parameters b, so the function $b(E, \theta)$ is not single-valued and consequently does not uniquely determine $u(r)$.

The correct quantum-mechanical formulation yields an oscillatory scattering intensity $I(E, \theta)$ which is much more complicated, but whose analysis can, in principle, yield more detailed information. Integration over all angles θ yields the total or integral elastic scattering cross section $Q(E)$,

$$Q(E) = 2\pi \int_0^{\pi} I(E, \theta) \sin\theta\, d\theta. \tag{2.10}$$

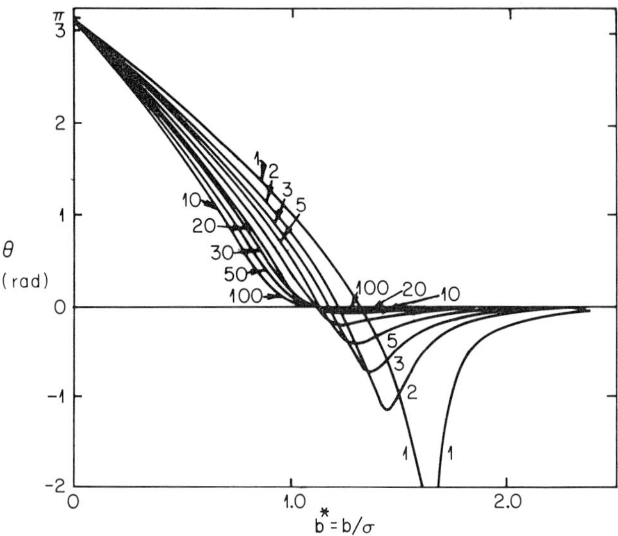

FIG. 5. Deflection angle θ for a Lennard-Jones (12, 6) pair energy versus $b^* = b/\sigma$ (according to classical mechanics); parameter $E^* = E/\varepsilon$. (From Bernstein and Muckerman, 1967.)

In the classical limit, the integration over angle may be replaced by integration over all impact parameters, replacing $I \sin\theta \, d\theta$ by $b \, db$, since

$$I_{\text{class}}(E, \theta) = b/(|\, d\theta/db \,| \sin\theta). \tag{2.11}$$

Experimental measurements of the intensity of scattering as a function of angle θ and of energy E can be made both for a molecular beam of molecules of a comparatively narrow spectrum of kinetic energies impinging upon a gas at thermal equilibrium and for the intersection of two such beams. In principle, such measurements should, if carried out over all energy ranges, yield information about $u(r)$ over the entire range of r. In practice, the experimental difficulties and the difficulties of interpreting the results have confined the studies to two distinct regions of $u(r)$.

(a) For large values of E, the collision cross section depends primarily upon the (nearly) hard-core repulsion and hardly at all upon the attraction (since the total energy E now greatly exceeds the depth ε of the pair energy well). With molecular beams collimated so that they represent a relatively narrow band of kinetic energies E, the scattering results (usually measured as the "total" cross section integrated from some angle θ_0 to 2π, θ_0 being chosen so that quantum-mechanical complications at small angles are largely excluded) can be interpreted in a straightforward way in terms of a repulsive $u(r) = Ae^{-Br}$ at short distances ($r \ll \sigma$). The most extensive studies on He–He scattering are now (Jordan and Amdur, 1967) in good agreement with theory for $0.6\ \text{Å} < r < 1.5\ \text{Å}$.

(b) Low-energy ("thermal") molecular beam scattering by a gas at thermal equilibrium yields information about the long-range attractive part of the pair energy, and if a simple form for $u(r)$ is assumed, about the intermediate range as well. A series of recent reviews (Ross, 1966; Mason and Monchick, 1967; Bernstein and Muckerman, 1967) cover this subject thoroughly, in both its theoretical and experimental aspects, so we cite only a few important features. Measurements of $I(\theta)$ as a function of θ and of $Q(E)$ as a function of E confirm the dependence of $u(r)$ upon the inverse-sixth power of the distance in the limit of large r [e.g., the r^{-6} term in Eq. (2.1)]. With this dependence verified, the total "effective" cross section yields approximate values (correct to within perhaps 10%) of the coefficient C_6.

3. *Determination from Transport Properties of the Dilute Gas*

It is well known that the "kinetic theory of gases" yields equations for various transport properties: the coefficient of diffusion D is the flux of molecules divided by the gradient in the number density; the coefficient of viscosity η is the flux (in the y direction) of the x component of momentum divided by the gradient (again in the y direction) in the x component of velocity; and the coefficient of thermal conductivity λ is the energy flux divided by the gradient in the temperature.

For dilute gases, the mechanism for such transport is essentially one of isolated collisions between molecules, the details of which depend upon the pair energy $u(r)$, as we have seen. These properties depend upon a set of "transport cross sections" $Q^{(l)}$ (Hirschfelder *et al.*, 1954),

$$Q^{(l)}(E) = 2\pi \int_0^\pi (1 - \cos^l \theta) I(E, \theta) \sin \theta \, d\theta. \tag{2.12}$$

Unlike the situation in Eq. (2.10), the small angle contributions to $I(E, \theta)$ (where quantum effects are important) are largely suppressed by the weighting function $(1 - \cos^l \theta)$, so a classical approximation utilizing Eqs. (2.9) and (2.11) is normally sufficient; for light molecules (e.g., H_2 or He) at very low temperatures, corrections are necessary.

When the distribution of kinetic energies in the system of colliding gas molecules is that of thermal equilibrium (i.e., the Maxwell–Boltzmann distribution), a further integration yields a set of "collision integrals" $\Omega^{(l,s)}(T)$,

$$\Omega^{(l,s)}(T) = (2kT/\pi\mu)^{1/2} \int_0^\infty e^{-E/kT} e^{s+2} Q^{(l)}(E)\, dE. \tag{2.13}$$

In a gas at thermal equilibrium only these collision integrals appear directly in the equations for the transport properties. Calculation of these for the dilute gas of nonattracting rigid spheres ("hard spheres") of diameter σ is relatively straightforward, so it is convenient to express the properties calculated for a more realistic pair energy in terms of the hard-sphere results, defining a set of reduced collision integrals

$$\Omega^{(l,s)*} = \Omega^{(l,s)} / \Omega^{(l,s)}_{\text{hard sphere}}$$

In the limit of low density and one-component systems, the appropriate equations are

$$D = (3/8)[(\pi mkT)^{1/2}/\pi\sigma^2\varrho] f_{\mathrm{D}} / \Omega^{(1,1)*}, \tag{2.14}$$

$$\eta = (5/16)[(\pi mkT)^{1/2}/\pi\sigma^2] f_\eta / \Omega^{(2,2)*}, \tag{2.15}$$

$$\lambda = (25/32)[(\pi mkT)^{1/2}/\pi\sigma^2](c_V/m) f_\lambda / \Omega^{(2,2)*}, \tag{2.16}$$

where m is the molecular mass, ϱ the density, and c_V the heat capacity per molecule. The factors f_{D}, f_η, and f_λ are factors involving the various integrals $\Omega^{(l,s)*}$ but in such a way that they are never much different from unity (usually within 1%).

A fourth transport property is thermal diffusion in mixtures. For mixtures of heavy isotopes (i.e., where the ratio of isotopic masses is not greatly different from unity), the thermal diffusion ratio α reduces, in the Kihara first approximation, to

$$\alpha = 15(6\Omega^{(1,2)*} - 5\Omega^{(1,1)*})/16\Omega^{(2,2)*}. \tag{2.17}$$

For hard spheres, all the collision integrals $\Omega^{(l,s)*}$ are unity and, of course, independent of temperature. For a realistic pair energy $u(r)$, they are functions of temperature, so the temperature dependence of the transport properties is another property to fit with the adjustable parameters of an assumed pair function.

The viscosity and the thermal conductivity depend upon the same collision integral $\Omega^{(2,2)*}$, so in principle yield the same information; they are especially useful in fitting the medium-energy repulsive region $[0 < u(r) < 10\varepsilon]$. Self-diffusion data tend to be less accurate and consequently somewhat less useful. The thermal diffusion ratio is in some respects the most sensitive quantity; it varies by as much as a factor of seven over the experimentally accessible region; it is especially sensitive at low temperature to changes in the outer wall of the energy well (Dymond and Alder, 1969).

4. *Determination from Thermodynamic Properties of Condensed Phases*

For molecules with intermolecular energies that are short-range, the thermodynamic properties (especially the energy and heat capacity) of dense phases will depend primarily upon the interactions of nearest neighbors, i.e., upon the details of the pair energy $u(r)$ near its minimum at $r = r_0$. Since our inability to cope with the lack of long-range order in liquids has hindered the development of simple and reliable molecular theories of liquids, attention has focused upon idealized models of the crystalline solid and there have been a series of attempts (Rice, 1941; Guggenheim and McGlashan, 1960a; McGlashan, 1965) to correlate crystal properties with parameters of a pair energy function expanded around $r = r_0$:

$$u(r) = -\varepsilon + \varkappa[(r - r_0)/r_0]^2 - \alpha[(r - r_0)/r_0]^3 + \cdots, \tag{2.18}$$

where ε is, as before, the depth of the energy well with respect to infinite separation of the molecules; $\varkappa$ is a measure of the curvature of the bottom of the well; and α (and coefficients of still higher-order terms if required) measures the departure from simple parabolic shape. Needless to say, the parameters ε, $\varkappa$, α, etc. can be evaluated for any analytic closed-form expression for $u(r)$.

If one assumes pairwise additivity, it is a relatively simple matter to formulate a good approximation to the total potential energy of the crystal (e.g., for a face-centered cubic lattice of the rare-gas solids with 12 nearest neighbors) by using any assumed form for $u(r)$, or by combining nearest-neighbor energies calculated from Eq. (2.18) with non-nearest-neighbor interactions calculated from the limiting $-C_6/r^6$ form for London attraction.

For such a crystal, vibrational energy (and entropy) must be calculated, possibly using the Einstein approximation, but better with the Debye approximation. Then calculated properties can be compared with experiment or, alternatively, the experimental data can be used to evaluate some adjustable parameters (e.g., ε, $\varkappa$, and α). Experimental data available include:

(a) The molar heat capacity $\tilde{C}_V$ of the crystal and hence, by integration from $T = 0$, the molar entropy $\tilde{S}$. In an idealized model, these depend exclusively upon the lattice vibrations (i.e., upon r_0, $\varkappa$, and α) and not at all upon ε.

(b) The molar lattice energy (i.e., the molar energy of sublimation to the ideal gas). At $T = 0$, this depends primarily upon ε, with corrections for the zero-point vibrational energy and for the difference between the nearest-neighbor lattice spacing and r_0 (both of which corrections depend upon r_0, $\varkappa$, and α).

(c) The lattice spacing (or, alternatively, the nearly equivalent molar volume of the crystal) at $T = 0$ and $p = 0$ and as a function of temperature and pressure, again functions of r_0, $\varkappa$, and α.

Unfortunately, the usefulness of this information depends crucially upon the assumption of pairwise additivity. If there are extra many-body contributions to the total energy, the parameters ε, r_0, $\varkappa$, α, etc. determined in this manner are, at best, those for an "effective" two-body energy in the dense phase and not those for the "true" pair energy appropriate for an isolated pair of molecules in the dilute gas. The failure of the lattice energy to fit parameters derived from the dilute gas has been used to estimate the magnitude of the many body energy (Munn and Smith, 1965); conversely, calculated values of the "triplet" energy have been used to correct the lattice energy for comparison with the pair $u(r)$ (Barker and Pompe, 1968).

D. Some Oversimplified Pair Energy Functions

First, we consider some simple forms for $u(r)$ and illustrate the insensitivity of thermodynamic functions to the details of the form by calculating the second virial coefficient B of the gas as a function of temperature.

1. *The "Square-Well" Pair Energy*

Probably the simplest pair energy function that can represent both an attractive and a repulsive region is that for rigid spheres of diameter σ with a uniform attractive energy $-\varepsilon$ for intermolecular distances between σ and $\alpha\sigma$ and no attraction or repulsion whatever for distances greater than $\alpha\sigma$ (see Fig. 6).

$$\begin{aligned} r < \sigma, &\qquad u = +\infty, \\ \sigma < r < \alpha\sigma, &\qquad u = -\varepsilon, \\ r > \alpha\sigma, &\qquad u = 0. \end{aligned} \tag{2.19}$$

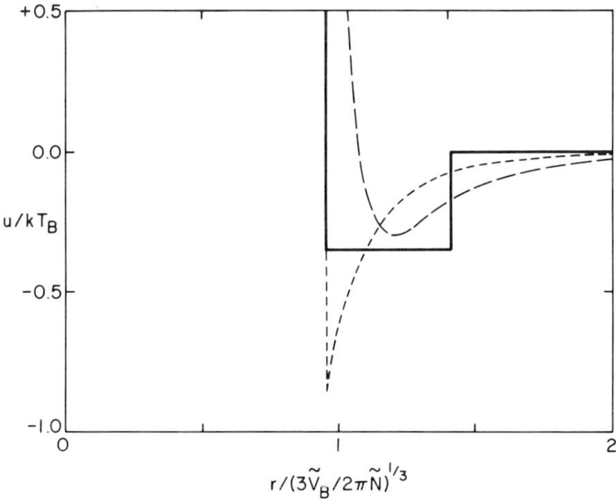

FIG. 6. Three oversimplified pair potential-energy functions. (——) square-well energy, Eq. (2.19) with $\alpha = 3/2$. (- - -) Sutherland (∞, 6) energy, Eq. (2.21). (— —) Lennard-Jones (12, 6) energy, Eq. (2.24). Note that the Boyle temperature T_B and the Boyle volume $\tilde{V}_B$ are used to convert u and r into reduced quantities. For the relations between (ε, σ) and (T_B, $\tilde{V}_B$), see Table II.

Substitution of this discontinuous function for $u(r)$ into Eq. (2.7) yields a simple closed expression for the second virial coefficient which, for ease of comparison with the results for other energy functions, is also shown as a power series in ε/kT:

$$\begin{aligned} B(T) &= (2\pi\tilde{N}\sigma^3/3)[1 - (\alpha^3 - 1)(e^{\varepsilon/kT} - 1)] \\ &= (2\pi\tilde{N}\sigma^3/3)[1 - \beta(\varepsilon/kT) - (\beta/2)(\varepsilon/kT)^2 - (\beta/6)(\varepsilon/kT)^3 - \cdots] \end{aligned} \tag{2.20}$$

where $\beta = \alpha^3 - 1$. For fixed β, there are two adjustable parameters, ε and σ.

2. *The Sutherland* (∞, 6) *Pair Energy*

A function for $u(r)$ that seems superficially somewhat more realistic combines the rigid-sphere infinite repulsion with an attraction term proportional to r^{-m}, where, in accord with the London theory for dispersion energy (Section IIA), m is almost invariably set equal to 6. If m is so fixed, there are again only two adjustable parameters, ε and σ (Fig. 6):

$$\begin{aligned} r < \sigma \qquad & u = +\infty, \\ r > \sigma \qquad & u = -\varepsilon(\sigma/r)^6. \end{aligned} \tag{2.21}$$

When Eq. (2.21) is substituted into Eq. (2.4), it is not possible to obtain the integral in closed form, but one may expand $\exp(-\varepsilon\sigma^6/kTr^6)$ in a power series and obtain the integrated result for $B(T)$ as a power series in ε/kT:

$$B(T) = (2\pi\tilde{N}\sigma^3/3)\{1 - (\varepsilon/kT) - [(\varepsilon/kT)^2/6] - [(\varepsilon/kT)^3/30] - \cdots\}. \tag{2.22}$$

Equations (2.20) and (2.22) can be made equivalent in the high-temperature limit if one sets the coefficients of the two $(1/kT)$ terms equal. Moreover, since there is no compelling reason to equate the quite different ε parameters in the two functions, one can in fact equate the coefficients of the $(1/kT)$ and $(1/kT)^2$ terms in Eqs. (2.20) and (2.22), so that these equations for $B(T)$ differ only in the $(1/kT)^3$ and higher terms. If this is done, one finds $\sigma_{\text{Sutherland}} = \sigma_{\text{square well}}$, $\varepsilon_{\text{Sutherland}} = 3\varepsilon_{\text{square well}}$, and $\alpha^3 - 1 = \beta = 3$ ($\alpha = 1.59$). In fact, Guggenheim (1953), in fitting $B(T)$ data for simple gases to the square-well $u(r)$, originally selected $\alpha = 1.50$, and, more recently (Guggenheim, 1966), $\alpha = 1.61$. It is evident that both Eqs. (2.20) and (2.22) yield a finite positive limit for $B(\infty)$ at high temperatures and negative B's decreasing toward $-\infty$ at low temperatures, and that only very precise experimental measurements over a wide range of temperatures (including quite low ones) could distinguish between the fit of one and that of the other.

3. *The Lennard-Jones (12, 6) Pair Energy*

Actually, since real molecules are not quite rigid spheres, the second virial coefficient B must slowly decrease at very high *relative* temperatures (kT/ε), as evidenced experimentally in the behavior of He and H_2 (Holborn and Otto, 1924; Tanner and Masson, 1930). This can be accounted for by the somewhat more realistic potential-energy function (Lennard-Jones, 1924, 1931)

$$u(r) = (j/r^n) - (k/r^6). \tag{2.23}$$

Values used for the repulsive exponent n have varied from 9 to 15 [the Sutherland $u(r)$ arises from setting $n = \infty$], but in most work n has been set equal to 12, not for any profound theoretical reason but for the mathematical simplifications which arise because $12 = 2 \times 6$. With $n = 12$, Eq. (2.23) can be rewritten in terms of the energy minimum $-\varepsilon$ and the distances σ or r_0 (see Fig. 6):

$$u(r) = 4\varepsilon[(\sigma/r)^{12} - (\sigma/r)^6] = \varepsilon[(r_0/r)^{12} - 2(r_0/r)^6], \qquad (2.24)$$

where $r_0 = 2^{1/6}\sigma = 1.122\sigma$. The pairs (ε, σ) or (ε, r_0) are now the only adjustable parameters.

The second virial coefficient $B(T)$ corresponding to this (12, 6) pair energy is again a function of $\tilde{N}\sigma^3$ and ε/kT. As with the Sutherland $(\infty, 6)$ pair energy, integration of Eq. (2.4) to a closed form is not possible, but $B(T)$ can be obtained as a series in half-integral powers of ε/kT:

$$B(T) = (2\pi\tilde{N}\sigma^3/3)(4\varepsilon/kT)^{1/4}\{\Gamma(3/4) - [(\varepsilon/kT)^{1/2}/2^{1/2}] - [(\varepsilon/kT)/2] - \cdots\}, \qquad (2.25)$$

where the gamma function $\Gamma(3/4) = 1.22541\ldots.$

4. *Comparisons of* $B(T)$

A plot of the reduced second virial coefficient $B/(2\pi\tilde{N}\sigma^3/3)$ as a function of kT/ε for each of these three pair energies might seem an appropriate way of comparing them. However, since the parameters ε and σ are to a considerable extent arbitrary, a more meaningful comparison will be obtained if we use for reducing parameters two "experimental" quantities (Kihara, 1953), the Boyle temperature T_B (at which $B = 0$) and the "Boyle volume" $\tilde{V}_B = T_B(dB/dT)_{T=T_B}$. Figure 7 shows the reduced second virial coefficient $B/\tilde{V}_B$ as a function of T/T_B for each of the three pair functions: square well (with $\alpha = 3/2$), Sutherland $(\infty, 6)$, and Lennard-Jones (12, 6).

The values of $B/\tilde{V}_B$ and its slope must coincide at $T = T_B$, so the different functions $u(r)$ will produce plots of $B/\tilde{V}_B$ versus T/T_B which differ at T_B only in curvature. Measurements over a short range of temperature determine B and dB/dT fairly easily, but these just determine two scale parameters ε and σ. The curvature d^2B/dT^2, determined directly from measurement of $B(T)$ over a wider range of temperature or indirectly from heat-capacity measurements [using the thermodynamic relation $\lim_{p\to 0}(\partial\tilde{C}_p/\partial p)_T = -Td^2B/dT^2$], permits fixing a third parameter, but

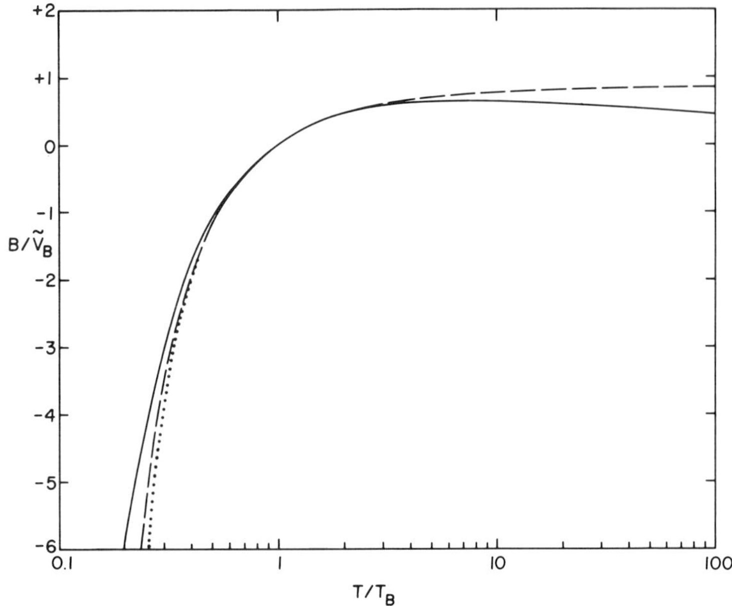

FIG. 7. Reduced second virial coefficient $B/\tilde{V}_B$ versus reduced temperature T/T_B for three pair energies: (——) Lennard-Jones (12, 6) energy; (— —) square-well energy (with $\alpha = 3/2$); (···) Sutherland (∞, 6) energy. The three curves must be coincident at $T = T_B$. At high temperatures, the square-well and Sutherland curves are virtually indistinguishable, so only the former is shown.

TABLE II

SECOND VIRIAL COEFFICIENTS FOR VARIOUS PAIR ENERGIES

	van der Waals Eq. (1.3)	Square well ($\alpha = 3/2$)	Sutherland (∞, 6)	Lennard-Jones[a] (12, 6)	Experiment (Argon[b])
kT_B/ε	—	2.846	1.171	3.418	—
$\tilde{V}_B/(2\pi\tilde{N}\sigma^3/3)$	—	1.186	1.174	0.811	—
T_{JT}/T_B	2	1.918	1.922	1.89	1.86
$T_B{}^2(d^2B/dT^2)_{T=T_B}/\tilde{V}_B$	-2	-2.351	-2.354	-2.29	-2.23
$B(T = 0.50T_B)/\tilde{V}_B$	-1	-1.198	-1.216	-1.13	-1.09
$B(T = 0.25T_B)/\tilde{V}_B$	-3	-5.32	-6.09	-4.10	-4.47

[a] Derived from calculations reported by Hirschfelder *et al.* (1954).
[b] Derived from data of Whalley *et al.* (1953) and Weir *et al.* (1967).

does not distinguish very well among Eqs. (2.20), (2.22), and (2.25) (see Table II). In effect, measurements of $B(T)$ can rarely do more than determine, for a particular *assumed form* of $u(r)$, a distance parameter σ, a representative well depth ε (not necessarily the "true" minimum), and a third parameter which is related to a kind of representative "width" of the energy well (e.g., the value of $r_2^3 - r_1^3$ for $u = -\varepsilon + k\bar{T}$, where $\bar{T}$ is representative of the range of temperatures covered by the measurements), and they can never do much more.

Table II compares the $B(T)$ curves at representative points for the three simple $u(r)$ functions with the experimental data for argon and with the especially simple equation $B = b - (a/RT)$ derived from the van der Waals equation (1.3). In the higher-temperature region, the ratio T_{JT}/T_B (where T_{JT} is the Joule–Thomson inversion temperature, at which $B = T\, dB/dT$) and the curvature $T_B^2(d^2B/dT^2)_{T=T_B}/\tilde{V}_B$ illustrate the fit, while the increasing divergence at lower temperatures is shown by comparing values of $B/\tilde{V}_B$ at $T/T_B = 0.50$ and at $T/T_B = 0.25$. (For argon, the critical temperature $T_c = 0.37T_B$, and the normal boiling point $T_b = 0.21T_B$).

In Figure 6, the energy functions $u(r)$ are appropriately scaled to fit T_B and $\tilde{V}_B$. While the collision diameters σ are similar for all three functions, the well depth ε varies enormously with the shape chosen for $u(r)$ (Sherwood and Prausnitz, 1964b). The fact that these radically different forms for $u(r)$ yield very similar functions for $B(T)$ illustrates the difficulty of obtaining information about the pair energy from the second virial coefficient alone.

E. The Pair Energy Function for Argon

To probe more deeply into the details of the shape and magnitude of the pair potential-energy function $u(r)$ requires a careful analysis and synthesis of all the available experimental data, utilizing properties which are sensitive to the details of $u(r)$ in different ranges of the intermolecular distance. This is really possible only for spherically symmetric molecules, i.e., the rare gases and the "effectively monatomic" CH_4, and in recent years much of the work has been concentrated on argon. Early work (Lennard-Jones, 1931) deduced from $B(T)$ data on Ar and the (12, 6) energy function [Eq. (2.24)] the parameters $\varepsilon/k = 120°K$ and $\sigma = 3.43$ Å. More recent evaluations, based upon newer data but still restricted to the same functional form, have not changed these very much; Sherwood and Prausnitz (1964b) report $\varepsilon/k = 118°K$ and $\sigma = 3.50$ Å.

However, as we have seen in the previous section, these numbers are strongly dependent upon the analytic form assumed, and when other expressions are substituted for Eq. (2.24) the uncertainties about the "true" depth ε of the energy well, the pair distance r_0 at the minimum, and the collision diameter σ can be very great indeed. Fitts (1966) has ably summarized the chaos in this field up to that time, and it must suffice here to summarize some of the most recent attempts to determine $u(r)$ for argon, all of which are relatively free from restrictive assumptions about analytic form.

1. *A "Piecewise" Pair Energy*

Guggenheim and McGlashan (1960a), after demonstrating once again the extremely arbitrary character of the Lennard-Jones (12, 6) function, suggested that a more realistic function could be built up by combining different analytic forms for $u(r)$ for different ranges of the distance r. In particular, they suggested the following:

(a) For large distances ($r > 1.5r_0$), the limiting form of Eq. (2.1) for the attractive energy, $u(r) = -C_6r^{-6}$, should be used, with the best value of C_6 deduced from theory.

(b) For distances near the minimum of $u(r)$ at r_0, Eq. (2.18) should be used, and the parameters ε, $\varkappa$, and α should be determined from the properties of crystalline argon, i.e., the energy of sublimation, compressibility, heat capacity, entropy, and vapor pressure. [These calculations were later refined by McGlashan (1965) using new experimental data]. In order to use these measurements, pairwise additivity had to be assumed, a procedure which subjected this work to much subsequent criticism.

(c) The collision diameter σ should be chosen to fit molecular beam results. This choice is important, but the details of $u(r)$ for $r < \sigma$ (where the energy is positive) are of little importance for the properties of liquids and gases at low temperatures, so a cut off [$u(r) = \infty$ for $r < \sigma$] was assumed.

On the grounds that the properties of interest were insensitive to the precise datails of $u(r)$ for other distances, Guggenheim and McGlashan sketched their function free-hand for values of r lying between these three well-defined regions and showed that the resulting second virial coefficient $B(T)$ was in good agreement with experiment. Their $u(r)$ is shown in Fig. 8, and the important parameters in Table III.

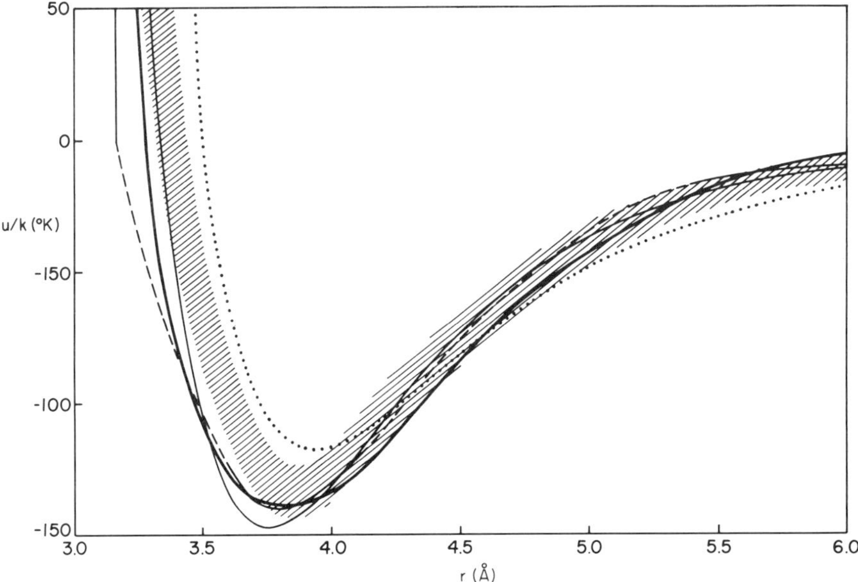

Fig. 8. The pair energy $u(r)$ for argon: (···) Lennard-Jones (12, 6) energy; (— - - —) Guggenheim and McGlashan piecewise energy (dashed parts sketched between important regions); (////) Mikolaj and Pings energy from X-ray scattering (shaded area indicates estimated error limits); solid curve, Barker and Pompe multiparameter analytic function; heavy solid curve, Dymond and Alder numerical function. (See text.)

2. *An "X-Ray" Pair Energy*

It can be shown that the appropriate Fourier transforms of X-ray scattering data yield not only the radial distribution function $g(r)$, but also a different function $C(r)$, the direct correlation function (see Chapters 2 and 4), which, unlike $g(r)$, is expected to be short-range (i.e., zero at large r). Moreover, the Percus–Yevick (PY) approximation to the molecular theory of fluids yields (Rowlinson, 1965a) a relation between the pair energy $u(r)$ and these two functions,

$$(u(r)/kT)_{\mathrm{PY}} = \ln\{[g(r, \varrho, T) - C(r, \varrho, T)]/g(r, \varrho, T)\}, \quad (2.26)$$

where the inclusion of the variables density ϱ and temperature T in the specification of $g(r)$ and $C(r)$ emphasize that these must be determined for the same state of the fluid. Mikolaj and Pings (1967) have evaluated $g(r)$ and $C(r)$ from X-ray diffraction data on 13 fluid states of argon; the direct

TABLE III

Comparison of Energy Functions

Function	(ε/k) (°K)	σ (Å)	r_0 (Å)	$(r_0^2/\varepsilon)(\partial^2 u/\partial r^2)_{r=r_0}$	C_6 (10^{-79} J m^6)
Lennard-Jones (12, 6) (Sherwood and Prausnitz, 1964b)	118	3.50	3.93	72	120
Guggenheim and McGlashan (1960a), McGlashan (1965)	140	3.165	3.80	66.5	60
Mikolaj and Pings (1967)	134 ± 10	3.38 ± 0.06	3.86 ± 0.03	—	—
Barker and Pompe (1968)	147	3.34	3.76	76.4	61.3
Dymond and Alder (1969)	138	3.28	3.81	—	66.9
Kihara core (Weir *et al.*, 1967)	164	3.15	3.47	99	30

correlation function so obtained is in fact short-range, but the pair energy obtained from Eq. (2.26) is not independent of T and ϱ, possibly because of the approximations inherent in the Percus–Yevick treatment (see Chapter 4), or because of the assumption of pairwise additivity. However the PY approximation is necessarily valid in the low-density limit, as is the assumption of pair additivity (there being substantially no contribution from triplets), so extrapolation of $[u(r)]_{\text{PY}}$ to the limit of zero density should yield the "true" $u(r)$ within the experimental error of the original X-ray measurements and the truncation error of the Fourier transforms.

The experimental slope of this plot of $[u(r)]_{\text{PY}}$ versus density is appreciably larger than can be accounted for by theory, even with the inclusion of triple-dipole effects (Copeland and Kestner, 1968; Barker *et al.*, 1969; Casanova *et al.*, 1970); this casts some doubt upon the correctness of the results obtained by extrapolation to zero density.

The $u(r)$ for argon estimated (as a band representing the probable error) by Mikolaj and Pings is shown in Fig. 8; the important parameters are listed in Table III.

3. *A Multiparameter Analytic Function*

Barker and Pompe (1968) have attempted to fit dilute-gas data from second virial coefficients (including new measurements down to 80°K), molecular beams, and transport properties to a flexible multiparameter function

$$u(R) = \varepsilon\{e^{-\alpha(R-1)}[A_0 + A_1(R-1) + A_2(R-1)^2 + A_3(R-1)^3] \\ -[\gamma_6/(\delta + R^6)] - [\gamma_8/(\delta + R^8)] - [\gamma_{10}/(\delta + R^{10})]\}, \qquad (2.27)$$

where $R = r/r_0$. While this function has in principle nine independently adjustable parameters, only three (α, ε, and r_0) were varied systematically. As is usual in modern treatments, γ_6 $(=C_6/\varepsilon r_0^{\,6})$ was taken from theory, and γ_8 and γ_{10} as well. The constant δ was set equal to 0.01 to avoid a spurious maximum at small values of R, while the repulsion parameters $A_0, \ldots, A_3$ were constrained (for given values of the others) to reproduce the molecular beam results at $r = 2$ Å. Then, for given α, the parameters ε and r_0 were varied to minimize the mean-square difference of the experimental and theoretical second virial coefficients. A range of five sets $\{\alpha, \varepsilon, r_0\}$ fit all the data (including gas transport properties) about equally well. These five pair energies were then used to calculate the lattice energy and the lattice spacing for crystalline argon at $T = 0$,

$p = 0$, using a calculated triple-dipole term (about 7% of the total) to allow for departures from pairwise additivity. None of these five sets yielded both the lattice energy and spacing correctly, but, by accepting a slightly poorer fit with $B(T)$, Barker and Pompe found a sixth set (α, ε, r_0) which reproduced the lattice properties very well. This "best" $u(r)$ is shown in Fig. 8 and Table III.

4. *A Numerical Function*

Dymond and Adler (1969) have abandoned the use of any specific analytic form for $u(r)$ and have examined systematically the effect upon dilute-gas properties (second virial coefficient and transport coefficients) of varying the long-range attraction, the outer wall of the energy well (bowl), the width of the well, and the repulsive energy. It was found that changing different regions of $u(r)$ simultaneously led to the same result as combining the separate changes, so that a smooth numerical function [values of $u(r)$ for 88 values of r/σ between 0.60 and 3.40] was combined with an analytical function $-C_6/r^6$ for $r/\sigma > 3.40$. The dilute-gas data are largely the same as those used by Barker and Pompe and the fit is equally good; the somewhat different $u(r)$ is shown in Fig. 8 and Table III. However, Dymond and Adler claim (1968) that, assuming pairwise additivity, their pair energy predicts the lattice energy and the lattice parameter to better than 1% and that, because of its value in the range of r for third nearest neighbors (Alder and Paulson, 1965), it stabilizes the face-centered cubic lattice by comparison with the hexagonal-close-packed lattice without extra three-body energies, something no analytic $u(r)$ has yet been able to do.

5. *Summary*

Table III summarizes the various values of ε, σ, r_0, C_6, and the reduced curvature $(r_0{}^2/\varepsilon)(d^2u/dr^2)_{r=r_0}$ at the minimum [i.e., $2\varkappa/\varepsilon$ in Eq. (2.18)] for these recently determined pair energy functions. Also shown are the results of a recent attempt (Weir *et al.*, 1967) to fit $B(T)$ to the Kihara spherical-core pair energy [see Eq. (2.31) of the next section].

This table illustrates the great importance of the asymptotic behavior of the $-C_6/r^6$ term in $u(r)$. The Lennard-Jones (12, 6) function overestimates C_6 and, in achieving the best fit to $B(T)$, obtains an ε which is certainly too small and values of σ and r_0 which are probably too large. Conversely, the latest fit to the Kihara core function underestimates C_6

and compensates with an ε which is probably too large and σ and r_0 which are surely too small.[†] The variations in ε, σ, and r_0 shown by the Mikolaj and Pings, Barker and Pompe, and Dymond and Alder studies probably represent a reasonable estimate of the remaining uncertainties in these quantities.

The uncertainty about the magnitude of triplet contributions to the thermodynamic functions of dense fluids has led to the recognition that the fitting of such properties to a simple pair energy function $u(r)$ determines at best an effective pair energy $u^*(r)$. There is some indication that the Lennard-Jones (12, 6) function is better for such an effective $u^*(r)$ than it is as a true pair function applicable to the dilute gas.

A first-order correction to the hypernetted-chain or Percus–Yevick approximations for fluids (see Chapter 4) leads (Rushbrooke and Silbert, 1967; Rowlinson, 1967; Rowlinson, 1969) to the equation

$$u^*(T,r) = u_{12}(r) - kTe^{\text{triplet}}(T,r), \tag{2.28}$$

where $e^{\text{triplet}}(T, r)$ is an integral which is probably negative in the dense fluid. If $e^{\text{triplet}}(T, r)$ is negative, the well depth for the effective energy function $u^*(r)$ would be shallower than that of the "true" pair function $u(r)$, in agreement with the results shown in Table III for the (12, 6) function and the better pair energies.

Because of the uncertainties in ε and σ, which are still considerable for Ar and which are surely greater for less-well-studied substances, it is extremely important, when correlating the properties of different substances, to use pair energy parameters derived in the same manner for comparable energy functions $u(r)$. This is illustrated in Table IV, where estimates of ε for Ar and Kr, derived by various methods for various analytic functions, are compared. Despite a variation of more than a factor of three in the ε's for one substance, the ratio $\varepsilon_{\text{Kr}}/\varepsilon_{\text{Ar}}$ has maximum variation from 1.31 to 1.50, and most are close to 1.40. The ratio of the experimental Boyle temperatures T_{B} is in fact 1.40, while that of the critical temperatures T_{c} is 1.39.

F. Pair Energy Functions for Polyatomic Molecules

Only a very few of the substances with which chemists and physicists ordinarily deal are monatomic: the rare "gases," He, Ne, Ar, Kr, Xe,

[†] Guggenheim and McGlashan (1960a) suggest that a fit to the lattice energy of a face-centered cubic crystal requires that $\varepsilon + (0.20C_6/r_0{}^6)$ be approximately constant.

TABLE IV

WELL DEPTH ε FOR Ar AND Kr

	(ε_{Ar}/k) (°K)	(ε_{Kr}/k) (°K)	$\varepsilon_{Kr}/\varepsilon_{Ar}$
Square well[a]	93	136	1.45
Lennard-Jones (12, 6)[a]	118	164	1.39
Morse[b]	133	200	1.50
Kihara[c]	138	196	1.42
Guggenheim–McGlashan[d,e]	140	191	1.37
Multiparameter[f]	147	208	1.41
Kihara[a]	147	216	1.47
Multiparameter[g]	153	216	1.41
Kihara[h]	164	214	1.31
Sutherland (∞, 6)[a]	305	422	1.38

[a] Sherwood and Prausnitz (1964b).
[b] Konowalow and Carra (1965).
[c] Rossi and Danon (1965).
[d] Guggenheim and McGlashan (1960b).
[e] McGlashan (1965).
[f] Dymond *et al.* (1965).
[g] Munn and Smith (1965).
[h] Weir *et al.* (1967).

etc., and metals, with their special interatomic forces. All the rest are polyatomic and necessarily lack complete spherical symmetry; for these, an intermolecular pair energy must depend upon other variables in addition to r, i.e., upon relative orientation, "shape," charge distribution, and other factors. For comparatively simple polyatomic molecules, three major complicating features must be distinguished as discussed in the following subsections.

1. "*Globularity*"

For molecules of tetrahedral or higher symmetry (e.g., CF_4, CCl_4, SF_6, P_4, etc.), the structure is globular or quasispherical, but the attractions and repulsions are not physically centered on the geometric center of the molecule but rather on the peripheral atoms. (An exception to this situation is CH_4, which—both in its intermolecular pair energy and the thermodynamic properties dependent thereon—seems to be virtually conformal with the truly monatomic rare gases. In general, hydrogen atoms can almost be ignored in considering the form of pair energy functions for polyatomic molecules; thus, ethane, CH_3CH_3, might be approximated by the isoelectronic F_2—corrected to the same internuclear distance.)

Thus, for CF_4 (and *a fortiori* for CCl_4), the intermolecular potential energy must be largely the sum of interactions between fluorine atoms (or chlorine atoms). Even if one assumes that these atom–atom interactions follow a simple pair energy function, such as those discussed in preceding sections, the resultant molecular pair energy will be more complex. It will of course be at least mildly dependent upon orientation, but the primary difference arises from the short-range character of the atom–atom interactions. Suppose, for example, we take a simple pair energy—the square well with $\alpha = 3/2$—for atom–atom interactions in CCl_4. Infinite repulsion occurs whenever two chlorine atoms attempt to get closer than the sum of their "van der Waals radii," i.e., less than an *atomic* collision diameter $\sigma_{\mathrm{Cl-Cl}}$, while the attractive energy is zero whenever all interatomic distances exceed $3\sigma_{\mathrm{Cl-Cl}}/2$. Translated into an *intermolecular* pair energy, measured from the molecular centers (i.e., the carbon atoms), each of the distances must be increased by something like twice the C—Cl bond distance. Expressed in terms of a molecular collision diameter, $\sigma = 2d_{\mathrm{C-Cl}} + \sigma_{\mathrm{Cl-Cl}}$, the outer wall of the energy well at $2d_{\mathrm{C-Cl}} + [3\sigma_{\mathrm{Cl-Cl}}/2]$ lies well inside $3\sigma/2$, i.e., for these globular polyatomic molecules, the width of the energy well will be narrower relative to overall molecular dimensions than for monatomic substances. Similar conclusions are drawn when more refined atom–atom energy functions are used.

Two useful energy functions have been proposed as refinements of the Lennard-Jones (12, 6) pair energy (Eq. 2.24), as follows.

The appropriate sums over atomic (12, 6) energy functions yield (Hamann and Lambert, 1954) a smooth function of the intermolecular distance r (with only mild angular variation) but with steeper attractive and repulsive branches than those that correspond to the exponents 12 and 6 in Eq. (2.24). The repulsion (being largely due to the nearest atoms) is affected more strongly than the attraction, as evidenced by the choice by Hamann and Lambert of the exponents 28 and 7,

$$u(r) = (4^{4/3}\varepsilon/3)[(\sigma/r)^{28} - (\sigma/r)^{7}] = \varepsilon[\tfrac{1}{3}(r_0/r)^{28} - \tfrac{4}{3}(r_0/r)^{7}]. \quad (2.29)$$

An alternative pair function, also spherically symmetric, is provided by the Kihara equation with a spherical core. Kihara (1953) assumed a

(12, 6) energy dependent not upon the distance between molecules, but upon the shortest distance between the edges of "cores" inside the molecule;

$$u(r) = \varepsilon[(\varrho_0/\varrho)^{12} - 2(\varrho_0/\varrho)^6], \tag{2.30}$$

where ϱ is the shortest distance between the cores of appropriate shape and ϱ_0 is its value at the energy minimum. For spherical cores of radius a, $\varrho = r - 2a$, and Eq. (2.30) can be rewritten in terms of $R = r/\sigma$ or $R' = r/r_0$ as

$$\begin{aligned} u(r) &= 4\varepsilon\{[(1-\gamma)/(R-\gamma)]^{12} - [(1-\gamma)/(R-\gamma)]^6\} \\ &= \varepsilon\{[(1-\gamma)/(R'-\gamma)]^{12} - 2[(1-\gamma)/(R'-\gamma)]^6\}, \end{aligned} \tag{2.31}$$

where $\gamma = 2a/\sigma$. Obviously, larger values of γ will be associated with greater globularity. For large values of R, Eq. (2.31) reduces to

$$u(r) = -4\varepsilon[(1-\gamma)\sigma/r]^6, \qquad R \gg 1, \tag{2.32}$$

giving the right asymptotic bahavior [Eq. (2.1)] with $C_6 = 4\varepsilon(1-\gamma)^6\sigma^6$, as the (28, 7) equation does not. In the region of the bottom of the potential-energy well, which is of most importance for liquids, the reduced curvature $(r_0{}^2/\varepsilon)(d^2u/dr^2)_{r=r_0}$ is 196 for the (28, 7) equation and $72(r_0/\varrho_0)^2 = 72\{1 + [(\gamma/2^{1/6})/(1-\gamma)]\}^2$ for the Kihara equation. These equations for globular molecules will be effectively equivalent in this region if $a = 0.197r_0 = 0.211\sigma$ ($\gamma = 0.422$) (Hildebrand and Scott, 1962).

Actually, the Kihara spherical core energy has been used for a three-parameter fit to second virial coefficients of monatomic substances. Thus, for argon (see also Tables III and IV), values of γ ranging from 0.08 to 0.16 have been proposed (Sherwood and Prausnitz, 1964b; Barker *et al.*, 1964; Rossi and Danon, 1965; Weir *et al.*, 1967).

2. *Multipole Contributions*

For most simple polyatomic molecules which lack the high symmetry of the quasispherical structures, departures from a simple, spherically symmetric potential-energy function are usually attributed to the orienting effect of an asymmetric distribution of charge, i.e., to dipole moments μ, quadrupole moments θ, and higher multipole moments. General formulations of these multipole–multipole interactions may be found elsewhere (Hirschfelder *et al.* 1954; Buckingham, 1959; Rowlinson, 1969); we confine ourselves here primarily to what is the principal multipole contribution in most liquids, the dipole–dipole interaction.

The electrostatic interaction energy between two permanent dipoles with moments μ_1 and μ_2 is

$$u_{\text{dipoles}}(r, \theta_1, \theta_2, \phi_1 - \phi_2) = -(\mu_1\mu_2/r^3)[2\cos\theta_1\cos\theta_2 - \sin\theta_1\sin\theta_2\cos(\phi_1 - \phi_2)] \quad (2.33)$$

where (θ_1, θ_2) and (ϕ_1, ϕ_2) are the usual azimuthal and equatorial polar coordinates specifying the orientation of the dipoles with respect to a line between their centers. Equation (2.33) is exact for "point dipoles," the hypothetical limit of very large charges separated by a very small distance; for real molecules, the equation must include higher terms corresponding to interactions of higher moments (multipoles) of the charge distribution.

The dipole–dipole interaction is, of course, only part of the total pair energy, so Stockmayer (1941) suggested that the pair energy be represented by the sum of a Lennard-Jones (12, 6) function and the dipole–dipole function of Eq. (2.33),

$$u(r, \theta_1, \theta_2, \phi_1 - \phi_2) = 4\varepsilon[(\sigma/r)^{12} - (\sigma/r)^6] - (\mu^2/r^3)f(\theta_1, \theta_2, \phi_1 - \phi_2), \quad (2.34)$$

where $\mu = \mu_1 = \mu_2$ and $f(\theta_1, \theta_2, \phi_1 - \phi_2)$ is the bracketed angle function in Eq. (2.33). Extensive calculations of the second virial coefficient B and other properties for this Stockmayer pair function have been reported (see Hirschfelder *et al.* (1954)).

In addition to the interactions between the permanent multipole moments, there are interactions between the permanent moment of one molecule and the moment which it induces in another. Here, the energy of the point dipole–induced dipole interaction may be written

$$u_{\text{dipole–induced dipole}}(r, \theta_1, \theta_2) = -\mu^2\alpha(3\cos^2\theta_1 + 3\cos^2\theta_2 + 2)/2r^6 \quad (2.35)$$

where, for simplicity, the polarizability α is assumed to be spherically symmetric. Normally, this energy is small by comparison with that between permanent dipoles [Eq. 2.33)] and the dispersion energy [Eq. (2.2)].

The total energy of the system is obviously related to the sum of pair potential energies averaged over all orientations. If this averaging is performed for an *isolated* pair separated by a fixed distance r between molecular centers, one obtains, with appropriate Boltzmann factor weighting $e^{-u/kT}$ (Keesom, 1921, 1922), from Eqs. (2.33) and (2.35) an average $\bar{u}(r, T)$,

$$\bar{u}_{\text{dipoles}}(r, T) = -(2\mu^4/3r^6kT) - (\mu^2\alpha/r^6) - 0(1/r^{12}). \tag{2.36}$$

It is to be noted that the dipole–dipole and dipole–induced dipole interactions average to leading terms proportional to r^{-6}, the same as the leading term for the London dispersion energy. At short distances, other orientationally dependent energies may be important, and when more than two molecules are present, the averaging is more complicated; in extreme cases (e.g., in the highly structured liquid water), Eq. (2.36) is completely useless even as a qualitative guide.

Higher-order multipole interactions may be averaged similarly. Thus, the dipole–quadrupole interaction, proportional to $\mu\theta/r^4$ and a function of angles, averages to an energy proportional to $-\mu^2\theta^2/r^8kT$; the quadrupole–quadrupole interaction averages to an energy proportional to $-\theta^4/r^{10}kT$; etc. From the nature of Boltzmann statistics, the averaged electrostatic energy for a pair of neutral molecules is necessarily negative (attractive).

3. "*Shape*" *Effects*

Many molecules (e.g., propane, $CH_3CH_2CH_3$), while having no large multipole moments, nonetheless have decidedly asymmetric pair energies simply because the molecule has a very nonspherical shape. The distribution of the atoms is so asymmetric that the attraction and repulsion, even if similar to that for argon on an atom-to-atom basis, is, when summed over all atoms, strongly dependent upon orientation.

Several approximate methods for handling this problem have been suggested. The most widely used is the Kihara core modification, Eq. (2.30), of the (12, 6) pair energy, now adapted to nonspherical cores. Thus, for ethane (CH_3CH_3), the core might be a thin cylinder, a strongly prolate ellipsoid, or even the carbon–carbon axis as a volumeless straight line. For any particular orientation, the distance ϱ is calculated as the distance between the nearest points on the two cores. Once the size and shape of the core is specified, together with ε and ϱ_0, the energy function u is complete; if desired, it can be averaged over all orientations to yield a $\bar{u}(r, T)$, and thermodynamic functions for dilute gases (e.g., the second virial coefficient) can be calculated as for other pair energies, but with proper allowance[†] for any temperature dependence of $\bar{u}$.

An alternative proposal (Pople, 1954) for representing the shape of molecules is to make the repulsive r^{-12} energy orientation-dependent. If the most natural form for this (a simple ellipticity) is chosen, averaging over all angles to obtain $\bar{u}$ merely alters the magnitude of the (temperature-independent) coefficient of r^{-12}, although a term proportional to r^{-24} with a temperature-dependent coefficient is introduced.

In principle, one could represent the pair energy $u(r, \theta_1, \theta_2, \phi_1 - \phi_2)$ of axially symmetric molecules (complete specification of nonlinear molecules requires two more angles) by the expression

$$u(r, \theta_1, \theta_2, \phi_1 - \phi_2) = \varepsilon_0 f(r/\sigma_0, \theta_1, \theta_2, \phi_1 - \phi_2, \gamma, \mu^2/\varepsilon_0\sigma_0{}^3, \theta^2/\varepsilon_0\sigma_0{}^5, \mu^2\bar{\alpha}/\varepsilon_0\sigma_0{}^6, \varkappa, \ldots), \tag{2.37}$$

but the dimensionless molecular parameters do not represent a demonstrably complete set unless one arbitrarily restricts oneself to the coefficients in the classical multipole expansion. Here, ε_0 and σ_0 are the scaling parameters of a simple, spherically symmetric pair energy $u_0(r)$, γ is a correction for globularity [e.g., the γ of the Kihara equation (2.31)], μ and θ are the dipole moment and quadrupole moment, respectively, $\bar{\alpha}$ is the average polarizability $(\alpha_\parallel + 2\alpha_\perp)/3$, and $\varkappa$ is the departure from symmetry of the polarizability ellipsoid $(\alpha_\parallel - 2\alpha_\perp)/3\bar{\alpha}$ (where $\alpha_\parallel$ and $\alpha_\perp$ are the polarizability parallel to and perpendicular to the principal axis).

III. The Principle of Corresponding States

A. Assumptions and "Derivation"

The idea that the equation of state of a fluid and the related thermodynamic functions could be expressed in a universal form using "reduced" variables (i.e., the actual variables multiplied by appropriate dimensional scaling factors constant for each substance) is an old one, starting with van der Waals (1873), and is inherent in almost all simple theories. However, the first general argument from statistical mechanics seems to be that given by Pitzer (1939). The assumptions and argument have been

[†] See footnote following Eq. (4.5), p. 58.

rephrased by others (Guggenheim, 1945; Rowlinson, 1969), but the basic features are common to all current formulations.

In order to compare polyatomic substances with monatomic ones, it is necessary to eliminate the effect of the internal degrees of freedom—rotation, vibration, excited electronic states. Fortunately, for many substances, these internal motions do not appear to be affected significantly by the presence or absence of neighboring molecules, so it seems appropriate to assume the following:

1. *The internal degrees of freedom are independent of density*, and consequently, for a given temperature, the same in the real fluid, liquid or gas, as in the reference ideal gas. This is equivalent to writing the total energy E_l of the lth quantum state of the system of N molecules in volume V as separable into the sum of an external contribution $E_{ext}(V, N)$ from the motion of the centers of mass of the molecules and an internal contribution E_{int}, the latter simply being the sum of the energies of the internal quantum states of the individual molecules, $\sum_{k=1}^{N}(\varepsilon_{int})_k$,

$$E_l = E_{ext}(V, N) + \sum_{k=1}^{N} (\varepsilon_{int})_k. \tag{3.1}$$

It follows that the canonical partition function $Z(T, V, N)$ may be written as

$$Z(T, V, N) = \sum_l e^{-E_l/kT} = (q_{int})^N \sum e^{-E_{ext}(V,N)/kT}, \tag{3.2}$$

where $q_{int} = \sum_i e^{-(\varepsilon_{int})_i/kT}$, the sum over all the internal quantum states of a single (isolated) molecule, thus independent of N and V and their ratio, the number density N/V.

Normally, the spacing of the energies E_{ext} of the external quantum states (essentially those for highly modified translations or, equivalently, very anharmonic vibrations) will be very small in comparison with kT, so we may safely assume the following:

2. *The external motion* ("translational" degrees of freedom) *of the centers of mass of the molecules is essentially classical.* This permits us to replace the quantum energy E_{ext} by the Hamiltonian function $\mathscr{H}\{\mathbf{p}, \mathbf{r}\}$ (classical sum of kinetic and potential energies) and the quantum sum by the integral over all positions $\mathbf{r}$ and momenta $\mathbf{p}$ of the N molecules, with division by $h^{3N}N!$ (where h is the Planck constant) to maintain the correspondence between classical and quantum statistical mechanics. Normally, the po-

tential energy U depends only upon the position variables $\{\mathbf{r}\}$, so we can write

$$\mathscr{H}\{\mathbf{p}, \mathbf{r}\} = \sum_{k=1}^{N} (\mathbf{p}_k^2/2m) + U\{\mathbf{r}\}; \tag{3.3}$$

m is the molecular mass and $U\{\mathbf{r}\}$ is shorthand for $U(\mathbf{r}_1, \mathbf{r}_2, \ldots, \mathbf{r}_N)$. If we now integrate with respect to the $d\mathbf{p}$'s, we obtain

$$\begin{aligned}
&\sum \exp[-E_{\text{ext}}(V, N)/kT] \\
&\quad = (1/h^{3N}N!) \int\int \cdots \int \exp[-\mathscr{H}\{\mathbf{p}, \mathbf{r}\}/kT]\, d\mathbf{p}_1 \cdots d\mathbf{p}_N\, d\mathbf{r}_1 \cdots d\mathbf{r}_N \\
&\quad = (1/h^{3N}N!) \left[\int_{-\infty}^{+\infty} \exp(-p^2/2mkT)\, dp \right]^{3N} \\
&\qquad \times \int\int \cdots \int \exp[-U\{\mathbf{r}\}/kT]\, d\mathbf{r}_1 \cdots d\mathbf{r}_N \\
&\quad = (2\pi mkT/h^2)^{3N/2}(1/N!) \int\int \cdots \int \exp[-U\{\mathbf{r}\}/kT]\, d\mathbf{r}_1 \cdots d\mathbf{r}_N.
\end{aligned} \tag{3.4}$$

If we define $\lambda = h/(2\pi mkT)^{1/2}$ and a "configuration integral" Q as

$$Q(T, V, N) = (1/N!) \int\int \cdots \int \exp[-U\{\mathbf{r}\}/kT]\, d\mathbf{r}_1 \cdots d\mathbf{r}_N, \tag{3.5}$$

we can rewrite the canonical partition function† [Eq. (3.2)] as

$$Z(T, V, N) = (q_{\text{int}}/\lambda)^{3N} Q. \tag{3.6}$$

The transformation to thermodynamic functions is straightforward, since the Helmholtz free energy A is directly related to Z,

$$\begin{aligned}
A(T, V, N) &= -kT \ln Z(T, V, N) \\
&= -3NkT[\ln(q_{\text{int}}/\lambda)] - kT \ln Q.
\end{aligned} \tag{3.7}$$

The last term on the right hand side is universally called the "configurational (Helmholtz) free energy,"

$$A^{\text{conf}}(T, V, N) = -\, kT \ln Q. \tag{3.8}$$

† The reader should be warned that there is no uniformly accepted symbolism for the canonical partition function and the configuration integral. That used here conforms to that in Volume II of this series, but others use Z and Q in other ways. For example, Hill (1960) and the present author, in other works, have used Z_N for the configuration integral and Q_N for the canonical partition function. Some define Q as excluding the factor $1/N!$

For completely independent molecules (i.e., those of an ideal gas), $U\{\mathbf{r}\} = 0$ and $Q = V^N/N!$,

$$A_{\text{id}}^{\text{conf}}(T, V, N) = -kT\ln(V^N/N!) = -NkT\ln(eV/N). \tag{3.9}$$

Following Rowlinson (1959), we define a "residual" property† (such as A^*, E^*, etc.) as the difference between the property of the real fluid and that of the ideal gas (subscript "id") in the same *volume* V and at the same T and N.

Thus, for fluids satisfying our assumptions 1 and 2,

$$\begin{aligned} A^*(T, V, N) &= A(T, V, N) - A_{\text{id}}(T, V, N) \\ &= A^{\text{conf}}(T, V, N) - A_{\text{id}}^{\text{conf}}(T, V, N) \\ &= A^{\text{conf}}(T, V, N) + NkT\ln(eV/N) \\ &= -kT\ln(QN!/V^N). \end{aligned} \tag{3.10}$$

The third assumption is usually that "the intermolecular potential energy is a function of the various intermolecular distances only" (Pitzer, 1939), but this seems to be implied by our first assumption and is certainly included in our formulation of Eq. (3.3), where U is a function of the $\mathbf{r}$'s only, since for all fluids, the choice of an origin from which $\mathbf{r}$ is measured is completely arbitrary. Our third assumption is then a statement of the universal form of U when expressed in a reduced form.

3. *The total (external) potential energy of a fluid may be expressed as the product of an energy parameter ε and a function of scaled (reduced) distances of separation r/σ between molecular centers, the same function for all conformal substances.*

$$\begin{aligned} U\{\mathbf{r}\} &= U(\mathbf{r}_1, \mathbf{r}_2, \mathbf{r}_3, \ldots, \mathbf{r}_N) \\ &= \varepsilon\Phi(r_{12}/\sigma, r_{13}/\sigma, r_{23}/\sigma, \ldots) = \varepsilon\Phi\{\mathbf{r}/\sigma\}. \end{aligned} \tag{3.11}$$

Until recently, this assumption was usually expressed in a more restrictive form, requiring an assumption of pairwise additivity:

3a. The total (external) potential energy of a fluid is the sum of the potential energies of the individual molecular pairs [Eq. (2.4)].

3b. The potential energy of a pair of molecules may be written $u(r) = \varepsilon\phi(r/\sigma)$, where ϕ is the same function for all conformal substances.

† These residual quantities A^*, S^*, E^*, etc. are identical with those which Hirschfelder *et al.* (1954) call the thermodynamic functions "of gas imperfection," A', S', etc.

These assumptions lead to a more restrictive pairwise additive form for $U\{\mathbf{r}\}$,

$$U\{\mathbf{r}\} = \varepsilon \sum_{i<j}\sum \phi(r_{ij}/\sigma). \tag{3.12}$$

For any pair function with only two adjustable constants, the parameters ε and σ can be defined and the function $\phi(r/\sigma)$ determined in principle, as we have seen in the preceding section; however, to obtain the principle of corresponding states, it is not necessary to specify the form of $\phi(r/\sigma)$.

It is not even necessary to assume pairwise additivity, although the assumption of the more general form of Eq. (3.11) imposes rather severe restrictions upon the parameters of the triplet energy, conditions which Barker *et al.* (1968) have discussed.

We may now formulate the configuration integral Q in the light of assumption 3;

$$\begin{aligned} Q(T, V, N) &= \frac{1}{N!}\int\int\cdots\int \exp\left[\frac{-U(\mathbf{r}_1, \mathbf{r}_2, \ldots, \mathbf{r}_N)}{kT}\right] d\mathbf{r}_1\, d\mathbf{r}_2 \cdots d\mathbf{r}_N \\ &= \frac{1}{N!}\int\int\cdots\int \exp\left[-\frac{\varepsilon}{kT}\,\Phi\left(\frac{r_{12}}{\sigma}, \frac{r_{13}}{\sigma}, \frac{r_{23}}{\sigma}, \ldots\right)\right] \\ &\quad\times d\mathbf{r}_1\, d\mathbf{r}_2 \cdots d\mathbf{r}_N \\ &= \frac{\sigma^{3N}}{N!}\int\int\cdots\int \exp\left[-\frac{\varepsilon}{kT}\,\Phi\left(\frac{r_{12}}{\sigma}, \frac{r_{13}}{\sigma}, \frac{r_{23}}{\sigma}, \ldots\right)\right] \\ &\quad\times \frac{d\mathbf{r}_1}{\sigma^3}\frac{d\mathbf{r}_2}{\sigma^3}\cdots\frac{d\mathbf{r}_N}{\sigma^3} = \frac{\sigma^{3N}}{N!}F\left(\frac{kT}{\varepsilon}, \frac{V}{\sigma^3}, N\right), \end{aligned} \tag{3.13}$$

where the function F indicates the functional dependence of the integral. Substitution into Eq. (3.10) yields

$$A^*(T, V, N) = -kT\ln[(\sigma^3/V)^N F(kT/\varepsilon, V/\sigma^3, N)]. \tag{3.14}$$

Thermodynamically, the Helmholtz free energy is an extensive property and must be proportional to N when T and V/N are held constant. If the equivalence of thermodynamic and statistical-mechanical results is to be maintained, the argument of the logarithm in Eq. (3.14) must be of the form

$$[(\sigma^3/V)^N F(kT/\varepsilon, V/\sigma^3, N)] = [\psi(kT/\varepsilon, V/N\sigma^3)]^N, \tag{3.15}$$

so that Eq. (3.14) becomes

$$A^*(T, V, N) = -NkT \ln \psi(kT/\varepsilon, V/N\sigma^3), \tag{3.16}$$

or, for the molar Helmholtz free energy $\tilde{A}^*$, which is a function of the temperature T and the volume per molecule V/N (or equivalently, the molar volume $\tilde{V} = \tilde{N}V/N$),

$$\tilde{A}^*(T, \tilde{V}) = -RT \ln \psi(kT/\varepsilon, \tilde{V}/\tilde{N}\sigma^3). \tag{3.17}$$

From Eq. (3.17), all the applications of the principle of corresponding states can be developed. In what follows, we shall restrict ourselves to the intensive molar quantities. There seems no advantage of maintaining the generality of Eq. (3.14); while the functional dependence of A^* upon T and V cannot be expressed explicitly, the dependence upon N is clear, as indicated in Eq. (3.15).

B. Equivalent Formulations of the Principle

Equation (3.17) expresses the principle of corresponding states in terms of a relation between molar residual Helmholtz free energy $\tilde{A}^*(T, \tilde{V})$ and reduced variables kT/ε and $\tilde{V}/\tilde{N}\sigma^3$. It is useful, however, to relate all the thermodynamic properties of a given fluid to those of a reference substance with parameters ε_0 and σ_0. Then, for another conformal substance with different ε_1 and σ_1, we may define ratios $f = \varepsilon_1/\varepsilon_0$ and $h = \sigma_1^3/\sigma_0^3$ and rewrite Eq. (3.17) as

$$\tilde{A}_1^*(T, \tilde{V}) = f\tilde{A}_0^*(T/f, \tilde{V}/h), \tag{3.18}$$

where the molar residual Helmholtz free energy $\tilde{A}_0^*$ of the reference substance is to be obtained at a different temperature T/f and a different molar volume $\tilde{V}/h$.

An even more elegant form of Eq. (3.18) is obtained by using the Massieu function $J = -A/T$. Then the prefactor f disappears and we have

$$\tilde{J}_1^*(T, \tilde{V}) = \tilde{J}_0^*(T/f, \tilde{V}/h). \tag{3.19}$$

Although the residual functions are more compact and more directly related to experimental quantities, most theoretical discussions of fluids and of the principle of corresponding states have tended to emphasize the slightly different configurational functions like A^{conf}. Equations (3.10) and (3.17) lead us to the configurational analog of Eq. (3.18):

$$\tilde{A}_1^{\mathrm{conf}}(T, \tilde{V}) = f\tilde{A}_0^{\mathrm{conf}}(T/f, \tilde{V}/h) - RT \ln h. \tag{3.20}$$

Experimentally, the more useful thermodynamic functions are those appropriate to processes at constant temperature and pressure, e.g., $\tilde{G}$, $\tilde{H}$, $\tilde{C}_p$, etc. This might tempt one to define pressure residual functions $\tilde{X}^{**}(T, p)$ as the difference between the property $\tilde{X}$ of the real fluid and that of the ideal gas at the same temperature and *pressure*. However, since most studies of liquids are carried out at or near zero pressure, where $-\tilde{G}_{\mathrm{id}}$ and $\tilde{S}_{\mathrm{id}}$ are approaching infinity, $\tilde{G}^{**}$ and $\tilde{S}^{**}$ are not convenient; the other pressure residuals ($\tilde{H}^{**}$, $\tilde{C}_p^{**}$, etc.) are the same as the ordinary residual functions ($\tilde{H}^*$, $\tilde{C}_p^*$, etc.). It is nonetheless useful to consider the residual functions $\tilde{G}^*$, $\tilde{S}^*$, etc. as functions of T and p, but the reference state is still the ideal gas in the same *volume*, although this can be expressed as an "ideal pressure" $p_{\mathrm{id}} = RT/\tilde{V}(T, p)$,

$$\tilde{G}^*(T, p) = \tilde{G}(T, p) - \tilde{G}_{\mathrm{id}}(T, p_{\mathrm{id}}) = \tilde{A}^*(T, \tilde{V}) + p^*/\tilde{V} \tag{3.21}$$

or, expressing $\tilde{G}_1^*(T, p)$ in terms of $\tilde{G}_0^*$ for a reference substance,

$$\tilde{G}_1^*(T, p) = f\tilde{G}_0^*(T/f, ph/f). \tag{3.22}$$

By appropriate differentiation and combination, all the other thermodynamic functions can be obtained from Eqs. (3.18), (3.19), or (3.22), but these derivations must be made with some care. For example, one way of obtaining the residual pressure $p^* = p(T, \tilde{V}) - p_{\mathrm{id}}(T, \tilde{V})$ is by differentiation of $\tilde{J}^*$:

$$\begin{aligned} p_1^*/T &= (\partial \tilde{J}_1^*/\partial \tilde{V}_1)_T = (\partial \tilde{J}_0^*/\partial \tilde{V}_1)_T = (\partial \tilde{J}_0^*/\partial \tilde{V}_0)_T(\partial \tilde{V}_0/\partial \tilde{V}_1)_T \\ &= [p_0^*(T/f, \tilde{V}/h)/T_0](1/h) = p_0^*(T/f, \tilde{V}/h)/(T/f)h \end{aligned}$$

or

$$p_1^* = (f/h)p_0^*(T/f, \tilde{V}/h), \tag{3.23}$$

or, expressed symmetrically, in terms of the actual pressure,

$$\phi_1(p, \tilde{V}, T) = \phi_0(ph/f, \tilde{V}/h, T/f) = 0. \tag{3.24}$$

Because the relations between the residual and configurational thermodynamic functions, while straightforward, are not always obvious on inspection, they are summarized in Table V. It should be noted that,

TABLE V

ALTERNATIVE THERMODYNAMIC FUNCTIONS[a]

"Residual"	"Configurational"
$\tilde{A}^*(T, \tilde{V})$	$\tilde{A}^{\text{conf}}(T, \tilde{V}) = \tilde{A}^* - RT\ln(e\tilde{V}/\tilde{N})$
$\tilde{S}^*(T, \tilde{V})$	$\tilde{S}^{\text{conf}}(T, \tilde{V}) = \tilde{S}^* + R\ln(e\tilde{V}/\tilde{N})$
$\tilde{E}^*(T, \tilde{V})$	$\tilde{E}^{\text{conf}}(T, \tilde{V}) = \tilde{E}^*$
$p^*(T, \tilde{V}) = p - (RT/\tilde{V})$	$p^{\text{conf}}(T, \tilde{V}) = p = p^* + (RT/\tilde{V})$
$\tilde{G}^* = \tilde{A}^* + p^*\tilde{V}$ $= \tilde{A}^* + p\tilde{V} - RT$	$\tilde{G}^{\text{conf}} = \tilde{A}^{\text{conf}} + p^{\text{conf}}\tilde{V}$ $= \tilde{G}^* + RT - RT\ln(e\tilde{V}/\tilde{N})$
$\tilde{H}^* = \tilde{E}^* + p^*\tilde{V}$ $= \tilde{E}^* + p\tilde{V} - RT$	$\tilde{H}^{\text{conf}} = \tilde{E}^{\text{conf}} + p^{\text{conf}}\tilde{V}$ $= \tilde{H}^* + RT$
$\tilde{C}_V^*(T, \tilde{V})$	$\tilde{C}_V^{\text{conf}} = \tilde{C}_V^*$
$\tilde{C}_p^* = \tilde{C}_V^* + (T\tilde{V}\alpha^2/\varkappa) - R$	$\tilde{C}_p^{\text{conf}} = \tilde{C}_V^{\text{conf}} + (T\tilde{V}\alpha^2/\varkappa)$ $= \tilde{C}_p^* + R$

[a] Here, $\alpha = (\partial \ln \tilde{V}/\partial T)_p$, the thermal expansivity, and $\varkappa = -(\partial \ln \tilde{V}/\partial p)_T$, the isothermal compressibility. No functional dependence is shown for $\tilde{G}^*$, $\tilde{H}^*$, etc. since $\tilde{G}^*(T, p) = \tilde{G}^*(T, \tilde{V})$, $\tilde{S}^*(T, p) = \tilde{S}^*(T, \tilde{V})$, as long as it is understood that p and $\tilde{V}$ refer to the same state of the fluid.

within each set, all the usual thermodynamic equations continue to apply (e.g., $\tilde{S}^* = -(\partial\tilde{A}^*/\partial T)_V$, $\tilde{G}^{\text{conf}} = \tilde{A}^{\text{conf}} + p^{\text{conf}}\tilde{V}$, etc.).

Still another way of expressing the principle of corresponding states is by expanding Eqs. (3.18) and (3.22) as Taylor series in powers of $(f - 1)$ and $(h - 1)$;

$$\begin{aligned}\tilde{A}^*(T, \tilde{V}) = {} & \tilde{A}_0^*(T, \tilde{V}) + \tilde{A}_f(f-1) + \tilde{A}_h(h-1) \\ & + \tfrac{1}{2}\tilde{A}_{ff}(f-1)^2 + \tilde{A}_{fh}(f-1)(h-1) \\ & + \tfrac{1}{2}\tilde{A}_{hh}(h-1)^2 + \cdots \end{aligned} \tag{3.25}$$

$$\begin{aligned}\tilde{G}^*(T, p) = {} & \tilde{G}_0^*(T, p) + \tilde{G}_f(f-1) + \tilde{G}_h(h-1) \\ & + \tfrac{1}{2}\tilde{G}_{ff}(f-1)^2 + \tilde{G}_{fh}(f-1)(h-1) \\ & + \tfrac{1}{2}\tilde{G}_{hh}(h-1)^2 + \cdots . \end{aligned} \tag{3.26}$$

This approach has proved especially useful in applications of the principle to mixtures. Table VI summarizes these coefficients as thermodynamic properties of the reference substance.

TABLE VI

COEFFICIENTS OF TAYLOR SERIES[a]

Equation (3.25)	Equation (3.26)
$\tilde{A}_f = \tilde{E}_0{}^*$	$\tilde{G}_f = \tilde{H}_0{}^* - p\tilde{V}_0{}^* = \tilde{E}_0{}^* = \tilde{A}_f$
$\tilde{A}_h = p_0{}^*\tilde{V} = p_0\tilde{V} - RT$	$\tilde{G}_h = p\tilde{V}_0{}^* = p_0\tilde{V} - RT = \tilde{A}_h$
$\tilde{A}_{ff} = -T\tilde{C}^*_{V0}$	$\tilde{G}_{ff} = -T\tilde{C}^*_{p0} + 2pT\tilde{V}_0\alpha_0 - p^2\tilde{V}_0\varkappa_0$ $= \tilde{A}_{ff} + RT - [\tilde{V}_0(\alpha_0 T - \varkappa_0 p)^2/\varkappa_0]$
$\tilde{A}_{fh} = p_0\tilde{V} - (T\tilde{V}\alpha_0/\varkappa_0)$	$\tilde{G}_{fh} = -pT\tilde{V}_0\alpha_0 + p^2\tilde{V}_0\varkappa_0$ $= \tilde{A}_{fh} + [\tilde{V}_0(\alpha_0 T - \varkappa_0 p)(1 - \varkappa_0 p)/\varkappa_0]$
$\tilde{A}_{hh} = -2p_0\tilde{V} + RT + (\tilde{V}/\varkappa_0)$	$\tilde{G}_{hh} = RT - p^2\tilde{V}_0\varkappa_0$ $= \tilde{A}_{hh} - [\tilde{V}_0(1 - \varkappa_0 p)^2/\varkappa_0]$

[a] It should be noted that expansions of $\tilde{A}^{\text{conf}}(T, V)$ and $\tilde{G}^{\text{conf}}(T, p)$ yield the same coefficients as those given for $\tilde{A}^*$ and $\tilde{G}^*$; only the initial terms ($\tilde{A}_0^{\text{conf}}$ and $\tilde{G}_0^{\text{conf}}$) are different.

C. EXPERIMENTAL PROPERTIES OF "SIMPLE FLUIDS"

Pitzer (1955) has suggested the name "simple fluid" for those substances (Ar, Kr, Xe, and CH_4) for which the assumptions which lead to the principle of corresponding states should be at least approximately valid. As a matter of fact, the experimental properties of these substances, when expressed in appropriate reduced form, do correspond very closely, as do those of a few slightly less simple substances (e.g., N_2, CO, O_2). As we shall see in the next section, the properties of many other fluids can be correlated in terms of small deviations from the principle of corresponding states, so it is important to have the properties of the reference simple liquids expressed explicitly in the form of graphs or tables or, where possible, analytic equations. Moreover, it is these properties which any truly satisfactory theory of liquids must account for in detail.

The material presented below does not represent any new critical review by the present author but is rather a synthesis of previously published analyses of experimental data (Pitzer *et al.*, 1955; Pitzer and Curl, 1957; Guggenheim, 1966; Rowlinson, 1969).

1. *The Equation of State*

We have seen [Eq. (3.23)] how an appropriately scaled pressure must be a universal function of a reduced volume and temperature, or

$$\phi(p\sigma^3/\varepsilon, \tilde{V}/\tilde{N}\sigma^3, kT/\varepsilon) = 0. \tag{3.27}$$

However, since the parameters ε and σ are not directly accessible (and, until the uncertainties about the pair energy are resolved, not even unambiguously related to any experimental measurements), we follow previous authors (starting with van der Waals) in using the pressure, molar volume, and temperature at the gas–liquid critical point as reducing parameters. Since the critical point is uniquely defined on the conformal $(p, \tilde{V}/\tilde{N}, T)$ surface by the conditions $(\partial p/\partial \tilde{V})_T = 0$ and $(\partial^2 p/\partial \tilde{V}^2)_T = 0$, it follows that the dimensionless ratios $p_c\sigma^3/\varepsilon$, $\tilde{V}_c/\tilde{N}\sigma^3$, and kT_c/ε should be the same set of numbers for all simple fluids. Hence, Eq. (3.27) can be replaced by

$$\phi(p/p_c, \tilde{V}/\tilde{V}_c, T/T_c) = 0. \tag{3.28}$$

At the critical point, the dimensionless compression factor $p_c\tilde{V}_c/RT_c$ should have the same value for all simple fluids, a prediction confirmed by experiment (see Table VII), which yields 0.291 for all the simple fluids, virtually within experimental error.

TABLE VII

Corresponding States of Fluids: The Equation of State

	T_c (°K)	p_c (atm)	$\tilde{V}_c$ (cm³ mole⁻¹)	$p_c\tilde{V}_c/RT_c$	T_B (°K)	T_B/T_c
Ne	44.5	26.9	41.7	0.308	122	2.74
Ar	150.7	48.4	74.6	0.291	411	2.73
Kr	209.4	54.3	92.2	0.291	575	2.74
Xe	289.8	58.0	118.8	0.290	768	2.65
Rn	377.5	62.4	140	0.28	—	—
N_2	126.0	33.5	90.2	0.292	327	2.59
CO	133.0	34.5	93.2	0.294	~345	2.6
O_2	154.3	49.7	74.5	0.292	406	2.63
CH_4	190.7	45.8	98.8	0.289	510	2.67
CF_4	227.7	37.2	147	0.292	518	2.28

Just as all equation-of-state data can be expressed in various ways (pressure–volume isotherms, pressure–temperature isochores, compression factors, etc.), so also can the reduced equation of state be expressed variously in terms of the appropriate reduced variables. Pitzer *et al.* (1955) give extensive tables of the compression factor $p\tilde{V}/RT$, itself a reduced quantity, for a large number of values of p/p_c and T/T_c. No simple analytic expression has been found adequate to express the reduced equation of state over the entire range of density and temperature appropriate to the fluid state.

In the dilute-gas region, the principle of corresponding states may be tested by examining the reduced virial coefficients $B/\tilde{V}_c = f_2(T/T_c)$,

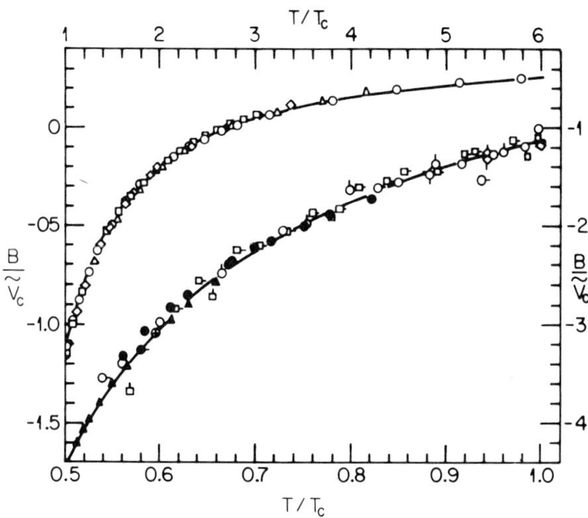

FIG. 9. The reduced second virial coefficient $B/\tilde{V}_c$ plotted against the reduced temperature T/T_c for four gases: (○) Ar; (△) Kr; (◇) Xe; (□) CH_4. The scales for the upper curve (for $T > T_c$) are at the top and on the left; the scales for the lower curve (for $T < T_c$) are at the bottom and on the right. The solid curves represent an empirical formula derived from the square-well, Eq. (2.20), with $\alpha = 1.61$: $B/\tilde{V}_c = 0.440 + 1.40[1 - \exp(0.75T_c/T)]$. (From Guggenheim, 1966.)

$C/\tilde{V}_c^2 = f_3(T/T_c)$, etc. For any pair energy function $u(r)$ with only two adjustable parameters (e.g., ε and σ, as in the fourth assumption of Section III, A), this is directly equivalent to the problem of $B(T)$ already discussed in Section II. Figure 9 (Guggenheim, 1966) shows $B/\tilde{V}_c$ for eight simple fluids; the curve represents a fit to the square-well function [Eq. (2.20)] with $\alpha = 1.61$.

Also shown in Table VII is the ratio of the Boyle temperature T_B to the critical temperature T_c. The constancy of this ratio is a measure of the adequacy of the principle of corresponding states to fit the situation of the dilute gas *and* that of the fairly dense gas at the critical point. The substance CF_4 is included for comparison; it does not satisfy this condition and cannot be regarded as a simple fluid. For the others, the ratio T_B/T_c is close to 2.7.

2. *Vapor–Liquid Equilibrium*

The properties of coexistent liquid and gas are even more sensitive tests of the principle of corresponding states. If the logarithm of the reduced vapor pressure (p/p_c) is plotted against the reciprocal of the reduced temperature (T_c/T), a very nearly straight line is obtained (Fig. 10)†

$$\ln(p/p_c) = 5.365[1 - (T_c/T)]. \tag{3.29}$$

Figure 11 shows (Rowlinson, 1959) the deviation of $\ln(p/p_c)$ for Ar, Kr, and Xe from that predicted by Eq. (3.29).

The molar entropy of vaporization $\Delta\tilde{S}^V = \Delta\tilde{H}^V/T$ is itself a reduced quantity, a function only of the reduced temperature T/T_c. It may be obtained from calorimetric measurements of the heat of vaporization $\Delta\tilde{H}^V$ or from the vapor-pressure curve with appropriate corrections for deviations from the ideal gas law (which are still not entirely negligible at the triple point of simple fluids). At the triple point $(T_t/T_c = 0.555)$, $\Delta\tilde{S}^V/R = 9.49$.

Still another property of the conjugate phase at equilibrium is the molar volume or its reciprocal, the concentration or density. Guggenheim (1945)

† Pitzer (1939) obtained from the vapor pressure of krypton the more nearly exact reduced equation

$$\ln(p/p_c) = -9.897(T_c/T) - 12.554\ln(T/T_c) + 1.454 + 8.443(T/T_c).$$

This can be reorganized to a form

$$\begin{aligned}\ln(p/p_c) = {} & 5.786[1 - (T_c/T)] + 12.554[\ln(T_c/T) - (T_c/T) + 1] \\ & + [8.443(T_c - T)^2/TT_c],\end{aligned}$$

which shows the relationship to Eq. (3.29). The small second and third terms on the right-hand side introduce a slight *S*-shaped curvature into the nearly straight line of Fig. 10, the curvature shown in Fig. 11.

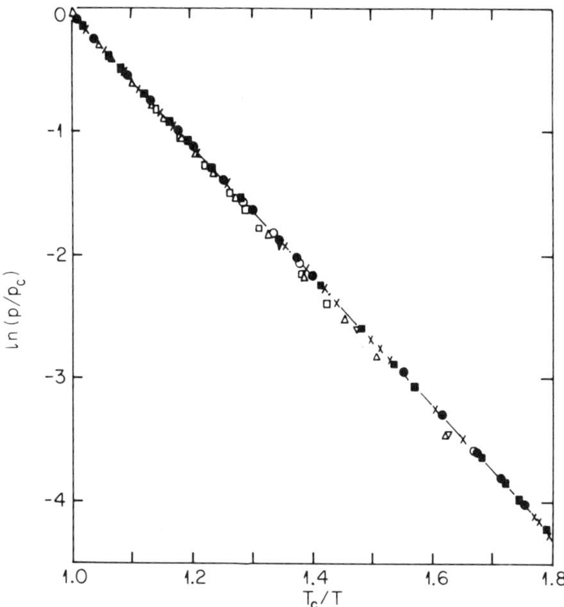

FIG. 10. Logarithm of reduced saturated vapor pressure $\ln(p/p_c)$ plotted against reciprocal reduced temperature T_c/T for seven liquids: (●) Ar, (▲) Kr, (×) Xe, (△) N_2, (▽) O_2, (□) CO, (○) CH_4. (From Guggenheim, 1966.)

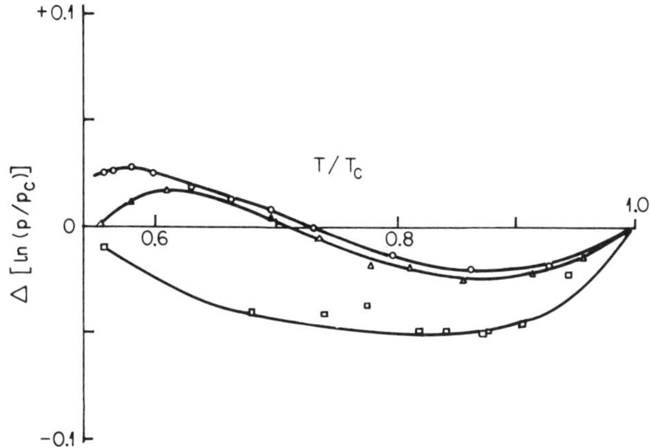

FIG. 11. The deviations of the logarithms of the reduced saturated vapor pressure from Eq. (3.29), $\Delta[\ln(p/p_c)]$ as a function of T/T_c. Liquids: (○) Ar; (△) Kr; (□) Xe. (From Rowlinson, 1959.)

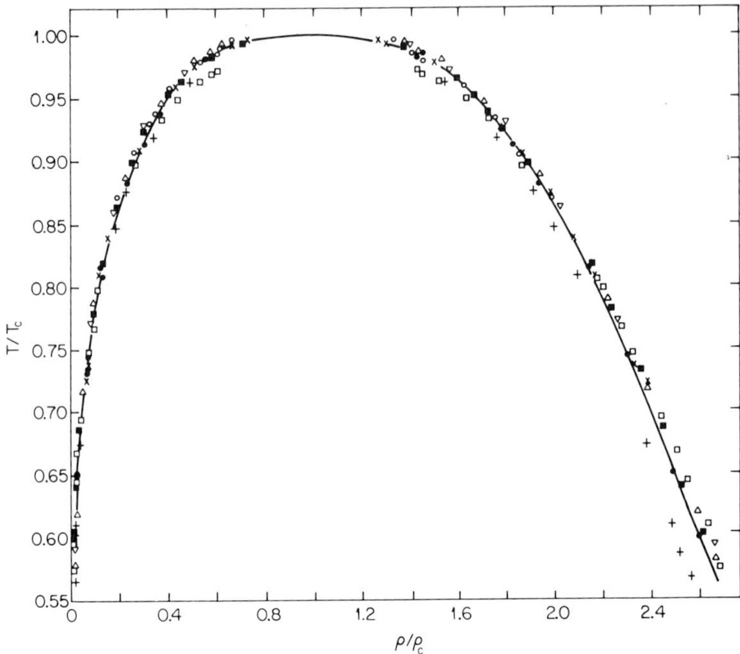

FIG. 12. Reduced temperature T/T_c plotted against reduced density ϱ/ϱ_c for coexisting vapor and liquid phases for eight fluids: (+) Ne, (●) Ar, (■) Kr, (×) Xe, (△) N_2, (▽) O_2, (□) CO, (○) CH_4. (From Guggenheim, 1966).

has shown that these densities are well represented by the equations†

$$\varrho_g/\varrho_c = \tilde{V}_c/\tilde{V}_g = 1 - 1.75[1 - (T/T_c)]^{1/3} + 0.75[1 - (T/T_c)], \quad (3.30)$$

$$\varrho_l/\varrho_c = \tilde{V}_c/\tilde{V}_l = 1 + 1.75[1 - (T/T_c)]^{1/3} + 0.75[1 - (T/T_c)]. \quad (3.31)$$

Combination of Eqs. (3.30) and (3.31) yields an equation for $\varrho_l + \varrho_g$ which is linear in T, in accord with the "law of the rectilinear diameter," while $\varrho_l - \varrho_g$ is proportional to $(T_c - T)^{1/3}$, in accord with current theories of critical phenomena (Kadanoff *et al.*, 1967; Fisher, 1967; Heller, 1967) (see Fig. 12).

Finally, from dimensional considerations, or by applying the assump-

† Equation (3.31) is remarkably successful in representing the liquid density. As the temperature is lowered, Eq. (3.30) correctly predicts a rapid decrease in the density of the saturated vapor, but, as ϱ_g becomes smaller and smaller, the percentage inaccuracy increases until, for $T/T_c < 0.564$, the predicted ϱ_g is actually negative. The percentage accuracy of $\varrho_l + \varrho_g$ and $\varrho_l - \varrho_g$ is always good.

tions of simple fluids to the liquid–vapor interface, one obtains a relation for the surface tension γ:

$$\gamma(\tilde{V}/\tilde{N})^{2/3}/kT = f(T/T_c). \tag{3.32}$$

Empirically, one finds

$$\gamma = \gamma_0[1 - (T/T_c)]^{11/9}, \tag{3.33}$$

where γ_0 is a constant which might be regarded as the surface tension of the (hypothetical) liquid in the limit $T \to 0$. It follows from Eqs. (3.32) and (3.33) that $\gamma_0(\tilde{V}_c/\tilde{N})^{2/3}/kT = \gamma_0 \tilde{V}_c^{2/3} \tilde{N}^{1/3}/RT_c$ should be the same dimensionless constant for all simple fluids; its value is about 4.5.

Table VIII summarizes these properties for a typical simple fluid.

TABLE VIII

VAPOR–LIQUID EQUILIBRIUM FOR SIMPLE FLUIDS

	T/T_c	p/p_c	$\Delta\tilde{S}^V/R$	ϱ_l/ϱ_c	ϱ_g/ϱ_c	γ/γ_0
Triple point	0.555	0.014	9.49	2.66	0.0079	0.37
	0.60	0.0294	8.52	2.60	0.0143	0.33
	0.65	0.0560	7.55	2.51	0.027	0.27
	0.70	0.1000	6.64	2.41	0.046	0.23
	0.75	0.165	5.80	2.30	0.075	—
	0.80	0.257	5.02	2.19	0.115	0.14
	0.85	0.380	4.28	2.05	0.175	—
	0.90	0.537	3.50	1.88	0.26	0.06
	0.95	0.742	2.52	1.67	0.40	—
Critical point	1.00	1.000	0.00	1.00	1.00	0.00

3. *Thermal Properties of the Liquid*

For simple fluids, the intramolecular vibrations and rotations, if any, must be the same as in the gas; therefore, only the intermolecular degrees of freedom concern us. At low temperatures (the normal liquid range), the molar heat capacity at constant pressure $\tilde{C}_p$ is about $5.05R$ for Ar, Kr, and Xe; since the corresponding value for the gas is $2.50R$, the residual molar heat capacity $(\tilde{C}_p{}^*) = 2.55R$. Values of $(\tilde{C}_p{}^*)$ at various tem-

TABLE IX

THERMAL PROPERTIES OF SIMPLE FLUIDS (LIQUID STATE AT SATURATION)

	T/T_c	$\tilde{C}_p^*/R$	$\alpha_p T_c$	$\gamma_V T_c/p_c$
Triple point	0.555	2.54	0.66	66
	0.60	2.59	0.69	58
	0.65	2.70	0.75	50
	0.70	2.91	0.83	44
	0.75	3.3	0.93	39
	0.80	3.8	1.20	33
	0.90	6.1	—	21
	1.00	∞	∞	6

peratures are shown in Table IX, together with values of the reduced thermal expansivity $\alpha_p T_c = T_c(\partial \ln \tilde{V}/\partial T)_p$, and the reduced thermal pressure coefficient $\gamma_V T_c/p_c = T_c(\partial p/\partial T)_V/p_c$. The reduced isothermal compressibility $\varkappa_T p_c = -p_c(\partial \ln \tilde{V}/\partial p)_T = \alpha_p T_c/(\gamma_V T_c/p_c)$ may be calculated from these. [It should be noted that, although these quantities are tabulated over a series of temperatures, they do not yield (without the use of additional information) temperature derivatives at either constant pressure or constant volume, since the properties are given along the saturation line, where both p and $\tilde{V}$ are changing with temperature.]

4. *Melting Properties*

The assumption that the internal degrees of freedom are the same as in the ideal gas seems inherently unlikely for a crystalline solid, so one would expect the principle of corresponding states to be applicable to the solid phase for monatomic substances only. The triple point, like the critical point, is a corresponding point, so Table X shows comparisons for the triple-point temperature T_t and pressure p_t, the molar entropy of fusion $\Delta\tilde{S}^F$, and the molar volume of fusion $\Delta\tilde{V}^F$. The data for N_2 and CH_4, also shown, demonstrate the inadequacy of the principle for these polyatomic solids, even though CH_4 is considered to be rotating in the crystal.

As seen in Tables VII and X and in Figs. 10–13, the three rare gases Ar, Kr, and Xe have reduced properties which are almost exactly coincident; as we shall see in the next section, the deviations for Ne are undoubtedly due primarily to quantum effects. Although there is good

TABLE X

CORRESPONDING STATES AND MELTING[a]

	T_t (°K)	T_t/T_c	p_t (atm)	$100p_t/p_c$	$\Delta\tilde{S}_F/R$	$\Delta\tilde{V}_F$ (cm³ mole⁻¹)	$100\Delta\tilde{V}_F/\tilde{V}_c$
Ne	24.56	0.549	0.428	1.58	1.64	—	—
Ar	83.76	0.556	0.679	1.41	1.69	3.52	4.67
Kr	115.94	0.555	0.721	1.33	1.69	4.49	4.86
Xe	161.36	0.558	0.805	1.39	1.71	5.59	4.71
Rn	202	0.54	0.646	1.04	1.6	—	—
N_2	63.15	0.502	0.124	0.37	1.37	2.01	2.22
CH_4	90.66	0.477	0.115	0.25	1.25	2.69	2.73

[a] T_c, p_c, and $\tilde{V}_c$ given in Table VII.

reason to believe that the potential energy function is not exactly conformal [i.e., does not satisfy Eq. (3.11)] for these fluids, it is not clear how much of the small variation in the observed reduced properties may be due to this lack of conformality, and how much may be due to experimental error.

IV. Deviations from the Principle of Corresponding States

Since most substances do not satisfy very closely the molecular assumptions necessary to "derive" the principle of corresponding states, it is necessary to consider the various sources of deviations, together with theoretical and semiempirical attempts to handle these problems.

A. Sources of the Deviations

One or more of the three assumptions which led to Eq. (3.17) may fail for a particular substance, so it is useful to consider them seriatim.

1. *Changes in Internal Degrees of Freedom*

It is inconceivable that the partition function for a dense fluid would be exactly separable into external factors ("translation") and internal factors ("rotation" and "intramolecular vibration"), with the latter completely independent of density. Hence, we can expect the first assumption to be strictly valid only for monatomic substances with no internal degrees of freedom whatever.

However, for many polyatomic substances, a high degree of almost free rotation may exist, and it is a well-established experimental fact that many vibration frequencies are shifted only slightly in the change from gas to liquid. Pitzer (1939) has discussed the qualitative changes in thermodynamic properties that are expected from restriction of rotation (decrease in liquid entropy) and changes in vibration frequencies (on the whole, decreased, producing increased entropy and enthalpy), but more recent work has tended to concentrate upon changes in the intermolecular pair energy.

The introduction of an angle-dependent perturbation to the pair energy allows, properly in principle, for *classical* restricted rotation and, although this is not theoretically appropriate for intramolecular motions, it may produce empirically similar effects upon the thermodynamic functions. Thus, empirical deviation parameters may derive in part from these intramolecular changes, even though they may be attributed to other causes.

2. *Quantum Effects upon Intermolecular "Translation"*

The quantum correction to the configurational partition function, and hence to the equation of state, of simple (monatomic) fluids was developed by de Boer (1948) and is discussed in detail by Hirschfelder *et al.* (1954). Basically, the classical integrand $e^{-U/kT}$ in Eq. (3.5) or (3.13) is replaced by a quantum-mechanical sum (the "Slater sum") over a complete set of orthonormal functions for the N-particle liquid; in the high-temperature limit this of course reduces to the classical form.

For pairwise additivity and a pair energy function of the form $u(r) = \varepsilon\phi(r/\sigma)$, it can be shown that the configuration integral, and hence the residual Helmholtz free energy, can be expressed as a function of three reduced parameters ε/kT, $\tilde{V}/\tilde{N}\sigma^3$, and Λ^*,

$$\tilde{A}^*/RT = F(kT/\varepsilon, V/N\sigma^3, \Lambda^*), \tag{4.1}$$

where the new dimensionless quantum-mechanical parameter Λ^*, a reduced de Broglie wavelength, is defined in terms of ε, σ, the Planck constant h, and the molecular mass m,

$$\Lambda^* = h/\sigma(m\varepsilon)^{1/2}. \tag{4.2}$$

When appropriate thermodynamic properties are expanded (around their classical values for $\Lambda^* = 0$) as a power series in Λ^*, the first correction term is proportional to $(\Lambda^*)^2$ and decreases in magnitude with increasing temperature; the difference between Bose–Einstein and Fermi–Dirac statistics is substantial only at very low temperatures and is proportional to $(\Lambda^*)^3$.

The evaluation of the parameter Λ^* depends upon obtaining good (or at least self-consistent) values of ε and σ; Table XI gives values of

TABLE XI

QUANTUM-MECHANICAL DEVIATION PARAMETERS

	Λ^*	Λ'		Λ^*	Λ'
^3He	3.08	2.80	CH_4	0.24	0.14
^4He	2.67	2.08	N_2	0.23	0.14
H_2	1.73	1.13	O_2	0.20	0.12
HD	1.41	0.89	Ar	0.19	0.11
D_2	1.22	0.76	Kr	0.10	0.06
Ne	0.59	0.36	Xe	0.06	0.04

Λ^* for the lighter gases (Hirschfelder *et al.*, 1954) based upon Lennard-Jones (12, 6) values of the parameters. An alternative procedure (Byk, 1921, 1922; Pitzer, 1939) used the product $M^{1/2}T_c^{1/2}\tilde{V}_c^{1/3}$ to assess the importance of quantum corrections; low values signaled potentially large corrections. The reciprocal of this quantity should be roughly proportional to Λ^*; transformed into a dimensionless product Λ', it is also shown in Table XI.

$$\begin{aligned} \Lambda' &= h/[(\tilde{V}_c/\tilde{N})^{1/3}(MkT_c/\tilde{N})^{1/2}] \\ &= \tilde{N}^{5/6}h/[M^{1/2}(kT_c)^{1/2}\tilde{V}_c^{1/3}]. \end{aligned} \tag{4.3}$$

Only for helium, the isotopes of hydrogen, and neon is Λ^* large enough to cause substantial deviations in the reduced properties of liquids. (Figure 13 shows some reduced-vapor-pressure curves.) For helium and hydrogen, these deviations are large; where adequate theoretical treatments exist (as in the case of the second virial coefficient of the gas), theory and experiment are in good agreement.

For neon, the deviations are small but unmistakable; by comparison

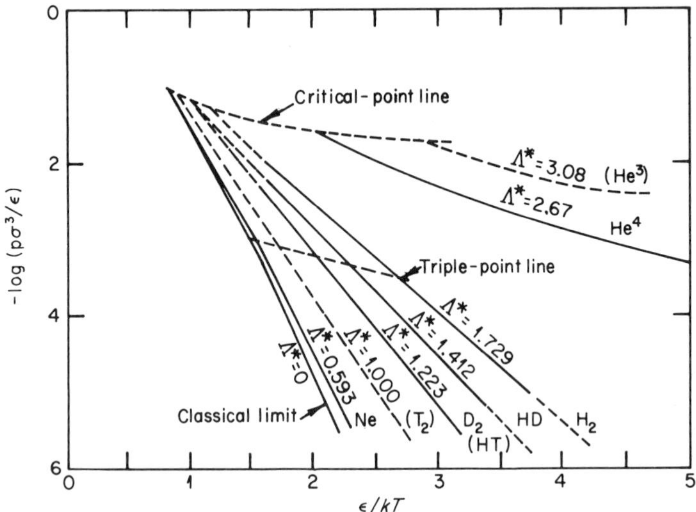

FIG. 13. Reduced vapor pressure $p\sigma^3/\varepsilon$ as a function of ε/kT for several gases. The figure shows that, for the lighter molecules He, H_2, and Ne, quantum deviations cannot be neglected, and that quantum effects become more appreciable for larger values of Λ^* (Lunbeck, 1950). (From Hirschfelder *et al.*, 1954.)

with the "simple fluids" at the triple point (Table X), the heat capacity $\tilde{C}_p/R$ is 0.9 too small, the entropy of vaporization $\Delta\tilde{S}^{\mathrm{V}}/R$ is 0.3 too small, and the volume $\tilde{V}/\tilde{V}_c$ is 0.01 too large, deviations in the right direction and of reasonable magnitude for $\Lambda^* = 0.6$.

Very small quantum-mechanical corrections would be expected for CH_4, Ar, and Kr [and such corrections are regularly applied to $B(T)$], but the small scatter of the reduced properties of the liquids shows no clear correlation with Λ^*; the variations may be due in large part either to experimental error or to the lack of perfect conformality of the pair energy $u(r)$.

3. *Nonconformality of the Pair Energy*

The major features of the deviations of the pair energy $u(r, \theta_1, \theta_2, \phi_1 - \phi_2)$ from the simple form $u(r)$ for a representative pair (e.g., two Ar atoms) were discussed in Section II. If one starts with the general equation for the pair energies of axially symmetric molecules [Eq. (2.37)], generalization of the "derivation" of Section III,A and the use of dimensional analysis [see, for example, Hakala (1967), or Leland and

Chappelear (1968)] yields for the compression factor

$$p\tilde{V}/RT = F(kT/\varepsilon_0, \tilde{V}/\tilde{N}\sigma_0^3, \gamma, \bar{\alpha}/\sigma_0^3, \varkappa, \Lambda^*, \mu^2/\varepsilon_0\sigma_0^3, \theta^2/\varepsilon_0\sigma_0^5, \ldots), \quad (4.4)$$

and the various reduced residual functions will depend upon the same dimensionless parameters. Here, $\bar{\alpha}/\sigma_0^3$ is introduced to allow for the possibility of "triple–dipole" contributions to $U\{\mathbf{r}\}$ and, in doing so, removes the need for the parameter $\mu^2\bar{\alpha}/\varepsilon_0\sigma_0^6$, which is now the product of two others; Λ^* is introduced to include quantum corrections explicitly.

A general theory which yields the unknown function F is not in sight, and even a perturbation treatment is not yet available except for a few special cases. In principle, each of the variables in Eq. (4.4) could be considered in a multivariable analysis of the experimental data for a large number of systems. This is not yet feasible because (a) the experimental data on liquids are neither sufficiently numerous nor sufficiently precise, (b) it is not certain that all the important variables are included in Eq. (4.4), and (c) the parameters ε_0, σ_0, γ, μ, θ, etc., are not sufficiently well established for most of the substances which would have to be considered.

The fact that the angle-averaged dipole–dipole energy has the same r^{-6} dependence as the principal source of attraction, the London dispersion energy, suggests that one might obtain an especially simple "effective" pair function by adding Eq. (2.36) to a Lennard-Jones (12, 6) function. However, the general statistical-mechanical equations (see Section III) involve the average $\langle\exp[-\sum u/kT]\rangle$, not $\langle\exp[-\sum\langle u\rangle/kT]\rangle$, where the pair energy has been preaveraged over all orientations, and such an expectation proves false in general. However, a perturbation treatment (Cook and Rowlinson, 1953; see also Rowlinson, 1969) shows that when the only angle-dependent part of u is of the form of Eq. (2.33), a transformation is correct to first-order terms in μ^4/kT. Moreover, a shape effect can be artificially simulated by an angle-dependent function proportional to r^{-6} which, after angle-averaging, becomes proportional to $T^{-1}r^{-12}$, conformal with the repulsive part of the Lennard-Jones pair energy.

The appropriate modifications to the configuration integral Q lead to a molar residual Helmholtz free energy $\tilde{A}^*$ which is, correct to terms containing $1/T$, equivalent to that for a simple fluid at temperature T with an effective $u(r)$,

$$u_{\text{eff}}(r, T) = 4\varepsilon_0\{(\sigma_0/r)^{12}[1 - 2\delta(T)] - (\sigma_0/r)^6[1 + 2\chi(T)]\}, \quad (4.5)$$

where the dipole–dipole factor† $\chi = \mu^4/24\varepsilon_0\sigma_0{}^6kT$. Equation (4.5) is conformal with the simple (12, 6) pair energy [Eq. (2.24)] with the substitution of temperature-dependent parameters ε_{eff} and σ_{eff},

$$\varepsilon_{\text{eff}}(T) = \varepsilon_0(1 + 2\chi)^2/(1 - 2\delta) \cong \varepsilon_0[1 + (2\delta + 4\chi)], \tag{4.6}$$

$$\sigma_{\text{eff}}(T) = \sigma_0[(1 - 2\delta)/(1 + 2\chi)]^{1/6} \cong \sigma_0[1 - \tfrac{1}{3}(\delta + \chi)], \tag{4.7}$$

where δ and χ are both inversely proportional to temperature. Thus, at low temperatures, $u_{\text{eff}}(r, T)$ has the same shape as $u_0(r)$ but the energy well is deeper and the collision diameter smaller. The essential equations for the principle of corresponding states follow directly;

$$\tilde{A}^*[T, \tilde{V}] = (1 + 2\delta + 4\chi)\tilde{A}_0{}^*[T(1 - 2\delta - 4\chi), \tilde{V}(1 + \delta + \chi)], \tag{4.8}$$

$$\phi[p, \tilde{V}, T] = \phi_0[p(1 - 3\delta - 5\chi), \tilde{V}(1 + \delta + \chi)], T(1 - 2\delta - 4\chi)], \tag{4.9}$$

where $\tilde{A}_0{}^*$ and ϕ_0 are the functions for the simple fluid with $\varepsilon = \varepsilon_0$ and $\sigma = \sigma_0$ and of course $\delta = \chi = 0$.

Equations (4.8) and (4.9) are correct only to the first power of $1/T$ in the correction terms δ and χ. To this extent, Eq. (4.9) leads to the conclusion that the compression factor at the critical point is independent of these perturbations.

$$(p_c\tilde{V}_c/RT_c) = (p_c\tilde{V}_c/RT_c)_0. \tag{4.10}$$

The critical compression factor is, in fact, not a universal constant; as the complexity of the substance increases, the value becomes distinctly smaller than 0.291.

Since the perturbations δ and χ affect the different parameters (ε and σ; or p, $\tilde{V}$, and T) differently, even this simple treatment must in principle require two more parameters.‡ In spite of this, however, it is found empirically that these and many other deviation parameters affect the equation of state in approximately similar ways. Thus, Pitzer (1955)

† This coefficient for the dipole–dipole contribution to $u_{\text{eff}}(r, T)$ is one-half that of the corresponding term in the angle-averaged energy $\bar{u}(r, T)$ [Eq. (2.36)]. Equation (4.5) is used to obtain the configuration integral Q and the Helmholtz free energy A, while Eq. (2.36) leads directly to the energy E. Since $E = [\partial(A/T)/\partial(1/T)]_V$, the coefficients (for A and E) of terms containing $(1/T)^n$ must differ by a factor $1/(n + 1)$.

‡ Cook and Rowlinson (1953) considered the modifications of the r^{-6} and r^{-12} terms separately and did not actually combine them. However, if the angular functions for the r^{-3} and r^{-6} terms before averaging are different (orthogonal) spherical harmonics, the two perturbations are separable and lead to Eqs. (4.5) to (4.9).

compared the reduced second virial coefficients $B/\tilde{V}_B$ at two reduced temperatures, $T/T_B = 0.286$ (near the triple point) and $T/T_B = 0.50$ (near the critical point), for three different pair energies, the Stockmayer function [Eq. (2.34)] for dipoles, the Kihara function for a spherical core, and the Kihara function for a linear core; he found (Fig. 14) the two Kihara results to be virtually indistinguishable and the results from the Stockmayer function to differ only slightly. (The dimensionless parameter $y = \mu^4/2\varepsilon_0{}^2\sigma_0{}^6$ could be as great as 0.7 before the difference from the Kihara deviation curve exceeded 1%; by this point, the dipole–dipole interactions are contributing more than a quarter of the total energy and the deviations from the simple fluid exceed 10%).

These similarities in behavior make it difficult to identify the specific sources of the deviations from the principle of corresponding states, but this theoretical difficulty becomes an empirical advantage. The convergent

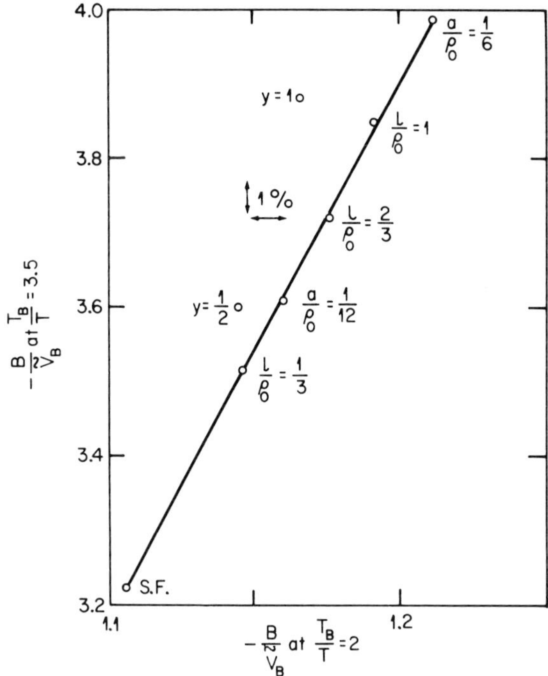

FIG. 14. Deviations of the reduced second virial coefficient $B/\tilde{V}_B$ from that of a simple fluid, calculated at $T_B/T = 3.5$ and $T_B/T = 2$ for the Kihara function [Eq. (2.30)] with a spherical core (various values of a/ϱ_0), for the Kihara function with a linear core (various values of l/ϱ_0), and for the Stockmayer function [Eq. (2.34)] (various values of y). (From Pitzer 1955).

behavior of different kinds of perturbations permits the use of a single empirical deviation parameter, or at most two, rather than the many indicated by theory.

B. Empirical Treatments of Deviations

1. *A Single Deviation Parameter—the Acentric Factor*

From Eqs. (4.8) and (4.9), other thermodynamic properties of a fluid (e.g., vapor pressure, density, entropy of vaporization, heat capacity, etc.) may be derived and used to estimate the magnitude of the deviation parameter δ (or, given the dipole moment μ, to test the equations with the deviation parameter χ). In particular, one can write $\delta(T) = \delta_c(T_c/T)$, where δ_c is the value of δ at the critical point. Cook and Rowlinson (1953; see also Rowlinson, 1954) have calculated δ_c from five different liquid properties of a number of substances and find for most a reasonable consistency between the values. Riedel (1954, 1955, 1956) has developed a similar set of empirical correlations in terms of a parameter α_k, the logarithmic slope of the vapor-pressure–temperature curve at the critical point.†

$$\alpha_k = (d \ln p/d \ln T)_{\sigma, T=T_c}. \tag{4.11}$$

(Here the subscript σ means that the derivative is taken along the saturation (or coexistence) curve.)

An extensive empirical correlation system (applied most thoroughly at temperatures above $T/T_c = 0.8$) was developed by Pitzer *et al.* (1955; see also Pitzer and Brewer, 1961) in terms of what they call the "acentric factor" ω, defined in terms of the reduced vapor pressure at a reduced temperature $T/T_c = 0.700$,

$$\omega = -\log_{10}[p(T/T_c = 0.700)/p_c] - 1.000. \tag{4.12}$$

This apparently arbitrary definition was chosen because of the ease and precision with which this quantity can be determined experimentally (as distinct from the limiting slope of the vapor-pressure curve at the critical point, which defines Riedel's α_k), and because at $T/T_c = 0.700$ the vapor pressure of a simple fluid is almost exactly $0.100p_c$, making the acentric factors of these monatomic substances essentially zero.

† Recent work (see Rowlinson, 1969) suggests that the curvature $(d^2p/dT^2)_\sigma$ of the vapor-pressure line may become infinite at $T = T_c$. If so, α_k as the true limiting value of $(d \ln p/d \ln T)_\sigma$ may be difficult to determine precisely.

With the aid of the experimental vapor-pressure curve for simple fluids, the correlation parameters δ_c, α_k, and ω can be related (Pitzer *et al.*, 1955; Rowlinson, 1969). If second-order perturbations are ignored, the following equation results:

$$\omega = 2.40\delta_c = 4.93(\alpha_k - 5.81). \tag{4.13}$$

For a series of substances, the deviations of a particular thermodynamic function from that for a single fluid at the same reduced temperature and pressure are plotted against ω, and first (and sometimes second) deviation coefficients are determined from the smoothed curve. Thus, for a reduced function $\tilde{Y}^*(T/T_c, p/p_c, \omega)$,

$$\tilde{Y}^*(T/T_c, p/p_c, \omega) = \tilde{Y}_0^*(T/T_c, p/p_c, 0) + (\partial\tilde{Y}^*/\partial\omega)_0\omega + \tfrac{1}{2}(\partial^2\tilde{Y}^*/\partial\omega^2)_0\omega^2 + \cdots \tag{4.14}$$

Tables of various functions with $\tilde{Y}_0^*$ and its ω-derivatives are given in the original literature. Here, it must suffice to reproduce a table of acentric factors for representative substances (Table XII) and to note qualitatively the direction and magnitude of the deviations in some of the important properties.

TABLE XII

ACENTRIC FACTORS FOR VARIOUS SUBSTANCES[a]

	ω		ω
Ne	(−0.026)	CF_4	0.18
Ar	−0.002	$C(CH_3)_4$	0.195
Kr	−0.002	n-C_4H_{10}	0.201
Xe	0.002	C_6H_6	0.215
CH_4	0.013	CO_2	0.225
O_2	0.021	NH_3	(0.250)
N_2	0.040	n-C_5H_{12}	0.252
H_2S	0.100	n-C_6H_{14}	0.301
C_2H_6	0.105	H_2O	(0.348)
C_3H_8	0.152	n-C_7H_{16}	0.352

[a] These parameters are calculated directly from the vapor pressures. It is not to be expected that the various thermodynamic properties of Ne (a quantum fluid) and NH_3 and H_2O (fluids with extensive hydrogen bonding) would all scale in accordance with these ω's. In particular, one does not expect to predict the properties of water from those of n-heptane even though they have almost identical acentric factors.

a. Equation of State. At the critical point, the critical compression factor $(p\tilde{V}/RT)_c$ decreases slowly[†] with increasing ω, from its simple fluid value of 0.291 to a value of about 0.27 for $\omega = 0.25$. At temperatures below the critical, the deviations of $(p\tilde{V}/RT)$ from the simple fluid values for the same reduced variables are invariably negative, increasingly so as one goes to higher pressures. Conversely, above the critical temperature, the deviations become increasingly positive as one goes to higher pressures and temperatures.[‡]

Pitzer and Curl (1957) have analyzed the data on second virial coefficients and have proposed the following empirical equation, linear in ω;

$$\begin{aligned} Bp_c/RT_c = {} & (0.1445 + 0.073\omega) - (0.330 - 0.46\omega)(T_c/T) \\ & - (0.1385 + 0.50\omega)(T_c/T)^2 - (0.0121 + 0.097\omega)(T_c/T)^3 \\ & - 0.0073\omega(T_c/T)^8. \end{aligned} \tag{4.15}$$

b. Vapor Pressure and Other Properties of Coexistent Phases. In effect, the acentric factor ω is a measure of the deviation of the slope of the vapor-pressure curve [which is nearly a straight line when $\ln(p/p_c)$ is plotted versus T_c/T, as was shown in Fig. 10] from that of the simple fluid. With increasing ω, this slope is steeper and the reduced vapor pressure at a given reduced temperature will be lower than that for the simple fluid. Moreover, since the reduced heat of vaporization $\Delta\tilde{H}^V/RT_c$ and the reduced entropy of vaporization $\Delta\tilde{S}^V/R$ are both closely related to this slope, it follows that these are greater than the corresponding simple fluid values by a factor which would be $(1 + \omega)$ if $\ln(p/p_c)$ were exactly linear in T_c/T, but which is in fact somewhat greater (as great as $1+1.5\omega$).

The reduced density of the liquid $\varrho_l/\varrho_c = \tilde{V}_c/\tilde{V}_l$, for a particular reduced temperature T/T_c, increases with increasing acentric factor, as evidenced by the increasing steepness of the slope of the rectilinear diameter $(\varrho_l + \varrho_g)/\varrho_c$ plotted against T/T_c (Rowlinson, 1954). This is a direct consequence of the decrease of the effective collision diameter σ_{eff} with decreasing temperature [Eq. (4.7)].

If the surface tension γ is fitted to the reduced temperature using Eq. (3.33), Curl and Pitzer (1958) find that the dimensionless constant

[†] The critical compression factor $(p\tilde{V}/RT)_c$ or its deviation from the simple fluid value was an early suggestion for a deviation parameter. The disadvantages of this quantity as an empirical parameter have been summarized by Leland and Chappelear (1968).

[‡] The deviations of $p\tilde{V}/RT$ are essentially zero along a line on the $(T/T_c, p/p_c)$ diagram commencing at (1, 0) and extending in a roughly linear way through the point (1.4, 8.5). At temperatures below this line, deviations are negative; and above it, positive.

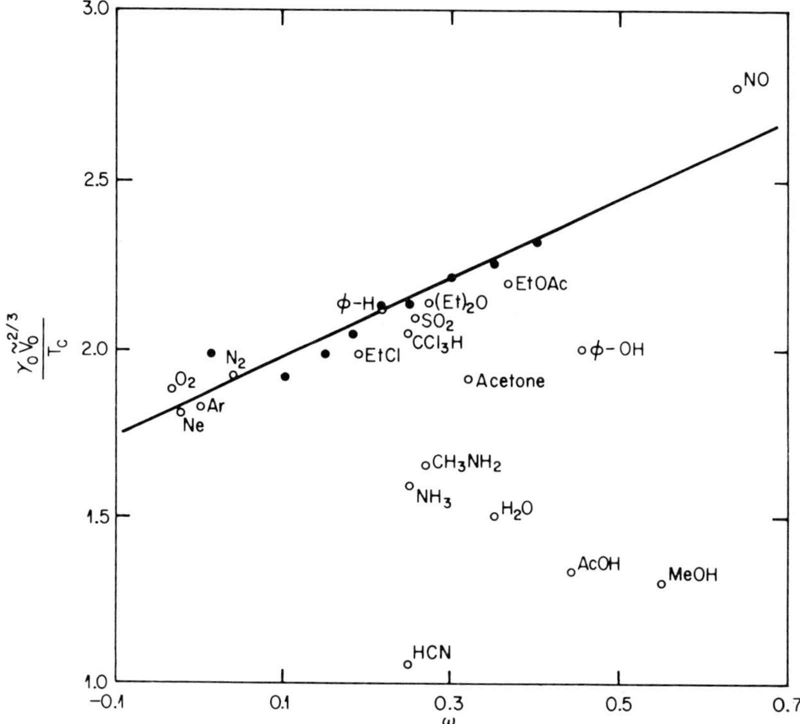

FIG. 15. The reduced-surface-tension parameter $\gamma_0 \tilde{V}_0^{2/3}/T_c$ [not made dimensionless as in Eq. (4.16)] as a function of the acentric factor ω; black dots indicate saturated hydrocarbons (Curl and Pitzer, 1958). (From Pitzer and Brewer, 1961.)

$\gamma_0(\tilde{V}_c/\tilde{N})^{2/3}/kT_c$ increases in a roughly linear manner with ω,[†]

$$\gamma_0(\tilde{V}_c/\tilde{N})^{2/3}/kT_c = 4.5(1 + 0.64\omega). \tag{4.16}$$

They suggest that the consistent fit of thermodynamic properties of liquids to the acentric factor formulation is a useful criterion for a "normal liquid," and that the surface tension, as evidenced by Eq. (4.16), is an especially sensitive test. Figure 15 shows this function for a series of liquids; the points for hydrogen-bonded liquids (e.g., NH_3, H_2O, HCN) lie far below the line for "normal liquids."

[†] Curl and Pitzer (1958) express Eq. (4.16) in terms of the parameter $\gamma_0 \tilde{V}_0^{2/3}/T_c$, where $\tilde{V}_0$ is the hypothetical molar volume at 0°K, approximately $0.28\tilde{V}_c$. Their equation is then $[\gamma_0\ (\text{dyn cm}^{-1})][\tilde{V}_0\ (\text{cm}^3\ \text{mole}^{-1})]^{2/3}/[T\ (°\text{K})] = 1.86 + 1.18\omega$, as shown in Fig. 15.

c. Thermal Properties. Of particular interest are the reduced heat capacities $\tilde{C}_V{}^*/R$ and $\tilde{C}_p{}^*/R$. Both increase markedly with increasing ω, especially the latter (Rowlinson, 1954).

Curl and Pitzer (1958; see also Pitzer and Brewer, 1961) tabulate $\log_{10}(f/p)$ (where f is the "fugacity"), $(\tilde{H}^\circ - \tilde{H})/RT_c$, and $(\tilde{S}^\circ - \tilde{S})/R$ and their first ω-derivatives for T/T_c from 0.8 to 4.0 and p/p_c from 0.2 to 9.0; unfortunately, this does not include much of the normal liquid range.

2. *Two Deviation Parameters—The Shape Factors*

For liquids with more complex intermolecular forces, indeed for almost any fluid if examined over wide ranges of temperature and pressure, a single perturbation parameter such as the acentric factor ω will prove inadequate, and the use of two or more has been suggested (Hirschfelder *et al.*, 1958). A more recent proposal of two "shape factors" by Leland and co-workers (Leach *et al.*, 1966; Leland and Chappelear, 1968) is an empirical outgrowth of the earlier work of Cook and Rowlinson (1953), as illustrated in Eqs. (4.5)–(4.9).

They suggest that one define a temperature-dependent pair energy function $u_{\text{eff}}(r, T)$ such that

$$u_{\text{eff}}(r, T) = \varepsilon(T)\phi[r/\sigma(T)], \tag{4.17}$$

and assume that a property of a fluid may be obtained by substituting reduced parameters $kT/\varepsilon(T)$ and $\tilde{V}/\tilde{N}[\sigma(T)]^3$ into the equations for reduced properties of a simple fluid, provided that this property does not require a temperature derivative of the configuration integral Q in its derivation. In principle, the functions $\varepsilon(T)$ and $\sigma(T)$ can be determined theoretically for the dilute gas by comparing the second virial coefficient $B(T)$ for a gas with an angle-dependent pair energy with that for a simple fluid with an assumed $u(r)$. This is not very useful, however, because of the uncertainties about $u(r)$ and about the contribution of triplet interactions at higher densities.

Leach *et al.* (1966) have developed shape factors θ_i and ϕ_i by simultaneous solution of the equations

$$(p\tilde{V}/RT)_i = f_0(T/\theta_i T_{ci}, \tilde{V}/\phi_i \tilde{V}_{ci}), \tag{4.18}$$

$$(f/p)_i = F_0(T/\theta_i T_{ci}, \tilde{V}/\phi_i \tilde{V}_{ci}), \tag{4.19}$$

where f_0 and F_0 are the reduced compression factor and fugacity coefficient for a reference fluid (for which $\theta = \phi = 0$), and the subscript i represents the substance whose properties are to be related to those of a reference fluid.

At moderate densities of the gas, solution of Eqs. (4.18) and (4.19) is equivalent to a simultaneous fit of the second and third virial coefficients $B(T)$ and $C(T)$ over a range of temperatures; such a fit includes the uncertain triplet interactions in $C(T)$. Application of the shape factors ϕ and θ so obtained to the fluid at higher densities shows that they are really somewhat density-dependent.

3. *Corresponding States for Chain Molecules*

The thermodynamic properties of a series of homologous chain molecules (e.g., the set of *n*-alkanes), while amenable to treatment by perturbation methods (e.g., the acentric factors† for the *n*-alkanes in Table XII), have inspired special treatment. One assumes a chain molecule made up of n segments whose interactions with neighboring segments of other molecules are characterized by parameters ε and σ analogous to those for interactions between isolated (monomer) segments. In the original lattice model (Prigogine *et al.*, 1953; see also Prigogine, 1957), the chain molecule occupied r sites in a lattice of coordination number z and could have a total of $qz = (z - 2)r + 2$ interactions with nearest-neighbor segments; the number of external degrees of freedom $3c$ introduced a third parameter (originally assumed to be 3 for $r = 1$, and approximately $3 + r$ for $r = 2$ and all higher polymers).

Later Hijmans (1961), in a more general but less rigorous formulation, suggested that r, q, and c be regarded as scaling parameters for obtaining the reduced volume, energy, and entropy for a chain-molecule fluid, $\tilde{V}/\tilde{N}r\sigma^3$, $\tilde{E}/\tilde{N}q\varepsilon$, and $\tilde{S}/\tilde{N}ck$. Dimensional considerations then lead to a reduced equation of state

$$\phi(pr/q, \tilde{V}/r, cT/q) = 0, \tag{4.20}$$

where ε and σ are taken as the same for all members of the series. (One does not in fact expect that each segment will have exactly the same ε and σ; any difference between end and middle segments, for example, is accounted for approximately by adjusting r and q, which no longer have to conform to the lattice model.)

Evidently, Eq. (4.20) can apply only to the dense fluid, for it yields a

† It is interesting to note that the acentric factors for the *n*-alkanes (see Table XII) increase linearly with the number of carbon atoms n. Except for methane, they fit very closely the equation $\omega = 0.05n$. This has led Guggenheim and Wormald (1965) to suggest for the single deviation parameter the number of carbon atoms n in the *n*-alkane whose reduced properties most closely resemble those of the substance in question.

compression factor $p\tilde{V}/RT$ which is proportional to the parameter c, a prediction which is manifestly incorrect at the critical point, where $(p\tilde{V}/RT)_c$ decreases only slowly with increasing chain length, and *a fortiori* for the ideal gas, where $p\tilde{V}/RT = 1$ for all substances.

Consequently, the parameters r, q, and c must be evaluated from the properties of the dense fluid only, without the aid of the critical constants. The equation of state at zero pressure (or that at saturation pressure, virtually equivalent at low temperatures) yields r and c/q relative to those for a reference fluid (for n-alkanes, usually n–C_7H_{16}). A third property of the dense fluid, e.g., the thermal pressure coefficient $\gamma_V = (\partial p/\partial T)_V$, which is proportional to c/r, completes the specification. The reduced enthalpy $\tilde{H}^*$, the reduced heat capacity $\tilde{C}_p{}^*$, etc. all involve relation to the properties of the ideal gas and are not appropriate for determining the scaling parameters. (For example, a difference in spatial configuration or restricted rotation will affect $\tilde{H}^*$ and $\tilde{C}_p{}^*$).

The properties of n-alkane liquids satisfy this extended theory of corresponding states remarkably well, particularly if the first members of the series (methane and ethane) are ignored. In fact, the parameters r, q, and c increase almost exactly linearly with the number of carbon atoms n [e.g., $r(n) = r_0 + r_1 n$, etc.]. This three-parameter corresponding-states formulation has been particularly fruitful in its application to binary mixtures. The properties of a mixture of homologs should depend only upon the mean values of r, q, and c, i.e., upon the mean value of n, $\bar{n}$, so all the properties of mixtures (or pure liquids) with the same $\bar{n}$ should coincide (except of course for the ideal entropy of mixing) at any fixed T and p; these are "congruent mixtures" and satisfy the "principle of congruence" (Brönsted and Koefoed, 1946). More generally, the concept of these three parameters for the equation of state has been adapted to the treatment of other pure fluids and binary mixtures (the latter with the addition of an empirical mixing energy) by Flory and his co-workers (Flory *et al.*, 1964; Flory, 1965; Orwoll and Flory, 1967).

V. The Equation of State for Dense Liquids

Of special interest, both for practical use and for theoretical interpretation, is the equation of state of the dense liquid at normal liquid temperatures. Accurate measurements of the volume as a function of temperature and pressure are available for a wide variety of substances over a temperature range from the triple point to $0.7T_c$ and higher and from zero pressure (or saturation pressure) up to pressures well in excess of p_c.

A. Experimental Results

The two mechanical properties of liquids which are easiest to measure are the orthobaric thermal expansivity α_σ and the (isochoric) thermal pressure coefficient γ_V. The former is simply calculated from the orthobaric density of the liquid under saturation pressure or from equivalent dilatometric measurements,

$$\alpha_\sigma = (\partial \tilde{V}/\partial T)_\sigma / \tilde{V} = (\partial \ln \tilde{V}/\partial T)_\sigma . \tag{5.1}$$

The closely related isobaric thermal expansivity α_p can be measured directly or can be calculated from α_σ with the aid of the isothermal compressibility $\varkappa_T$ and the slope of the vapor pressure curve $\gamma_\sigma = (\partial p/\partial T)_\sigma$,

$$\alpha_p = (\partial \tilde{V}/\partial T)_p / \tilde{V} = \alpha_\sigma + \varkappa_T \gamma_\sigma . \tag{5.2}$$

The thermal pressure coefficient $\gamma_V = (\partial p/\partial T)_V$ is easily measured in a pressure bomb acting essentially as a constant-volume thermometer; the technique has been developed and exploited by Hildebrand and coworkers (Westwater *et al.*, 1928; Hildebrand and Carter, 1932; Alder *et al.*, 1954). The derivative $(\partial \gamma_V/\partial T)_V = (\partial^2 p/\partial T^2)_V$ is very small, so a straight-line plot of p against T yields an accurate value of γ_V.

The determination of the isothermal compressibility

$$\varkappa_T = -(\partial \ln \tilde{V}/\partial p)_T$$

is more difficult, especially if one wants the value near the saturation pressure of the liquid. Although there are extensive measurements of the compressibility of liquids to very high pressures, it is usually better to calculate $\varkappa_T$ for low pressures from the thermodynamic identity

$$\varkappa_T = \alpha_p / \gamma_V . \tag{5.3}$$

In comparing results for different fluids, we follow the procedure suggested by the extended corresponding-states treatment of chain molecules (see the preceding section) and avoid using constants from the dilute gas or the critical point as reduction parameters. Useful reduced quantities are then

$$\alpha_p T = (\partial \ln \tilde{V}/\partial \ln T)_p ,$$

$$T(\alpha_p T)' = T[\partial(\alpha_p T)/\partial T]_p = [\partial^2(\ln \tilde{V})/(\partial \ln T)^2]_p ,$$

and $\gamma_V \tilde{V}/R$, all determined at (or extrapolated to) zero pressure. Table XIII shows these quantities for Ar, CCl_4, n-C_6H_{14}, and n-C_6F_{14}.

TABLE XIII

EQUATION-OF-STATE PROPERTIES FOR VARIOUS LIQUIDS[a]

	T (°K)	T/T_c	$\tilde{V}/\tilde{V}_c$	$\alpha_p T$	$T(\alpha_p T)'$	$\gamma_V \tilde{V}/R$
Ar	83.31 (T_t)	0.556	0.378	0.368	0.57	7.32
	90	0.598	0.389	0.412	0.73	6.7_4
	100	0.664	0.407	0.511	1.15	5.9_1
CCl_4	250.2 (T_t)	0.449	0.332	0.29	—	16.8
	273	0.491	0.341	0.32	0.5	15.0
	298	0.536	0.352	0.367	0.6	13.4
	323	0.581	0.363	0.42	0.7	11.6
n-C_6H_{14}	273	0.538	0.346	0.359	0.49	14.68
	298	0.587	0.358	0.415	0.76	12.88
	323	0.636	0.371	0.489	1.10	11.31
	348	0.685	0.387	0.591	1.64	9.86
n-C_6F_{14}	273	0.160	—	0.445	0.93	18.0
	298	0.666	—	0.538	1.18	15.0
	323	0.722	—	0.656	1.5	12.1

[a] All data corrected to zero pressure. Sources: Rowlinson (1969) (Ar); Benninga and Scott (1955) (CCl_4); Dunlap and Scott (1962) (C_6H_{14}, C_6F_{14}); Orwoll and Flory (1967) (C_6H_{14}).

The magnitude and temperature dependence of these quantities must be accounted for by any satisfactory equation of state, whether empirical or theoretically derived.

B. ENERGY–ENTROPY–VOLUME RELATIONS; THE INTERNAL PRESSURE

Measurement of the thermal pressure coefficient γ_V yields direct information concerning the dependence of the entropy of a fluid upon density, since

$$\gamma_V = (\partial p/\partial T)_V = (\partial \tilde{S}/\partial \tilde{V})_T. \tag{5.4}$$

Moreover, the use of the so-called "thermodynamic equation of state" leads to the internal pressure $(\partial \tilde{E}/\partial \tilde{V})_T$ as well,

$$p = T(\partial \tilde{S}/\partial \tilde{V})_T - (\partial \tilde{E}/\partial \tilde{V})_T, \tag{5.5}$$

or

$$(\partial \tilde{E}/\partial \tilde{V})_T = T\gamma_V - p. \tag{5.6}$$

For the liquid under its saturation vapor pressure, the internal pressure is essentially just $T\gamma_V$ (exactly so for zero pressure.)

Measurements of γ_V as a function of T and p offer information concerning two approximations in van der Waals-like theories of fluids, (a) the assumption that the molar residual energy $\tilde{E}^*$ and entropy $\tilde{S}^*$ are functions of density only and not of temperature, and (b) the assumption that the residual energy is proportional to the first power of the density [i.e., inversely proportional to molar volume, Eq. (1.6)]. Needless to say, neither can be exactly valid for any realistic model of a fluid, but they may prove to be useful approximations.

In the dilute-gas region, the first of these assumptions requires that all the virial coefficients be of the form $C_1 + (C_2/T)$, where the constant C_1 arises from an entropy contribution and C_2 from an energy. While the van der Waals form $b - (a/RT)$ is a useful zeroth approximation for the second virial coefficient, the temperature dependence of the higher virial coefficients is quite different. However, the approximation seems to be more nearly appropriate for the dense liquid and may be tested by examining the temperature dependence of the thermal pressure coefficient, since we have seen that, for this assumption to be true [Eq. (1.8)], $(\partial\gamma_V/\partial T)_V$ must be zero. In fact, for many liquids, the isochore (p versus T for constant V) is an extraordinarily good straight line, and careful measurements must be made over a very wide range of pressure to establish the magnitude (or even the sign) of $(\partial^2 p/\partial T^2)_V$. It is now clear (Rowlinson, 1969) that for most, if not all, liquids, it is small and negative; in taking a liquid from zero pressure to 1000 atm along an isochore, γ_V might decrease by 10%. Unfortunately, one must be careful in transferring these conclusions about the smallness of $(\partial\gamma_V/\partial T)_V$ into predictions of other properties with other units; thus, $T(\partial\gamma_V/\partial T)_V$, which has the same dimensions as γ_V itself, has about one third its magnitude, hardly negligible. However, over the range of liquid densities, assumption (a) is a useful approximation, but it must not be extended over the whole range of densities to the dilute gas.

The second assumption (b) can be examined in various ways. If we assume as a general form

$$\tilde{E}^* = \tilde{E}(T, \tilde{V}) - \tilde{E}^\circ_{\text{id}}(T) = -\Delta\tilde{E}^{\text{V}} = -a/\tilde{V}^n, \tag{5.7}$$

where $\Delta\tilde{E}^{\text{V}}$ is the energy of vaporization to the ideal gas, then

$$(\partial\tilde{E}/\partial\tilde{V})_T = (\partial\tilde{E}^*/\partial\tilde{V})_T = na/\tilde{V}^{n+1}, \tag{5.8}$$

and the ratio of the internal pressure to the cohesive energy density $(-\tilde{E}^*/\tilde{V})$ determines the exponent n,

$$n = (\partial\tilde{E}/\partial\tilde{V})_T/(-\tilde{E}^*/\tilde{V}) = TV\gamma_{V,p=0}/(\Delta\tilde{E}^{\mathrm{V}}/\tilde{V}). \qquad (5.9)$$

Values of n derived in this way from thermal pressure measurements are summarized in Table XIV. For many liquids, n is, in fact, approximately

TABLE XIV

EXPONENTS FOR THE EQUATION[a] $\tilde{E}^* = -a/\tilde{V}^n$

	T (°K)	n	$n' = m - 1$
Ar	85	0.86	0.5
Hg	298	0.33	—
CS_2	298	0.89	—
C_6H_6	298	1.05	—
CCl_4	273	1.08	1.4
	298	1.09	
	323	1.08	
n-C_6H_{14}	273	1.08	1.3
	298	1.09	
	323	1.10	
n-C_6F_{14}	273	1.32	2.1
	298	1.31	
	323	1.29	

[a] Sources: Hildebrand and Scott (1950, 1962), and references to Table XIII.

1; even when it is not, it is surprisingly nearly independent of temperature, an experimental result not consistent with some sophisticated theories of liquids (Benninga and Scott, 1955).

Unfortunately, unless one assumes that Eq. (5.7) holds exactly at all densities, the exponent n deduced from Eq. (5.9) is a kind of average value obtained as one integrates Eq. (5.8) over the range of volume from the dense liquid to the ideal gas, and is not necessarily the exponent appropriate to the differential quantity $(\partial\tilde{E}/\partial\tilde{V})_T$ at liquid densities.

As an alternative, one may try to examine in detail the change in $(\partial\tilde{E}/\partial\tilde{V})_T$ with small changes of $\tilde{V}$ at constant temperature,

$$-\tilde{V}(\partial^2\tilde{E}/\partial\tilde{V}^2)_T/(\partial\tilde{E}/\partial\tilde{V})_T \equiv m = n' + 1. \qquad (5.10)$$

Only if n is effectively constant at all densities will $m - 1 \equiv n'$ be equal to n.

Equation (5.10) can be transformed into an expression involving more directly measured quantities,

$$m = -\{T(\partial\gamma_V/\partial T)_p - T(\partial\gamma_V/\partial T)_V + \gamma_V\}/T\alpha_p\gamma_V. \qquad (5.11)$$

Since $(\partial\gamma_V/\partial T)_V$ is not usually available, it has been set equal to zero in the calculations reported in Table XIV. If, in fact, it is normally small and negative, the "true" values of n' will be somewhat smaller. For CCl_4, this might bring n and n' into agreement, but for Ar, n' is already too small.

A more complex treatment of the internal pressure was suggested by Hildebrand (1929) to account for the values of $(\partial\tilde{E}^*/\partial\tilde{V})_T$ at high densities (see Fig. 16). Noting that, at high pressures, the liquid is compressed beyond its "close-packed" density, he suggested that there must be at least a small contribution from the repulsive branch of $u(r)$. Empirically, the experimental data in Fig. 16 fit the equation

$$\begin{aligned}\tilde{E}^* &= -(a/\tilde{V}) + (a'/\tilde{V}^9) = -(a/\tilde{V})[1 - (a'/a\tilde{V}^8)] \\ &= -(a/\tilde{V})(1 - \xi), \end{aligned} \qquad (5.12)$$

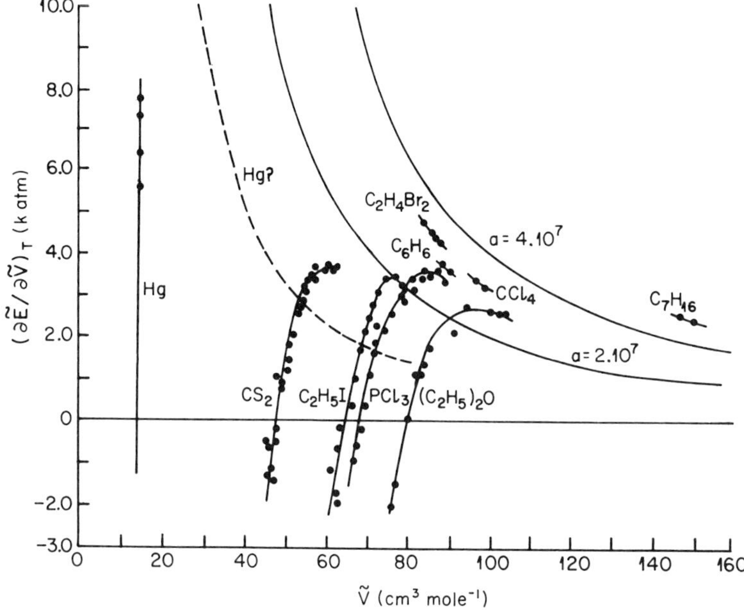

FIG. 16. The internal pressure $(\partial\tilde{E}/\partial\tilde{V})_T$ as a function of molar volume $\tilde{V}$ for several liquids. (From Hildebrand and Scott, 1962.)

where ξ is the fraction of the cohesive energy arising from the repulsion term $a'/\tilde{V}^9$. Differentiation yields

$$(\partial\tilde{E}^*/\partial\tilde{V})_T = (a/\tilde{V}^2) - (9a'/\tilde{V}^{10}) = (a/\tilde{V}^2)(1 - 9\xi), \tag{5.13}$$

$$(\partial^2\tilde{E}^*/\partial\tilde{V}^2)_T = -(2a/\tilde{V}^3) + (90a'/\tilde{V}^{11}) = -(2a/\tilde{V}^3)(1 - 45\xi), \tag{5.14}$$

or

$$n = (1 - 9\xi)/(1 - \xi) \approx 1 - 8\xi, \tag{5.15}$$

and

$$n' = m - 1 = (1 - 81\xi)/(1 - 9\xi) \approx 1 - 72\xi, \tag{5.16}$$

Thus, even for $\xi = 0.01$, n and n' are not unity, but 0.92 and 0.21, respectively.

C. Simple Analytic Equations of State

Recent work on the hard-sphere equation of state and on perturbations thereof using attractive potential energies has reawakened interest in explicit analytic equations of the van der Waals type. Most of the background for these extensive theoretical treatments will be reviewed in subsequent chapters, so only a brief summary of the basic conclusions will be given here:

(a) Monte Carlo (Wood and Jacobson, 1957) and molecular-dynamics (Alder and Wainwright, 1960) calculations upon finite systems of hard spheres have yielded an equation of state ($p\tilde{V}/RT$ as a function of $\tilde{V}/\tilde{V}_0$, where $\tilde{V}_0$ is the volume of spheres in a close-packed lattice) which is reasonably reliable for all values of $\tilde{V}/\tilde{V}_0$ from 1.5 to ∞. These results are in complete agreement with the calculated first seven virial coefficients (Ree and Hoover, 1967) and in good agreement with the results of calculations using the Percus–Yevick approximation (Thiele, 1963; Wertheim, 1963, 1964).

(b) When the attractive energies in a fluid are treated as a perturbation upon the hard-sphere infinite repulsion (Zwanzig, 1954; Smith and Alder, 1959; Frisch *et al.*, 1966; Barker and Henderson, 1967, 1968; Kozak and Rice, 1968; Rasaiah and Stell, 1970), it is found that the leading term of the correction to the hard sphere pressure p_{hs} is proportional to the second power of the density and independent of the temperature (i.e., equivalent to the van der Waals $-a/\tilde{V}^2$) and that, for liquid densities, the higher terms are an order of magnitude smaller and only weakly temperature-dependent.

In effect, the structure of the liquid, and hence the entropy, is determined almost entirely by the steep repulsive forces, in this case those between hard spheres. The addition of an attractive perturbation merely lowers the level of the "smoothed potential" energy in which the molecules move.

When the computer results became available, Longuet-Higgins and Widom (1964) proposed as a new and simple equation of state the explicit combination[†] of the hard-sphere results and the van der Waals energy $-a/\tilde{V}$, yielding an expression for the pressure

$$p = p_{\mathrm{hs}} - (a/\tilde{V}^2). \tag{5.17}$$

They showed that the use of this equation with the two branches (solid and fluid) of the molecular-dynamics results for hard spheres gave a good representation of the melting of argon. In that part of their work, they used the computer-derived curves directly, but in comparing the predictions of Eq. (5.17) with the experimental properties of dense liquids, they used an analytic form for $(p\tilde{V}/RT)_{\mathrm{hs}}$ obtained from a "scaled-particle" approach by Reiss *et al.* (1959), a form identical to that obtained later (Thiele, 1963) from the Percus–Yevick "compressibility" approximation [form III for $\phi(y)$ in Table XV]. Other slightly variant forms for $(p\tilde{V}/RT)_{\mathrm{hs}}$ have been proposed and examined by Guggenheim (1965).

Table XV summarizes the various analytic forms which have been proposed for $(p\tilde{V}/RT)_{\mathrm{hs}} = \phi(y)$; it is convenient to express these in terms of the density variable y,

$$y = b/4\tilde{V} = (2\pi\tilde{N}\sigma^3/3)/4\tilde{V}. \tag{5.18}$$

(It should be noted that the largest physically meaningful value of y for a hard-sphere system is $b/4\tilde{V}_0 = (2\pi\tilde{N}\sigma^3/3)/(4\tilde{N}\sigma^3/\sqrt{2}) = \pi/3\sqrt{2} = 0.74$, corresponding to close-packing in a face-centered-cubic lattice.) Equation (5.17) can then be rewritten as

$$p\tilde{V}/RT = \phi(y) - (4ay/RTb). \tag{5.19}$$

It is evident from Table XV that the exact virial expansion for hard spheres lies between the two Percus–Yevick functions II and III, and rather closer to the latter; this is confirmed by the machine calculations

[†] Stell (1970) has pointed out that Eq. (5.17) has a long history, starting with Boltzmann (1898) and Happel (1906).

TABLE XV

HARD-SPHERE COMPRESSION FACTORS $(p\tilde{V}/RT)_{hs} = \phi(y)$

$\phi(y)$	Expansion in powers of y
Exact	$1 + 4y + 10y^2 + 18.365y^3 + 28.24y^4 + 39.5y^5 + 56.5y^6 + \cdots$
I (van der Waals) $1/(1 - 4y)$	$1 + 4y + 16y^2 + 64y^3 + 256y^4 + 1024y^5 + 4096y^6 + \cdots$
II (PY-pressure) $(1 + 2y + 3y^2)/(1 - y)^2$	$1 + 4y + 10y^2 + 16y^3 + 22y^4 + 28y^5 + 34y^6 + \cdots$
III [scaled particle (PY-compressibility)] $(1 + y + y^2)/(1 - y)^3$	$1 + 4y + 10y^2 + 19y^3 + 31y^4 + 46y^5 + 64y^6 + \cdots$
IV (Guggenheim) $1/(1 - y)^4$	$1 + 4y + 10y^2 + 20y^3 + 35y^4 + 56y^5 + 84y^6 + \cdots$
V $(1 + 2y)/(1 - 2y)$	$1 + 4y + 8y^2 + 16y^3 + 32y^4 + 64y^5 + 128y^6 + \cdots$
VI $(1 + 3y + 4y^2)/(1 - 2y)(1 + y)$	$1 + 4y + 10y^2 + 18y^3 + 38y^4 + 74y^5 + 150y^6 + \cdots$
VII (Flory) $1/(1 - y^{1/3})$	—

right up to $\tilde{V}/\tilde{V}_0 = 1.5$ ($y = 0.5$). Also shown in Table XV are the original van der Waals function (I), which diverges greatly from the exact hard-sphere function at high densities, the function suggested by Guggenheim (IV), two new functions (V and VI), and one (VII) recently popularized by Flory† (1965; see also Flory *et al.*, 1964).

There are various ways in which Eq. (5.19) could be compared with experiment (e.g., with the properties of liquid argon), not all equivalent. In effect (although not necessarily in an explicit fashion), one selects two properties of a fluid to evaluate the a and b parameters and then compares the predicted values of other properties with their experimental values.

As we have seen in Section II, the parameters a and b will be proportional to $\varepsilon\sigma^3$ and σ^3, respectively, where ε and σ are the energy and distance scaling parameters in a pair energy function $u(r)$. They can be evaluated in a variety of ways, including, among others: (a) from the second virial coefficient B, which, for this equation of state [all forms for $\phi(y)$ except VII], is simply $b - (a/RT)$, so the Boyle temperature $T_B = a/Rb$ and the Boyle volume $\tilde{V}_B = b$; (b) from the critical constants T_c, $\tilde{V}_c$, and p_c, which can be related to a and b for each choice of $\phi(y)$; or (c) exclusively from properties of the dense liquid by fitting two properties [e.g., $\tilde{V}$, $\alpha_p T = (\partial \ln V/\partial \ln T)_p$, $\gamma_V \tilde{V}/R$, $\Delta\tilde{E}^V/RT$, etc.). The rationale for Eq. (5.19) suggests that it is most appropriate for high densities, so one should expect method (c) to be most successful. However, if Eq. (5.19) should be approximately applicable at lower densities, this is an unexpected but useful dividend, and we explore this possibility first.

Table XVI compares the predictions of Eq. (5.19) for the critical constants, given each of the seven different $\phi(y)$. The most striking feature is the failure of any of the forms to fit the critical compression factor $(p\tilde{V}/RT)_c$ very well. All the different functions $\phi(y)$, I–VI, yield values of this which are much closer to each other than to the experimental 0.291. Indeed, it can be shown that, if all the virial coefficients for hard spheres are positive [as they are for all these $\phi(y)$], Eq. (5.19) yields $(p\tilde{V}/RT)_c > 1/3$.

† The Flory equation of state was first introduced by Eyring and Hirschfelder (1937) and derived from a kind of cell model. In principle, it is more nearly appropriate to a solid than to a liquid, and its use at densities far lower than those of normal dense liquids was never contemplated by Flory. Needless to say, it cannot yield a virial expansion and hence values of T_B and $\tilde{V}_B$ from an expression for the second virial coefficient. It is not really expected to apply even at critical densities, so the "critical constants" derived in Table XVI and used in Tables XVII and XVIII are potentially misleading.

TABLE XVI

CRITICAL CONSTANTS FOR VARIOUS EQUATIONS OF STATE

Equation of state[a]	T_c/T_B	$\varrho_c/\varrho_B = \tilde{V}_B/\tilde{V}_c$	$(p\tilde{V}/RT)_c$
I (van der Waals)	0.296	0.333	0.375
II [$(PY)_p$]	0.382	0.538	0.357
III [Scaled particle-$(PY)_c$]	0.376	0.515	0.360
IV (Guggenheim)	0.373	0.506	0.361
V	0.392	0.536	0.366
VI	0.374	0.505	0.363
VII (Flory)	—	—	(0.536)[b]
Argon	0.368	0.55	0.291

[a] Roman numerals correspond to the equations given in Table XV.

[b] This number is presented only for completeness, since the Flory equation is not expected to apply at the critical point. The Flory equation does not yield a second virial coefficient, so b is not expected to equal $\tilde{V}_B$. At the critical point, the Flory equation yields $y_c = 0.279$, and $RT_cb/4a = 0.119$.

For functions II–VI, the agreement of T_c/T_B and ϱ_c/ϱ_B with experiment is much better, so the principal failure must lie in the poor fit of the pressure variable p_cb^2/a. The small differences in T_c/T_B and ϱ_c/ϱ_B mean, however, that, in comparing predicted and experimental properties of the dense liquid, the fit will vary somewhat depending upon whether one chooses the set $(T_B, \tilde{V}_B)$ or the set $(T_c, \tilde{V}_c)$ as reference.

Another set of properties for comparison are those of the dense liquid at a representative temperature, low enough that the vapor pressure is virtually zero. Then we may rewrite Eq. (5.19) as

$$\begin{aligned}\phi(y) = 4ay/RTb &= (T_B/T)(\tilde{V}_B/\tilde{V}) \\ &= -\tilde{E}^*/RT = \Delta\tilde{E}^V/RT \\ &= \tilde{V}(\partial\tilde{E}/\partial\tilde{V})_T/RT = \gamma_V\tilde{V}/R. \end{aligned} \tag{5.20}$$

Unfortunately, the three quantities on the right-hand side of Eq. (5.20), all of which should equal $4ay/RTb$, are not themselves the same. At the triple point of argon, 83.81°K, $T_t/T_B = 0.205$ and $\varrho_t/\varrho_B = \tilde{V}_B/\tilde{V}_t = 1.46$ yielding $(T_B/T)(\tilde{V}_B/\tilde{V}) = 7.12$ to compare with $\gamma_V\tilde{V}/R = 7.32$ and $\Delta\tilde{E}^V/RT = 8.52$, the ratio of the latter two being the $n = 0.86$ which we have already discussed.

Specification of *one* of these dimensionless ratios [or of others such as T_t/T_c or $\alpha_p T_t = (\partial \ln \tilde{V}/\partial \ln T)_{p,T=T_t}$] by the experimental values at the triple point of argon permits the use of Eq. (5.20) to deduce y_t and all the other properties at this point for comparison with experiment. Tables XVII–XIX show these comparisons for four different procedures: fitting T_t/T_c, fitting $\Delta\tilde{E}^V/RT$ (the method used by Longuet-Higgins and Widom (1964) and by Guggenheim (1965)), fitting $\gamma_V\tilde{V}/R$, and fitting $\alpha_p T$.

Properties compared with experiment in these tables are T/T_B, $\varrho/\varrho_B = \tilde{V}_B/\tilde{V}$, $\alpha_p T$, $T(\alpha_p T)' = [\partial^2(\ln \tilde{V})/(\partial \ln T)^2]_p$, $\gamma_V\tilde{V}/R$, and the vapor pressure expressed as $\ln(p\tilde{V}/RT)$ derived from the equation, valid at low temperatures and liquid densities,

$$\ln(p\tilde{V}/RT) = \left(\int_0^{y_l} \{[\phi(y) - 1]/y\}\, dy\right) - 1 - (4ay/RTb). \qquad (5.21)$$

It is interesting to note that, although the functions II–IV fit the actual hard-sphere system better than any of the others, it is functions V–VII which yield better results for the density–temperature curve of a real fluid. This is shown by the much better fit for $\alpha_p T$ and for $T(\alpha_p T)'$

TABLE XVII

LIQUID PROPERTIES DEFINED BY $T/T_c = 0.556$

Equation of state[a]	T/T_B	ϱ/ϱ_B	$\alpha_p T$	$T(\alpha_p T)'$	$\gamma_V\tilde{V}/R$	$\ln(p\tilde{V}/RT)$
I (van der Waals)	0.165	0.792	0.356	0.826	4.81	−4.24
II [$(PY)_p$]	0.212	1.828	0.651	1.02	8.61	−5.78
III [Scaled particle-$(PY)_c$]	0.209	1.664	0.619	0.97	7.98	−5.54
IV (Guggenheim)	0.207	1.600	0.600	0.96	7.71	−5.43
V	0.218	1.489	0.428	0.91	6.83	−5.11
VI	0.208	1.450	0.457	0.96	6.98	−5.17
VII (Flory)	—	([b])	0.398	0.639	11.54	−5.20
Argon (triple point)	0.205	1.46	0.368	0.57	7.32 (8.52)[c]	−5.88

[a] Roman numerals correspond to the equations given in Table XV.

[b] $y/y_c = 2.73$, compared with the experimental value $y_t/y_c = 2.64$ for Ar.

[c] This is the experimental value of $\Delta\tilde{E}^V/RT$ for argon at its triple point; all the equations I–VII yield the same value for $\Delta\tilde{E}^V/RT$ as for $\gamma_V\tilde{V}/R$.

TABLE XVIII

LIQUID PROPERTIES DEFINED BY $\Delta\tilde{E}^V/RT = 8.52$ OR BY $\gamma_V\tilde{V}/R = 7.32$[a]

Equation of state[b]	T/T_B	$\varrho/\varrho_B = \tilde{V}_B/\tilde{V}$	$\alpha_p T$	$T(\alpha_p T)'$	$\ln(p\tilde{V}/RT)$
I (van der Waals)	0.104	0.883	0.153	0.230	−7.38
	0.118	*0.863*	*0.188*	*0.307*	*−6.33*
II [$(PY)_p$]	0.214	1.82	0.658	1.05	−5.72
	0.233	*1.71*	*0.767*	*1.45*	*−4.96*
III [Scaled particle-$(PY)_c$]	0.200	1.706	0.581	0.86	−5.91
	0.220	*1.608*	*0.675*	*1.16*	*−5.11*
IV (Guggenheim)	0.195	1.659	0.545	0.784	−5.99
	0.214	*1.568*	*0.633*	*1.08*	*−5.17*
V	0.186	1.580	0.312	0.55	−6.40
	0.208	*1.519*	*0.386*	*0.77*	*−5.47*
VI	0.180	1.535	0.348	0.62	−6.33
	0.201	*1.472*	*0.426*	*0.85*	*−5.42*
VII (Flory)	([c])	([c])	0.664	2.08	−3.09
			0.694	*2.26*	*−2.35*
Argon (triple point)	0.205	1.46	0.368	0.57	−5.88

[a] Values given by the latter are given in italics.

[b] Roman numerals correspond to the equations given in Table XV.

[c] For $\Delta\tilde{E}^V/RT = 8.52$, the Flory equation of state yields $T/T_c = 0.680$, compared with the experimental value $T_t/T_c = 0.556$ for Ar, and $y/y_c = 2.46$, compared with the experimental value $y_t/y_c = 2.64$. For $\gamma_V\tilde{V}/R = 7.32$, the Flory equation of state yields $T/T_c = 0.740$ and $y/y_c = 2.31$.

and by the graphs of ϱ/ϱ_c versus T/T_c shown in Fig. 17. While interesting and useful, this better fit is unlikely to be of any fundamental significance; the prediction of second derivatives of the free energy [and in the case of $T(\alpha_p T)'$, a third derivative] is notoriously difficult and any success here is probably fortuitous.

Another second derivative is the configurational heat capacity C_V^{conf}, which Eq. (5.19) requires to be zero [regardless of the form of $\phi(y)$ so long as it is temperature-independent]. Here, all the equations I–VII are all equally unsatisfactory; for argon at its triple point, $\tilde{C}_V/R = 2.35$, leaving $\tilde{C}_V^{\mathrm{conf}}/R = 0.85$.

TABLE XIX

LIQUID PROPERTIES DEFINED BY $\alpha_p T = 0.368$

Equation of state[a]	T/T_B	ϱ/ϱ_B	$T(\alpha_p T)'$	$\gamma_V \tilde{V}/R$	$\ln(p\tilde{V}/RT)$
I (van der Waals)	0.167	0.79	0.874	4.72	− 4.17
II $[(PY)_p]$	0.132	2.30	0.35	17.5	−12.08
III [Scaled particle-$(PY)_c$]	0.135	2.04	0.35	15.1	−10.66
IV (Guggenheim)	0.139	1.93	0.357	13.86	−12.79
V	0.203	1.53	0.71	7.57	− 5.66
VI	0.186	1.52	0.68	8.16	− 6.06
VII (Flory)	([b])	([b])	0.750	12.15	− 5.66
Argon (triple point)	0.205	1.46	0.57	7.32 (8.52)[c]	− 5.88

[a] Roman numerals correspond to the equations given in Table XV.

[b] For $\alpha T = 0.368$, the Flory equation of state yields $T/T_c = 0.536$, compared with the experimental value $T_t/T_c = 0.556$ for Ar, and $y/y_c = 2.77$, compared with the experimental value $y_t/y_c = 2.64$.

[c] This is the experimental value of $\Delta\tilde{E}^V/RT$ for argon at its triple point; all the equations I–VII yield the same value for $\Delta\tilde{E}^V/RT$ as for $\gamma_V\tilde{V}/R$.

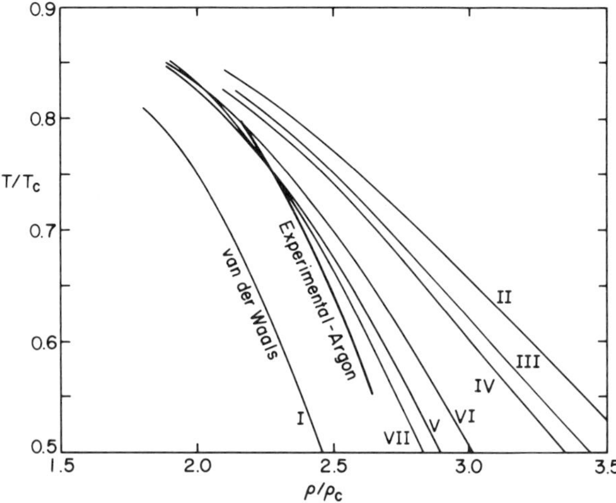

FIG. 17. Reduced temperature T/T_c versus reduced liquid density ϱ/ϱ_c for various forms of Eq. (5.19). (The various functions I–VII are identified in Table XV.) Note the comparison with the experimental values for liquid argon. The reduced density ϱ/ϱ_c is evaluated for $p = 0$, so does not continue to the critical point.

The success of the Flory equation of state (function VII) is especially interesting. Its $\phi(y)$ is theoretically more nearly appropriate for a solid, but since its $\tilde{V}_0$ is a freely adjustable parameter, it gives a very good representation of the variation of liquid density with temperature. It does not predict the pressure variables (e.g., $\gamma_V \tilde{V}/R$) nearly so well, but in practice these are scaled by a third adjustable parameter (essentially the Hijmans c of Section IV, B, 3) which corrects for this.

Different ones of these simple equations of state yield better fits for different properties, so which is the "best" empirical equation to use will remain a matter of opinion. What is important is that (with the exception of VII) they give a reasonably good qualitative representation of the properties of fluids over the entire range of liquid densities and over quite a range of temperatures. Many equations more theoretically sophisticated and obtained with superficially more rigor have proved less satisfactory.

Acknowledgments

The major part of this chapter was written while I was a visiting professor at the University of Otago, Dunedin, New Zealand, and I wish to thank the University of Otago and the U.S. Educational Foundation in New Zealand for their hospitality and support. I wish to thank Dr. D. V. Fenby and Profs. C. M. Knobler, M. L. McGlashan, J. S. Rowlinson, and B. Widom for their criticism and suggestions.

General References

Faraday Soc. (1965). General discussion on intermolecular forces. *Discuss. Faraday Soc.* **40.**

Hirschfelder, J. O., Curtiss, C. F., and Bird, R. B. (1954). "Molecular Theory of Gases and Liquids." Wiley, New York.

Hirschfelder, J. O. ed. (1967). Intermolecular forces. *Advan. Chem. Phys.* **12.**

Leland, Jr., T. W., and Chappelear, P. S. (1968). *Ind. Eng. Chem.* **60** (7), 15.

Rowlinson, J. S. (1969). "Liquids and Liquid Mixtures," 2nd ed. Plenum Press, New York.

Special References

Alder, B. J., and Paulson, R. H. (1965). *J. Chem. Phys.* **43**, 4172.

Alder, B. J., and Wainwright, T. E. (1960). *J. Chem. Phys.* **33**, 1439.

Alder, B. J., Haycock, E. W., Hildebrand, J. H., and Watts, H. (1954). *J. Chem. Phys.* **22**, 1060.

Axilrod, B. M. (1951). *J. Chem. Phys.* **19**, 719.

Axilrod, B. M., and Teller, E. (1943). *J. Chem. Phys.* **11**, 299.

Barker, J. A., and Henderson, D. (1967). *J. Chem. Phys.* **47**, 2856, 4718.

Barker, J. A., and Henderson, D. (1968). *J. Chem. Educ.* **45**, 2.

BARKER, J. A., and LEONARD, P. J. (1964). *Phys. Lett.* **13**, 127.

BARKER, J. A., and POMPE, A. (1968). *Aust. J. Chem.* **21**, 1683.

BARKER, J. A., FOCK, W., and SMITH, F. (1964). *Phys. Fluids* **7**, 897.

BARKER, J. A., HENDERSON, D., and SMITH, W. R. (1968). *Phys. Rev. Lett.* **21**, 134.

BELL, R. J. (1965). *Proc. Phys. Soc. (London)* **86**, 17.

BENNINGA, H., and SCOTT, R. L. (1955). *J. Chem. Phys.* **23**, 1911.

BERNSTEIN, R. B., and MUCKERMAN, J. T. (1967). *Advan. Chem. Phys.* **12**, 389.

BOLTZMANN, L. (1898). "Vorlesung über Gastheorie," Part II. Barth, Leipzig. [English Transl. "Lectures on Gas Theory" (S. G. Brush, transl.). Univ. California Press, Berkeley and Los Angeles, 1964.]

BRÖNSTED, J. N., and KOEFOED, J. (1946). *Kgl. Dan. Vidensk. Selsk. Mat. Fys. Skr.* **22**, No. 17.

BUCKINGHAM, A. D. (1959). *Quart. Rev. Chem. Soc.* **13**, 183.

BYK, A. (1921). *Ann. Phys. (Leipzig)* [**4**] **66**, 157.

BYK, A. (1922). *Ann. Phys. (Leipzig)* [**4**] **69**, 161.

CASIMIR, H. G. B., and POLDER, D. (1938). *Phys. Rev.* **73**, 360.

COOK, D., and ROWLINSON, J. S. (1953). *Proc. Roy. Soc. Ser. A* **219**, 405.

COPELAND, D. A., and KESTNER, N. R. (1968). *J. Chem. Phys.* **49**, 5214.

CURL, Jr., R. F., and PITZER, K. S. (1958). *Ind. Eng. Chem.* **50**, 265.

DALGARNO, A., and KINGSTON, A. E. (1961). *Proc. Phys. Soc. (London)* **78**, 607.

DALGARNO, A., MORRISON, I. M., and PENGELLY, R. M. (1967). *Int. J. Quantum Chem.* **1**, 161.

DEBOER, J. (1948). *Physica (Utrecht)* **14**, 139.

DUNLAP, R. D., and SCOTT, R. L. (1962), *J. Phys. Chem.* **66**, 631.

DYMOND, J. H., and ALDER, B. J. (1968). *Chem. Phys. Lett.* **2**, 54.

DYMOND, J. H., and ALDER, B. J. (1969). *J. Chem. Phys.* **51**, 308.

DYMOND, J. H., RIGBY, M., and SMITH, E. B. (1965). *J. Chem. Phys.* **42**, 2801.

EYRING, H., and HIRSCHFELDER, J. O. (1937). *J. Phys. Chem.* **41**, 249.

FARADAY SOC. (1965). General discussion on intermolecular forces. *Discuss. Faraday Soc.* **40**.

FISHER, M. E. (1967). *Rep. Progr. Phys.* **92**, 615.

FITTS, D. D. (1966). *Annu. Rev. Phys. Chem.* **17**, 69.

FLORY, P. J. (1965). *J. Amer. Chem. Soc.* **87**, 1833.

FLORY, P. J., ORWOLL, R. A., and VRIJ, A. (1964). *J. Amer. Chem. Soc.* **86**, 3507, 3515.

FRISCH, H., KATZ, J. L., PRAESTGARD, E., and LEBOWITZ, J. L. (1966). *J. Phys. Chem.* **70**, 2016.

GORDON, R. G. (1968). *J. Chem. Phys.* **48**, 3929.

GUGGENHEIM, E. A. (1945). *J. Chem. Phys.* **13**, 253.

GUGGENHEIM, E. A. (1953). *Rev. Pure Appl. Chem.* **3**, 1.

GUGGENHEIM, E. A. (1965). *Mol. Phys.* **9**, 43, 199.

GUGGENHEIM, E. A. (1966). "Applications of Statistical Mechanics." Oxford Univ. Press (Clarendon), London and New York.

GUGGENHEIM, E. A., and MCGLASHAN, M. L. (1960a). *Proc. Roy. Soc. Ser. A* **255**, 456.

GUGGENHEIM, E. A., and MCGLASHAN, M. L. (1960b). *Mol. Phys.* **3**, 563.

GUGGENHEIM, E. A., and WORMALD, C. J. (1965). *J. Chem. Phys.* **42**, 3775.

HAKALA, R. W. (1967). *J. Phys. Chem.* **71**, 1880.

HAMANN, S. D., and LAMBERT, J. A. (1954). *Aust. J. Chem.* **7**, 1.

HAPPEL, H. (1906). *Ann. Phys. (Leipzig)* [4], **21**, 342.

HELLER, P. (1967). *Rep. Progr. Phys.* **92**, 731.

HIJMANS, J. (1961). *Physica (Utrecht)* **27**, 433.

HILDEBRAND, J. H. (1929). *Phys. Rev.* **34**, 649, 984.

HILDEBRAND, J. H., and CARTER, J. M. (1932). *J. Amer. Chem. Soc.* **54**, 3592.

HILDEBRAND, J. H., and SCOTT, R. L. (1950). "Solubility of Nonelectrolytes" 3d ed. Reinhold, New York. Reprinted as a paperback with new preface and corrections, Dover, New York, 1964.

HILDEBRAND, J. H., and SCOTT, R. L. (1962). "Regular Solutions," Prentice-Hall, Englewood Cliffs, New Jersey.

HILL, T. L. (1960). "An Introduction to Statistical Thermodynamics." Addison-Wesley, Reading, Massachusetts.

HIRSCHFELDER, J. O., ed. (1967). Intermolecular forces. *Advan. Chem. Phys.* **12.**

HIRSCHFELDER, J. O., CURTISS, C. F., and BIRD, R. B. (1954). "Molecular Theory of Gases and Liquids." Wiley, New York.

HIRSCHFELDER, J. O., BUEHLER, R. J., McGEE, Jr., H. A., and SUTTON, J. R. (1958). *Ind. Eng. Chem.* **50**, 375, 386.

HOLBORN, L., and OTTO, J. (1924). *Z. Phys.* **30**, 320.

JORDAN, J. E., and AMDUR, I. (1967). *J. Chem. Phys.* **46**, 165.

KAC, M., UHLENBECK, G. E., and HEMMER, P. C. (1963). *J. Math. Phys. (N.Y.)* **4**, 216.

KADANOFF, L. P., GOTZE, W., HAMBLEN, D., HECHT, R., LEWIS, E. A. S., PALCIAUSKAS, V. V., RAYL, M., SWIFT, J., ASPNES, D., and KANE, J. (1967). *Rev. Mod. Phys.* **39**, 395.

KARPLUS, M., and KOLKER, H. J. (1964). *J. Chem. Phys.* **41**, 3955.

KEESOM, W. H. (1921). *Phys. Z.* **22**, 126, 643.

KEESOM, W. H. (1922). *Phys. Z.* **23**, 225.

KELLER, J. B., and ZUMINO, B. (1959). *J. Chem. Phys.* **30**, 1351.

KIHARA, T. (1953). *Rev. Mod. Phys.* **25**, 831.

KONOWALOW, D. D., and CARRA, S. (1965). *Phys. Fluids* **8**, 1585.

KOZAK, J. J., and RICE, S. A. (1968). *J. Chem. Phys.* **48**, 1226.

LEACH, J. W., CHAPPELEAR, P. S., and LELAND, T. W. (1966). *Proc. Amer. Petrol. Inst. (Sect. III)* **46**, 223.

LELAND, Jr., T. W., and CHAPPELEAR, P. S. (1968). *Ind. Eng. Chem.* **60** (7), 15.

LENNARD-JONES, J. E. (1924). *Proc. Roy. Soc. Ser. A* **106**, 463.

LENNARD-JONES, J. E. (1931). *Proc. Phys. Soc. (London)* **43**, 461.

LONDON, F. (1930). *Z. Phys. Chem. Abt. B* **11**, 222.

LONDON, F. (1937). *Trans. Faraday Soc.* **33**, 8.

LONGUET-HIGGINS, H. C., and WIDOM, B. (1964). *Mol. Phys.* **8**, 549.

LUNBECK, R. J. (1950). Ph. D. Thesis, Amsterdam.

McGLASHAN, M. L. (1965). *Discuss. Faraday Soc.* **40**, 59.

MARGENAU, H. (1938). *J. Chem. Phys.* **6**, 897.

MARGENAU, H., and KESTNER, N. (1967). "Theory of Intermolecular Forces." Pergamon Press, New York.

MASON, E. A., and MONCHICK, L. (1967). *Advan. Chem. Phys.* **12**, 329.

MASON, E. A., and SPURLING, T. H. (1968). "The Virial Equation of State," Pergamon Press, New York.

MIKOLAJ, P. G., and PINGS, C. J. (1967). *J. Chem. Phys.* **46**, 1412.

MUNN, R. J., and SMITH, F. J. (1965). *J. Chem. Phys.* **43**, 3998.

ORWOLL, R. A., and FLORY, P. J. (1967). *J. Amer. Chem. Soc.* **89**, 6814, 6822.

PITZER, K. S. (1939). *J. Chem. Phys.* **7**, 583.
PITZER, K. S. (1955). *J. Amer. Chem. Soc.* **77**, 3427.
PITZER, K. S., and BREWER, L. (1961). "Thermodynamics," 2nd ed. (original book by G. N. Lewis and M. Randall). McGraw-Hill, New York.
PITZER, K. S., and CURL, Jr., R. F. (1957). *J. Amer. Chem. Soc.* **79**, 2369.
PITZER, K. S., LIPPMANN, D. Z., CURL, Jr., R. F., HUGGINS, C. M., and PETERSEN, D. E. (1955). *J. Amer. Chem. Soc.* **77**, 3433.
POPLE, J. A. (1954). *Proc. Roy. Soc. Ser. A* **221**, 498, 508.
PRIGOGINE, I. (1957). "The Molecular Theory of Solutions." North-Holland Publ., Amsterdam.
PRIGOGINE, I., TRAPPENIERS, N., and MATHOT, V. (1953). *Discuss. Faraday Soc.* **15**, 93.
RASAIAH, J., and STELL, G. (1970). *Mol. Phys.* **18**, 249.
REE, F. H., and HOOVER, W. G. (1967). *J. Chem. Phys.* **46**, 4181.
REISS, H., FRISCH, H. L., and LEBOWITZ, J. L. (1959). *J. Chem. Phys.* **31**, 369.
RICE, O. K. (1941). *J. Amer. Chem. Soc.* **63**, 3.
RIEDEL, L. (1954). *Chem.-Ing. Tech. Z.* **26**, 83, 259, 679.
RIEDEL, L. (1955). *Chem.-Ing. Tech. Z.* **27**, 209, 475.
RIEDEL, L. (1956). *Chem.-Ing. Tech. Z.* **28**, 557.
ROSS, J., ed. (1966). Molecular beams. *Advan. Chem. Phys.* **10**.
ROSSI, J. C., and DANON, F. (1965). *Discuss. Faraday Soc.* **40**, 97.
ROTHE, E. W., and NEYNABER, R. H. (1965). *J. Chem. Phys.* **43**, 4177.
ROWLINSON, J. S. (1954). *Trans. Faraday Soc.* **50**, 647.
ROWLINSON, J. S. (1959). "Liquids and Liquid Mixtures," 1st ed. Butterworth, London.
ROWLINSON, J. S. (1965a). *Mol. Phys.* **9**, 217.
ROWLINSON, J. S. (1965b). *Discuss. Faraday Soc.* **40**, 19.
ROWLINSON, J. S. (1967). *Mol. Phys.* **12**, 513.
ROWLINSON, J. S. (1969). "Liquids and Liquid Mixtures," 2nd ed. Plenum Press, New York.
RUSHBROOKE, G. S., and SILBERT, M. (1967). *Mol. Phys.* **12**, 505.
SHERWOOD, A. E., and PRAUSNITZ, J. M. (1964a). *J. Chem. Phys.* **41**, 413.
SHERWOOD, A. E., and PRAUSNITZ, J. M. (1964b). *J. Chem. Phys.* **41**, 429.
SLATER, J. C., and KIRKWOOD, J. G. (1931). *Phys. Rev.* **37**, 682.
SMITH, E. B., and ALDER, B. J. (1959). *J. Chem. Phys.* **30**, 1190.
STELL, G. (1970). Personal Communication.
STOCKMAYER, W. H. (1941). *J. Chem. Phys.* **9**, 398.
TANNER, C. C., and MASSON, I. (1930). *Proc. Roy. Soc. Ser. A* **126**, 268.
THIELE, E. (1963). *J. Chem. Phys.* **39**, 474.
VAN DER WAALS, J. D. (1873). On the continuity of the gaseous and liquid states. Ph. D. Thesis, Univ. of Leiden.
WEIR, R. D., WYNN JONES, I., ROWLINSON, J. S., and SAVILLE, G. (1967). *Trans. Faraday Soc.* **63**, 1320.
WERTHEIM, M. S. (1963). *Phys. Rev. Lett.* **10**, 321.
WERTHEIM, M. S. (1964). *J. Math. Phys. (N.Y.)* **5**, 643.
WESTWATER, W., FRANTZ, H. W., and HILDEBRAND, J. H. (1928). *Phys. Rev.* **31**, 135.
WHALLEY, E., LUPIEN, Y., and SCHNEIDER, W. G. (1953). *Can. J. Chem.* **31**, 722.
WOOD, W. W., and JACOBSON, J. D. (1957). *J. Chem. Phys.* **27**, 1207.
ZWANZIG, R. W. (1954). *J. Chem. Phys.* **22**, 1420.

Chapter 2

Structure of Liquids

SOW-HSIN CHEN

I. Characteristics of the Liquid State

An equilibrium state of matter can be specified by giving its pressure p, specific volume v (or density ϱ_m), and temperature T. There exists for every real substance a relation between p, v, and T, called the equation of state, which defines a surface in the p–v–T state space. There are certain regions of this surface in which the substance can exist in a single phase

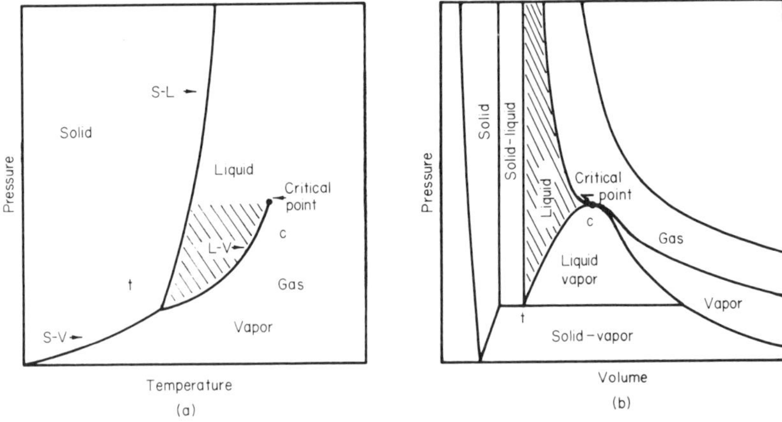

FIG. 1. (a) Projection of a typical p–v–T surface of simple liquids onto the p–T plane. Here, t is the triple point and c the critical point. The shaded area corresponds to the liquid phase. (b) Projection of the same p–v–T surface onto the p–v plane. The shaded area corresponds to the liquid phase.

only. Figure 1 shows two perpendicular projections of a typical p–v–T surface onto the p–T and p–v planes for a simple substance like argon. It is seen that the liquid phase exists over an extended region of the state space confined between the "vapor-pressure curve" (marked L–V) and the "melting curve" (marked S–L) shown in fig. 1(a). The triple point t and the critical point c mark the lowest and the highest temperatures at which existence of the liquid phase is still possible, and at the critical point, the liquid and gaseous phases are indistinguishable. Both the temperature and the density at the two points can be appreciably different. Near the triple point, the density of a liquid is only slighlty smaller than that of the solid, while at the critical point, the density is closer to that of the gas. As we shall show later, the structure of a liquid is strongly dependent on its density; therefore discussions of the structure of a liquid would not be meaningful without a detailed specification of its thermodynamic state.

The existence of three states of matter can be understood physically in terms of the relative importance of thermal energy and potential energy of a constituent atom (or molecule) in determining the state of the matter. At low temperatures, where the thermal energies of atoms are small, a substance exists in a solid state in which atoms arrange themselves in a close-packed crystalline form, each subjected to the influence of the potential energies of all other atoms. The thermal energy causes only a small perturbation to an otherwise perfect lattice arrangement in which, on the average, a long-range order exists. The solid state is therefore characterized by its anisotropy in structure. On the other hand, at high temperatures, the increased thermal energies of atoms disrupt the regular crystalline binding and the substance transforms into a gaseous state. In the gaseous state, the average distance between atoms is much larger (about ten times) than that in the solid state and the chaotic nature of thermal motions completely destroys the long-range order. In the gaseous state, correlation between positions of atoms is negligible and the state is characterized by chaotic thermal motions with occasional binary encounters of atoms. The liquid state is somewhere in between these two extreme states of matter and thus both the thermal motions and the atomic bindings are important. Besides executing the random thermal motions, an atom in a liquid is constantly interacting with many of its neighbors and therefore correlations between positions of an atom and its near neighbors are important. Thus, in liquids, there exists a short-range order, the precise meaning of which we shall elaborate in subsequent sections. Much of the short-range order in a liquid is caused by the finite size of atoms, and the degree of the short-range order is strongly dependent on the density of the liquid. Perhaps the only universal feature of the structure of liquid in its whole range of existence is its "isotropy", which means that the order is independent of the direction of observation.

It is important to observe that liquids have a unique ability to form a free surface, a property which is absent even for a compressed gas, and which shows that the cohesion between atoms plays an important role in determination of the liquid structure. Another characteristic of the liquid state is its lack of rigidity and comparatively low resistance to flow. This is due to the comparatively large diffusion coefficient in liquids. In liquids, the thermal energy of an atom is almost large enough to overcome the potential barrier of neighboring atoms and after a number of collisions an atom may jump to a nearby vacant site. Most simple liquids have diffusion coefficients D of the order 10^{-5} cm^2/sec and intermolecular separations σ of the order 3–4 Å. If we assume an atom to undergo a random walk in liquids, we can estimate the average time it takes to jump a step by $\tau = \sigma^2/6D$, which is about 10^{-11} sec. We may also regard τ as approximately the relaxation time of the local structure. In this sense, we can say that, if one applies an external perturbation of the frequency higher than 10^{11} Hz, then the local structure does not have time to relax and a liquid will respond much like a solid. This point is, as we shall see in Section IV, to some extent supported by the neutron inelastic scattering experiment.

A. Qualitative Description of the Short-Range Order in Liquids

On the basis of thermodynamic data alone, one can argue with equal validity that liquids are either gaslike or solidlike. Following the success of the van der Waals theory in explaining the continuous transition from the gaseous to the liquid phases across the critical point, it is natural to assume that near the critical point a liquid has a completely disordered structure similar to a gas. Experimentally, one knows that for a large number of monatomic substances the average entropy of fusion is 2.2 cal $^\circ K^{-1}$ mol^{-1}, which is close to the value of the so-called communal entropy 2.0 cal $^\circ K^{-1}$ mol^{-1}. It can be shown by consideration of the partition function that for a system of N noninteracting particles the increase in entropy in going from a completely ordered state to a completely disordered state is Nk_B (where k_B is the Boltzmann constant), which is just 2.0 cal $^\circ K^{-1}$ mol^{-1}. This consideration seems to indicate that a solid completely loses its order in transforming to a liquid. On the other hand, one can find equally convincing evidence which points to the similarities between properties of liquids and solids at the melting point. Table I, which is taken from Egelstaff (1967), lists some properties of liquids and the corresponding solids for two simple substances, argon and sodium. One notices that the density changes only slightly upon melting, which implies that the atomic separations and hence the interactions do not change significantly from solids to liquids; also, the fact that the specific heat C_v has a value roughly equal to the harmonic oscillator value $3Nk_B$ in both solids and liquids indicates that the nature of thermal motions in the two states cannot be too different.

However, the fact that solids lose their long-range order upon melting does not exclude the possibility for liquids to have short-range order. Moreover, we shall see later that the short-range order depends strongly on the density of the liquid, which would then explain why liquids and solids share many common properties near the melting curve, while near the critical point, liquids have similar properties to gases. What is important in a theory of liquids is therefore to find a quantitative way of defining the concept of short-range order and to be able to measure it experimentally for a wide range of conditions.

Our present information regarding the structure of liquids has been mostly furnished by the results of radiation scattering experiments. Historically, it was Zernike and Prins (1927) and Debye (1931) who first established the relation between the angular intensity distribution of X-ray scattering and the local density distribution of particles in liquids.

TABLE I[a]

Property	Argon		Sodium	
	Solid	Liquid	Solid	Liquid
Melting and boiling points at 1 atm (°K)	84 (mp)	87.5 (bp)	371 (mp)	1150 (bp)
Liquid range at 1 atm (°K)	—	3	—	782
Liquid range critical point/triple point	—	1.8	—	7.5
Density (g cm^{-3})	1.636	1.407	0.951	0.927
Latent heat of vaporization (cal g mol^{-1})	1850 (0°K)	1357.5 (87.5°K)	24,000	23,000
Heat capacity per atom				
C_v (ergs °K^{-1})	$2.89k_B$	$2.32k_B$	$3.1k_B$	$3.4k_B$
C_p (ergs °K^{-1})	$3.89k_B$	$5.06k_B$	$3.3k_B$	$3.8k_B$
Compressibility (cm^2 dyn^{-1})				
isothermal	0.95×10^{-10}	2.00×10^{-10}	1.7×10^{-11}	1.9×10^{-11}
adiabatic	0.55×10^{-10}	0.92×10^{-10}	1.6×10^{-11}	1.8×10^{-11}
Viscosity (P)	—	2.8×10^{-3}	—	6.8×10^{-3}
Diffusion coefficient (cm^2 sec^{-1})	10^{-9}	1.6×10^{-5}	1.9×10^{-7}	4.3×10^{-5}
Heat conductivity (cal sec^{-1} cm^{-1} °K^{-1})	7.1×10^{-4}	2.9×10^{-4}	0.322	0.203

[a] From Egelstaff (1967).

From the interpretation of the X-ray Laue photograph, it was found that the distribution of particles within a small region about any fixed particle was partially ordered. However, it was also found that this partial ordering of near-neighbor particles was fundamentally different from the long-range crystalline order in solids. The diffraction pattern of a single crystal consists of symmetrically placed points of high intensity which are the image of the reciprocal lattice points. The diffraction pattern for an amorphous solid or very fine powder of crystallites would consist of a series of sharp rings concentric with the incident beam, with the intensity decreasing rapidly toward outer rings. On the other hand, for a liquid, although the ring pattern is similar to the amorphous solid or the powder, except for the first one, the rings are rather diffuse, with intensity decreasing rapidly toward outer rings. The diffuseness of the rings is evidence that the order in liquids is not a sharp one, and the rapid decrease of the intensity precludes the existence of long-range order. The existence of a ring pattern instead of symmetrically placed points also shows that liquids are structurally isotropic.

As will be shown in Section IV, X-ray diffraction only gives information on the time-averaged or static, structure, of a system. The advent of the neutron inelastic scattering technique made it possible for the time-dependent, or dynamic, structure of liquids to be studied. The essential point is that an inelastically scattered neutron spectrum can be analyzed in terms of the space–time correlations of particles in the system. The theoretical foundation of the interpretation was laid down by Van Hove in 1954 and the first experimental investigation in water was performed by Brockhouse in 1958. This aspect of the time-dependent structure of liquids will be discussed in Section III.

It is the intention of the present chapter to show that the structural information, both static and dynamic, in liquids can be used as a starting point for a theory of liquids. In Section II, we will outline a well-known procedure by which the equilibrium thermodynamic properties of liquids can be obtained through the use of the static structural information. This is the so-called pair theory of liquids, in which the starting point is the assumption of the existence of an effective pair potential between atoms. Therefore, in the subsequent sections, we briefly discuss the form of the effective pair potential for two types of simple liquids, namely the monatomic dielectric liquid and the liquid metal. It is rather plausible that, if the equilibrium properties of liquids can be obtained through the static structural information, the nonequilibrium properties such as the transport coefficients can also be obtained through the dynamic structure information. Unfortunately, the theoretical program is not yet completed in this respect, and in Section III we will only present some rudimentary results obtained in the recent literature.

The viewpoint we adopt in this chapter is sometimes called the "distribution-function" approach to the theory of liquids. This is because, as we shall see in the following sections, we base our discussions of the structure of liquids on a series of atomic (or molecular) distribution functions specifying the probability of finding sets of molecules in particular configurations. This approach has a strength that the lower-order distribution functions are experimentally observable quantities and that within the approximation of the pair theory the statistical mechanics gives all the equilibrium properties of liquids in terms of these distributions functions. The pair theory of liquids and the theory of transport coefficients in liquids are both undergoing vigorous development at present. We will try to present some most recent experimental efforts in this direction and to shown their theoretical relevance. The experimental results we shall quote include both scattering experiments and the so-called "molecular-dynamics" calculations, which can properly be viewed as a form of experiment.

There are other approaches to the theory of liquids, based primarily on analogy between the solid and the liquid states. The starting point of these "lattice"-type theories (such as the cell theory the hole theory, and the tunnel theory of liquids) is usually a physical assumption about the static structure of liquids. The purpose of these theories is to calculate the equilibrium properties of liquids starting from these structural assumptions. Whatever the success or the failure of the theories, they are of no concern to us in this chapter because of two reasons: our purpose here is to discuss the structure of liquids, both static and dynamic, in general and rigorous terms; and the lattice-type theories are completely unsuited for the discussion of the dynamic structure and transport properties of liquids. However, readers who are interested in the lattice theory of liquids should consult an excellent exposition of these topics in a recent monograph by Barker (1963).

B. The Pair Potential in Dielectric Liquids

In the pair theory of liquids, one assumes that the total potential energy of a liquid can be written as a sum of two-particle interactions,

$$U\{N\} = \sum_{i>j}^{N} u(r_{ij}). \tag{1.1}$$

In general, in a dense system like a liquid there should be terms involving three particles, four particles, etc. However, the recent success of molecular-dynamic calculations performed in simple liquid argon using this approximation suggests that the many-body forces, if they are at all important, behave so as to realize an effective pair interaction which is state-independent to a good approximation. There are some limits to be placed upon $u(r)$ by the requirement that the system be thermodynamically stable. The requirement for the total potential is

$$U\{N\} \geq -NB, \tag{1.2}$$

where B is a positive constant. One set of conditions for Eq. (1.2) to be satisfied, when working in a three-dimensional space and when $u(r)$ is bounded from below, is

$$\begin{aligned} u(r) &> C_1/r^{3+\delta} \qquad \text{for} \quad r \to 0 \\ |u(r)| &< C_2/r^{3+\delta} \qquad \text{for} \quad r \to \infty, \end{aligned} \tag{1.3}$$

where C_1 and C_2 are positive constants. The first of these conditions ensures that the core of the potential is sufficiently repulsive that the system does not collapse and the second ensures that the potential is sufficiently short-ranged that the partition functions remain bounded. In practice, with the potentials discussed below, conditions (1.3) are satisfied.

For a dielectric liquid with spherical molecules, some choices of relatively simple analytical forms for $u(r)$ are:

1. The Lennard-Jones potential,

$$u(r) = 4\varepsilon[(\sigma/r)^{12} - (\sigma/r)^6], \qquad \varepsilon > 0, \quad \sigma > 0. \tag{1.4}$$

2. The modified Buckingham potential,

$$\begin{aligned} u(r) &= \frac{\varepsilon}{1-(6/\alpha)}\left[\frac{6}{\alpha}\exp\alpha\left(1-\frac{r}{r_{\mathrm{m}}}\right)-\left(\frac{r_{\mathrm{m}}}{r}\right)^6\right]_{r>r_{\max}} \\ &= +\infty \qquad \text{for} \quad r < r_{\max}. \end{aligned} \tag{1.5}$$

3. The hard-sphere potential,

$$\begin{aligned} u(r) &= +\infty \qquad \text{for} \quad r < \sigma \\ &= \quad 0 \qquad \text{for} \quad r \geq \sigma. \end{aligned} \tag{1.6}$$

These potentials are illustrated in Fig. 2.

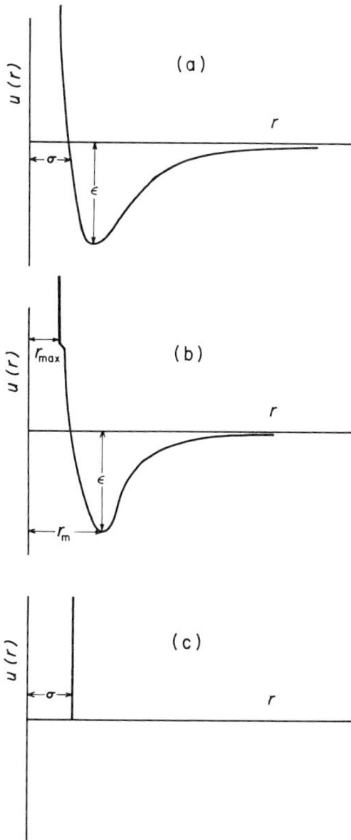

FIG. 2. (a) The Lennard-Jones potential. (b) The modified Buckingham potential. (c) The hard-sphere potential.

In general, the potential consists of a steep repulsive term and a weak attractive term. The repulsive term occurs from the fact that the Pauli exclusion principle gives rise to an effective repulsion when two atoms are sufficiently close together that their electron shells overlap. Quantum-mechanical calculations show that the repulsive term should be of the exponential form as in the modified Buckingham potential, but in practice the r^{-12} form as in the Lennard-Jones potential or an infinite potential wall as in the hard-sphere potential are substitued for convenience. The attractive term arises from the van der Waals force between two dipoles and can be shown to have a form $\sim r^{-6}$. In any event, except for the simplest atom, like argon, the force between a pair of molecules is a complicated one, and one should regard the above potentials as purely phenomenological for the purpose of computations. The recent molecular-dynamic calculations seem to show that the Lennard-Jones potential is a good representation for argon. In practice, one could determine the parameters σ, the core diameter, and ε, the well depth, in the L-J potential by calculating the second virial coefficient from the potential and fitting it to the temperature dependence of the coefficient determined experimentally in the gaseous state. In this way, one obtains the values listed in Table II of the parameters for some simple atoms and molecules. A fuller discussion of these topics may be found in Chapter 1.

TABLE II

LENNARD-JONES PARAMETERS OBTAINED FROM SECOND VIRIAL COEFFICIENTS[a]

Gas	ε/k_B (°K)	σ (Å)
Ne	34.9	2.78
A	119.8	3.405
Kr	171	3.60
CH_4	148.2	3.817

[a] From Egelstaff (1967).

C. THE PAIR POTENTIAL IN METALLIC LIQUIDS

In considering atomic interactions in liquid metals, one should begin by asking two basic questions: (1) Can the total potential energy of the liquid be written as a sum of effective central pair potentials as in (1.1)? (2) Can one write down a general form of the effective potential assuming question 1 to be answered affirmatively?

Due to the existence of free electrons in a metallic system, the effective ion–ion interaction is much more subtle than any of the forms we wrote down for insulators. In the case of insulators, we implicitly assume that the basic atom–atom interaction in a simple system is the same in gaseous, liquid, and solid states. The fact that metallic properties only appear when atoms are brought together to form a liquid or a solid indicates that the effective interaction in a condensed system is quite different from interaction between two atoms in a vacuum. Despite this complication, the recent approximate theory of simple metals answers question 1 in the affirmative (Harrison 1966). In a simple metal such as Na, Al, or Mg where the ion core is small, one can picture the system as consisting of point metal ions immersed in a uniform electron gas. As an ion moves around, electrons follow like a cloud around it and screen off most of the strong electrostatic interactions. Thus one can imagine a metallic liquid as a collection of weakly interacting neutral "pseudoatoms." According to this simple theory, interaction between a pair of pseudoatoms consists of two terms: first, a direct interaction which is the Coulomb interaction between ions with a modified charge Z^*; second, an indirect interaction between ions via electrons. The second contribution is the more complicated one and is called the band-structure energy, which takes into account the screening effect of electrons. Thus

$$\begin{aligned} u(r) &= u_{\mathrm{d}}(r) + u_{\mathrm{in}}(r) \\ &= (Z^{*2}e^2/r) + (v/\pi^2) \int_0^\infty F(q)(\mathrm{Sin}\, qr/qr) q^2 \, dq, \end{aligned} \tag{1.7}$$

where v is the volume per atom and $F(q)$ the energy–wave number characteristic tabulated by Harrison (1966) for most simple metals. Figure 3 shows the effective pair interaction in aluminum calculated by Harrison from the theory. We see, besides the steep repulsion at short distance due to the finite size of the ionic core, oscillations at larger distances. This characteristic oscillation can be shown to be due to the sharpness of the Fermi surface in metal. In fact, the asymptotic behavior

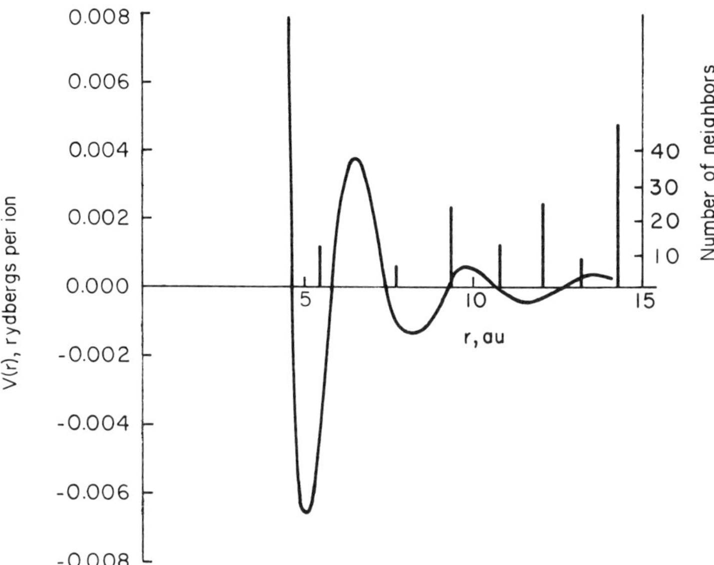

Fig. 3. The effective pair potential in aluminum (a calculation by Harrison, 1966).

of (1.7) at large distance can be shown to be

$$u(r) \simeq [9\pi Z^2[W(2k_F)]^2/E_F]\left[\frac{\cos 2k_F r}{(2k_F r)^3}\right], \tag{1.8}$$

where Z is the ionic charge, E_F and k_F the Fermi energy and momentum, and $W(q)$ the so-called pseudopotential. In a real metal, the Fermi surface is not perfectly sharp and one expects the oscillation to be damped exponentially. This damping is necessary to make the potential (1.7) satisfy the stability criteria (1.3).

II. Static Structure of Liquids

A. Static Pair Correlation Function and Its Relation to Thermodynamic Properties of Liquids

A quantitative description of the static structure of a liquid can be made through a series of molecular distribution functions defined in the following. Consider a system of N particles contained in a fixed volume at a temperature T. The Hamiltonian of the system can be written as

$$H = \sum_{i=1}^{N} \frac{1}{2M} \mathbf{P}_i^2 + U\{N\}, \tag{2.1}$$

where M is the molecular mass, $\mathbf{P}_i$, the momentum of the ith molecule, and U the potential energy of an assembly of N particles, assumed to depend on the coordinates of the N molecules only.

We define, in the canonical ensemble of the system, the n-particle distribution function as

$$n_N^{(n)}(1, 2, \ldots, n) = (N!/(N-n)!)(1/Q_N) \int_V \cdots \int \exp(-U\{N\}/k_B T)\, d\{N-n\}, \tag{2.2}$$

where

$$Q_N = \int_V \cdots \int \exp(-U\{N\}/k_B T)\, d\{N\} \tag{2.3}$$

is the configurational partition function. According to the definition (2.2), the n-particle distribution function consists of two factors. The first factor,

$$(1/Q_N) \int_V \cdots \int \exp(-U\{N\}/k_B T)\, d\{N-n\},$$

expresses the probability, when we observe the configuration of the system of N particles with given U, that particle 1 will be observed in $d\mathbf{r}_1$ at $\mathbf{r}_1$, particle 2 in $d\mathbf{r}_2$ at $\mathbf{r}_2, \ldots,$ and particle n in dr_n at $\mathbf{r}_n$, irrespective of the configuration of the remaining $N - n$ molecules. The second factor takes into account that molecules are indistinguishable and there are N choice for the molecule in $d\mathbf{r}_1$, $N - 1$ for $d\mathbf{r}_2, \ldots,$ and $(N - n + 1)$ for $d\mathbf{r}_n$, or a total of

$$N(N-1) \cdots (N-n+1) = N!/(N-n)!$$

possibilities of taking n molecules from N to put into the given configuration. Therefore $n_N^{(n)}(1, 2, \ldots, n)\, d1\, d2 \cdots dn$ is the probability that, if the configuration of the system of N molecules is observed, a molecule (not necessarily molecule 1) will be found in $d\mathbf{r}_1$ at $\mathbf{r}_1$, and another in $d\mathbf{r}_2$ at $\mathbf{r}_2, \ldots,$ and yet another in $d\mathbf{r}_n$ at $\mathbf{r}_n$. It is important to note that, by definition (2.2), all distribution functions are functions of two intensive variables, v (atomic volume) and temperature, besides the particle coordinates.

From the definition (2.2), we have the normalization condition

$$\int_V \cdots \int n_N^{(n)}(1, 2, \ldots, n)\, d\{n\} = N!/(N-n)! \tag{2.4}$$

and, in particular,

$$\int_V n_N^{(1)}(1)\, d1 = N, \tag{2.5}$$

$$\int_V \int n_N^{(2)}(1, 2)\, d1\, d2 = N(N-1). \tag{2.6}$$

In a homogeneous, isotropic system such as a liquid, $n_N^{(1)}(1)$ must be a constant independent of the position and thus is equal to the number density, N/V, following from (2.5). Likewise, in a liquid, $n_N^{(2)}(1, 2)$ must depend on the relative positions of particles 1 and 2, $\mathbf{r}_{12}$, only, and therefore, from (2.6),

$$\int_V n_N^{(2)}(12)\, d(12) = (1/V)N(N-1) = n_N^{(1)}(1)(N-1). \tag{2.7}$$

If the distribution of molecules is completely random, the probability that molecule 1 is in $d\mathbf{r}_1$, 2 in $d\mathbf{r}_2$, . . . , and n in $d\mathbf{r}_n$ is

$$\frac{d1}{V}\frac{d2}{V}\cdots\frac{dn}{V}=\frac{1}{V^n}\,d\{n\}.$$

Then

$$n_N^{(n)}(1, 2, \ldots, n)=\frac{1}{V^n}\frac{N!}{(N-n)!}=\left(\frac{N}{V}\right)^n\left[1+O\left(\frac{1}{N}\right)\right] \quad (2.8)$$

This random distribution is realized in the limit $T\to\infty$ or in a gas of infinite dilution, where the correlation between molecules is completely lost. We therefore define an n-particle correlation function $g_N^{(n)}(1, 2, \ldots, n)$ such that

$$n_N^{(n)}(1, 2, \ldots, n)=(N/V)^n g_N^{(n)}(1, 2, \ldots, n). \quad (2.9)$$

For a liquid, (2.5) and (2.9) give

$$g_N^{(1)}(1)=1. \quad (2.10)$$

Also, $g_N^{(2)}(1, 2)=g_N^{(2)}(12)$, and therefore (2.6) and (2.7) lead to

$$(N/V)\int_V g_N^{(2)}(12)\,d(12)=N-1. \quad (2.11)$$

Instead of a closed system with a fixed N, we can also consider an open system with variable N. Imagine a fixed volume element within the original canonical system. The number of particles N in the volume element is subject to change due to thermal fluctuations of the system. The statistical behavior of the subsystem is then described by the grand canonical ensemble.

Define the n-particle distribution function in the ground canonical ensemble as

$$n^{(n)}(1, 2, \ldots, n)=\sum_{N\geq n} n_N^{(n)}(1, 2, \ldots, n)P_N(V, T) \quad (2.12)$$

where

$$P_n(V, T)=Z^N Q_N/\Xi N! \quad (2.13)$$

is the probability of finding N particles in the sybsystem and Ξ is the grand partition function,

$$\Xi=\sum_N Z^N Q_N/N!, \quad (2.14)$$

and $Z = \exp(\mu/kT)$ is the activity (or fugacity). From (2.12), we immediately get

$$\int_V n^{(n)}(1, 2, \ldots, n)\, d\{n\} = \langle N!/(N! - n)!\rangle_{\text{all } N}, \tag{2.15}$$

which, in a liquid and in the special cases of $n = 1, 2$, reduces to

$$n^{(1)}(1) = \langle N\rangle/V = \varrho \tag{2.16}$$

$$\int_V \int n^{(2)}(1, 2)\, d1\, d2 = V \int_V n^{(2)}(12)\, d(12) = \langle N^2\rangle - \langle N\rangle. \tag{2.17}$$

We have denoted the number density by ϱ. Equation (2.17) is important in the discussion of thermodynamic properties, as will be seen below.

In general, the difference between a relation derived from the canonical ensemble and one from the grand canonical ensemble is that, in the latter, N is replaced by $\langle N\rangle$. In the limit that the system becomes large (or of macroscopic size), that is, $V \to \infty$, $N \to \infty$, while $(N/V) = \varrho = \text{const}$, the probability distribution for N becomes very sharply peaked about $\langle N\rangle$, so that measurements of N would effectively give $\langle N\rangle$ and the results obtained from the two ensembles coincide. In this limit, (2.15) becomes

$$\int_V \cdots \int n^{(n)}(1, 2, \ldots, n)\, d\{n\} = \langle N^n\rangle[1 + O(1/\langle N\rangle)], \tag{2.15a}$$

and one can similarly define the correlation function by

$$n^{(n)}(1, 2, \ldots, n) = (\langle N\rangle/V)^n g^{(n)}(1, 2, \ldots, n). \tag{2.18}$$

For a macroscopic system, $g^{(n)}$ and $g_N^{(n)}$ are usually indistinguishable.†

In the following, we shall be exclusively concerned with $g^{(2)}(1, 2)$ which, for a liquid, will be denoted simply as $g(r)$ and is called the "static pair correlation function." The pair correlation function is a quantity of fundamental importance in the theory of the liquid state because it can be measured rather directly in the scattering experiment and also because of the fact that the thermal and caloric equations of state as well as some other thermodynamic quantities of liquids are expressible entirely in terms of it and knowledge of the pair potential between molecules. The function $g(r)$ is sometimes called the "radial distribution function." The reason can be seen as follows. We see from the definition that $n^{(2)}(1, 2)\, d1\, d2$ is a measure of the joint probability of finding a

† The situations in which $g^{(n)}$ and $g_N^{(n)}$ must be distinguished do not arise in this chapter

molecule in $d1$ and another in $d2$. Therefore the conditional probability of having a specific molecule in $d1$ and another in $d2$ is then proportional to $n^{(2)}(1, 2)\, d1\, d2/n^{(1)}(1)\, d1$. Since in a liquid $n^{(1)}(1) = \varrho$ and $n^{(2)}(1, 2)\, d1\, d2 = \varrho^2 g(r_{12})\, d\mathbf{r}_{12}\, d\mathbf{r}_1$, we have then that $g(r_{12})\, d\mathbf{r}_{12}$ is proportional to the probability of having a molecule in $d\mathbf{r}_2$ at $\mathbf{r}_{12}$, knowing a specified molecule at $\mathbf{r}_1$. It follows from dimensional considerations that $g(r)4\pi r^2\, dr$ is the average number of molecules at a distance between r and $r + dr$ from a specified molecule at the origin. Thus $g(r)$ is a dimensionless quantity describing the distribution of molecules in the radial direction viewing from a molecule located at the origin. In a liquid, owing to finite size of a molecule, $g(r)$ is zero for $r < \sigma$, where σ is the hard-core diameter of the molecule. As $r = |\mathbf{r}_1 - \mathbf{r}_2|$ becomes very large, we expect that there would be no correlations between a molecule at $\mathbf{r}_1$ and at $\mathbf{r}_2$. This means $n^{(2)}(1, 2) = n^{(1)}(1)n^{(1)}(2) = \varrho^2$. Where correlations between molecules are appreciable, $g(r)$ oscillates around the asymptotic value unity. Figure 4 shows the radial distribution function of liquid argon at three different temperatures obtained from the computer experiment of Verlet (1968).

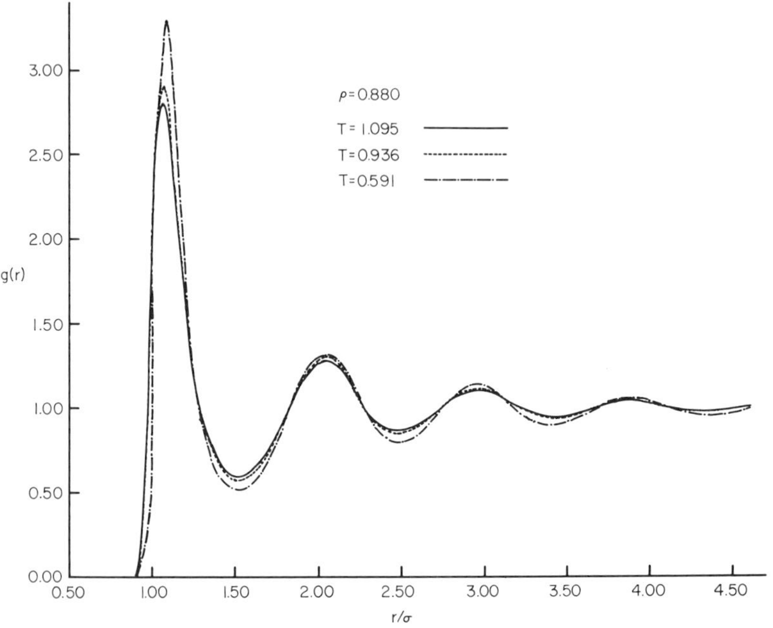

FIG. 4. The radial distribution function for liquid argon from molecular-dynamic computer experiment (Verlet, 1968).

1. *Ornstein–Zernike Fluctuation Theorem*

Equation (2.17), which relates integral of the pair distribution function to the fluctuation in number of molecules in the subsystem, can be rewritten in terms of the pair correlation function. Since

$$\int_V \int n^{(1)}(1)n^{(1)}(2)\, d1\, d2 = \langle N\rangle^2, \tag{2.19}$$

substract (2.19) from (2.17) to obtain

$$\int_V \int [n^{(2)}(1, 2) - n^{(1)}(1)n^{(1)}(2)]\, d1\, d2 = \langle N^2\rangle - \langle N\rangle^2 - \langle N\rangle \tag{2.20}$$

or, using (2.18),

$$V(\langle N\rangle/V)^2 \int_V [g(12) - 1]\, d(12) = \langle N^2\rangle - \langle N\rangle^2 - \langle N\rangle. \tag{2.21}$$

On the other hand, thermodynamic fluctuation theory gives

$$\langle N^2\rangle - \langle N\rangle^2 = \langle N\rangle^2 V k_B T \chi_T, \tag{2.22}$$

where χ_T is the isothermal compressibility. Substituting (2.22) into the right-hand side of (2.21), we finally obtain

$$\varrho \int_0^\infty (g(r) - 1)4\pi r^2\, dr = \varrho k_B T \chi_T - 1. \tag{2.23}$$

The upper limit of the integration can be extended to infinity in usual situations where the dimensions of the volume element are much larger than the range of correlations of the liquids. This relation, besides being one of the most fundamental relations connecting $g(r)$ to the thermodynamic property of a liquid, has a significant implication on the limiting behavior of the liquid structure to be discussed in Section IV. We also notice from the derivation that the fluctuation theorem (2.23) is quite general and does not depend on the pair approximation of liquids.

2. *The Caloric Equation of State*

To make further progress in deriving thermodynamic quantities from the pair correlation function, we need to invoke the pair approximation. Recall that the canonical partition function is given in terms of the configurational partition function (2.3) as

$$Z_N = Q_N/(N!\,\Lambda^{3N}), \tag{2.24}$$

where Λ is the thermal de Broglie wavelength,

$$\Lambda = h/(2\pi M k_B T)^{1/2}. \tag{2.25}$$

In terms of the partition function Z_N, we can evaluate the internal energy E according to

$$\begin{aligned} E &= k_B T^2 \left(\frac{\partial \ln Z_N}{\partial T} \right)_{N,V} \\ &= \frac{3}{2} N k_B T + k_B T^2 \left(\frac{\partial \ln Q_N}{\partial T} \right)_{N,V} \\ &= \frac{3}{2} N k_B T + \bar{U}, \end{aligned} \tag{2.26}$$

and where $\bar{U}$ is given by

$$\begin{aligned} \bar{U} &= (1/Q_N) \int_V \cdots \int U \exp[-U/k_B T]\, d1\, d2 \cdots dN \\ &= \tfrac{1}{2} N(N-1)(1/Q_N) \int_V \cdots \int u(12) \exp[-U/k_B T]\, d1 \cdots dN \\ &= \tfrac{1}{2} \int_V \int u(12) \\ &\quad \times \left[[N(N-1)/Q_N] \int_V \cdots \int \exp[-U/k_B T]\, d3 \cdots dN \right] d1\, d2 \\ &= \tfrac{1}{2} \int_V \int u(12) n_N^{(2)}(1,2)\, d1\, d2 \\ &= \tfrac{1}{2} (N^2/V) \int_0^\infty u(r) g(r) 4\pi r^2\, dr. \end{aligned} \tag{2.27}$$

The upper limit of infinity is permissible since $u(r) \to 0$ faster than r^{-3} as $r \to \infty$. Thus the internal energy per unit volume is given by

$$E/V = \tfrac{3}{2} \varrho k_B T + \tfrac{1}{2} \varrho^2 \int_0^\infty u(r) g(r) 4\pi r^2\, dr. \tag{2.28}$$

This equation gives E/V as a function of temperature and density, since $g(r)$ is also a function of T and ϱ, and is called the caloric equation of state.

3. *The Thermal Equation of State*

We use the "virial theorem," which states that for a system in equilibrium the average value of the virial of the external and internal forces exerted on the molecular system is equal to minus twice the average value of the kinetic energy of the system. If this theorem is applied to a fluid enclosed in a vessel of colume V under an external pressure p, the virial of the external force is $-3pV$, and the virial of the intermolecular force is equal to

$$\sum_i \mathbf{r}_i \cdot \mathbf{F}_i = -\tfrac{1}{2} \sum_i \sum_j r_{ij}(du/dr_{ij}).$$

Therefore

$$-2 \sum_i (1/2M)P_i^2 = -3pV - \tfrac{1}{2}\left\langle \sum_i \sum_j r_{ij}(du/dr_{ij}) \right\rangle_{\text{av}}.$$

Solving this equation for pV and expressing the average in terms of the canonical ensemble and noting that the average value of the kinetic energy is $(3/2)k_{\text{B}}NT$, we obtain

$$\begin{aligned} pV &= \tfrac{2}{3}\,\tfrac{3}{2}\,Nk_{\text{B}}T + \tfrac{1}{6}(1/Q_N) \int_V \sum_{i,j} r_{ij}(du/dr_{ij}) \exp[-U/k_{\text{B}}T]\, d\{n\} \\ &= Nk_{\text{B}}\, T + \tfrac{1}{6} \int\!\!\int_V r_{12}(du/dr_{12}) \\ &\quad \times \left[\{N(N-1)/Q_N\} \int_V \exp[-U/k_{\text{B}}T]\, d3 \cdots dN \right] d1\, d2 \\ &= Nk_{\text{B}}T + \tfrac{1}{6} \int\!\!\int_V r_{12}(du/dr_{12}) n_N^{(2)}(1, 2)\, d1\, d2, \end{aligned}$$

or, dividing through by volume, we finally get

$$p = \varrho k_{\text{B}}T + \tfrac{1}{6}\varrho^2 \int_0^\infty r(du/dr)g(r)4\pi r^2\, dr. \tag{2.29}$$

This equation gives p as a function of ϱ and T and is called the thermal equation of state for an obvious reason.

In order to get all other thermodynamic quantities besides the pressure and internal energy, we proceed as follows. One can get, for example, the free energy A by integrating

$$d(A/N) = -p\, dv \qquad \text{at} \quad T = \text{const} \tag{2.30}$$

$$d(A/T) = E\, d(1/T) \qquad \text{at} \quad v = \text{const} \tag{2.31}$$

where v is $1/\varrho$, the atomic volume. It is seen that the knowledge of the function $g(r)$ at a particular set of values of (ϱ', T') suffices, according to (2.28) and (2.29), to compute E/V and p at (ϱ', T'). On the other hand, in order to calculate a complete set of thermodynamic functions at $(\varrho'\ T')$ through use of (2.30) or (2.31), one needs to know $g(r)$ over a range of values of ϱ and T. This means that it is very important experimentally to measure $g(r, \varrho, T)$ under a wide range of conditions in order to compute various thermodynamic quantities.

4. *Elastic Moduli of Liquids*

As we mentioned in Section I, when a mechanical force is applied suddenly to a liquid, the liquid responds elastically at first, just as if it were a solid body. The initial response may be described by two quantities, the high-frequency limit G_∞ of the shear modulus (or modulus of rigidity), and the high-frequency limit K_∞ of the bulk modulus (or modulus of compression). Zwanzig and Mountain (1965) recently derived rigorous expressions of these two elastic moduli in terms of the pair correlation function in the pair approximation. The expressions they gave are

$$G_\infty = \varrho k_B T + (12\pi/15)\varrho^2 \int_0^\infty g(r)(d/dr)(r^4\, du/dr)\, dr \tag{2.32}$$

$$K_\infty = \tfrac{2}{3}\varrho k_B T + p + (2\pi/9)\varrho^2 \int_0^\infty g(r) r^3 (d/dr)(r\, du/dr)\, dr. \tag{2.33}$$

These two expressions together with the pressure expression (2.29) give a generalized Cauchy identity

$$K_\infty = \tfrac{5}{3} G_\infty + 2(p - \varrho k_B T). \tag{2.34}$$

The Cauchy relation is valid when an isotropic system is in equilibrium under two-body central forces.

If one further assumes the two-body central force to be the Lennard-Jones form as given in (1.4), one can further relate G_∞ and K_∞ to the internal energy E, i.e.,

$$G_\infty = (26/5)\varrho k_B T + 3p - (24/5)\varrho E \tag{2.35}$$

$$K_\infty = (20/3)\varrho k_B T + 7p - 8\varrho E. \tag{2.36}$$

These two relations allow one to get numerical data on G_∞ and K_∞ from the experimental equation of state and internal energy data. The calculated shear modulus G_∞ for argon has the value 10^9 dyn cm^{-2} at the critical point and 4×10^9 dyn cm^{-2} at the triple point.

B. Direct Correlation Function and Approximate Relations between the Correlation Function and the Pair Potential

So far, we have regarded the pair correlation function $g(r)$ as a given quantity, together with the pair potential $u(r)$, it constitutes a basic input to the calculation of equilibrium properties of liquids. However, it is apparent from the definition (2.2) that $g(r)$ is a derived quantity in terms of the more basic total potential of a liquid. In practice, it is very difficult to integrate (2.2) to obtain $g(r)$, but if we limit ourselves to the pair approximation, there exist approximate integral equations which connect $g(r)$ and $u(r)$.

For the purpose of the theoretical development, it is useful to introduce the "net correlation function" $h(r)$ defined by

$$h(r) = g(r) - 1. \tag{2.37}$$

$h(r)$ expressions the net correlation from the asymptotic value 1 and can take negative values. In terms of $h(r)$ we can then define the "direct correlation function" $c(r)$ as

$$h(r) = c(r) + \varrho \int c(|\mathbf{r} - \mathbf{r}'|)h(r')\, d\mathbf{r}'. \tag{2.38}$$

The function $c(r)$ was introduced originally by Ornstein and Zernike (1914) to take into account the fact that the net correlation between two particles is a sum of a direct effect, $c(r)$, of particle 1 on particle 2 plus an indirect effect of all other particles, expressed by the convolution of $\varrho h(r)$ and $c(r)$. From this physical picture, it can be argued that $c(r)$ should be short-ranged in comparison with $h(r)$. We shall see in Section IV that $c(r)$ is also an experimentally measurable quantity like $h(r)$ and, for a simple liquid, argon, it is indeed short-ranged as, was shown by a recent X-ray measurements of Mikolaj and Pings (1967).

There are currently three approximate relations between $h(r)$ and $u(r)$ in use. They are first-order theories in the sense that all of them assume smallness of $h(r) - c(r)$, which, according to (2.38), is equal to

$$h(r) - c(r) = \varrho \int c(|\mathbf{r} - \mathbf{r}'|)h(r')\, d\mathbf{r}'. \tag{2.38a}$$

They are, accordingly, exact in the limit of $\varrho \to 0$, i.e., dilute-gas limit. A fuller discussion of these theories may be found in Chapter 4.

1. *Born–Green Equation*

$$u(r)/k_B T = -\ln[1 + h(r)] + \varrho \int E(|\mathbf{r} - \mathbf{r}'|)h(r')\, d\mathbf{r}' \qquad (2.39)$$

where

$$E(r) = \int_r^\infty (u'(x)/k_B T)[1 + h(x)]\, dx$$
$$\to u(r)/k_B T \qquad \text{for large} \quad r. \qquad (2.40)$$

This equation has been used by Johnson *et al.* (1964) to obtain $u(r)$ for liquid metals from measured $h(r)$.

2. *Hyperchain Equations*

$$u(r)/k_B T = -\ln[1 + h(r)] + \varrho \int c(|\mathbf{r} - \mathbf{r}'|)h(r')\, d\mathbf{r}. \qquad (2.41)$$
$$= -\ln[1 + h(r)] + [h(r) - c(r)]. \qquad (2.42)$$

These two equations together determine $c(r)$ and $h(r)$ in terms of $u(r)$. It is known that they give poor numerical values of the calculated thermodynamic properties of liquids.

3. *Percus–Yevick Equation*

$$u(r)/k_B T = -\ln[1 + h(r)] + \ln[1 + h(r) - c(r)]; \qquad (2.43)$$

together with (2.38a), this equation determines $h(r)$ and $c(r)$. These equations are known to give fairly good thermodynamic data even at high densities and have an additional advantage that an exact analytical solution is obtainable for the case of the hard-sphere potential.

Both the hyperchain (HC) and the Percus–Yevick (PY) theories can be cast into a form in which they make definitive assumptions regarding the direct correlation function:

$$\begin{aligned} \text{HC:} \quad & c(r) \simeq -(u(r)/k_B T) + h(r) - \ln[1 + h(r)] \\ \text{PY:} \quad & c(r) \simeq g(r)[1 - \exp(u(r)/k_B T)]. \end{aligned} \qquad (2.44)$$

Consider both $h(r)$ and $u(r)/k_B T$ as small quantities. We can expand the logarithm and the exponential term to get

$$\begin{aligned} \text{HC:} \quad & c(r) \simeq -(u(r)/k_B T) - h^2(r)/2 + \cdots \\ \text{PY:} \quad & c(r) \simeq -(u(r)/k_B T) - h(r)u(r)/k_B T + \cdots. \end{aligned} \qquad (2.45)$$

It is clearly seen that the form of $c(r)$ is closely related to the form of the pair potential and that, in principle, the determination of $c(r)$ experimentally would give direct information on the form of the pair potential. Recently, Mikolaj and Pings (1967) were able to obtain good data for $h(r)$ and $c(r)$ for argon at various temperatures and densities using X-ray scattering. From (2.43), we see that, in the range of densities for which the PY approximation is good, the right-hand side of

$$u(r) = k_B T \ln\left[\frac{1 + h(r, \varrho, T) - c(r, \varrho, T)}{1 + h(r, \varrho, T)}\right]$$

[from (2.43)] should be independent of the density ϱ and equal to the pair potential $u(r)$. It is normally accepted that the Lennard-Jones potential with $\sigma = 3.405$ Å and $\varepsilon/k_B = 119.8$°K is a good representation for the pair potential in argon. Figure 5 shows the comparison of the L-J potential and the right-hand side of (2.43) at three different densities. It can perhaps be concluded that the PY approximation is better at lower densities.

Before we end this section, let us note a useful relation between the direct correlation function and the compressibility. By integrating both

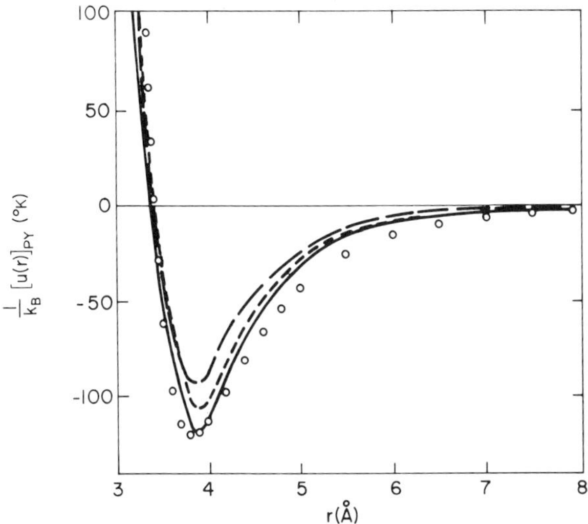

FIG. 5. Pair potential functions of argon predicted by the PY equation (2.43) for three densities along the 163°K isotherm. The open circles represent the Lennard-Jones potential with parameters $\sigma = 3.405$ Å and $\epsilon/k_B = 119.8$°K; (— —) $\varrho_m = 0.780$ g cm^{-3}; (- - -) $\varrho_m = 0.536$ g cm^{-3}; (——) $\varrho_m = 0.280$ g cm^{-3} (Mikolaj and Pings, 1967).

sides of (2.38) with respect to $\mathbf{r}$ and using the fluctuation theorem (2.23), we easily get

$$1 - \varrho \int c(r)\, d\mathbf{r} = (\varrho k_{\mathrm{B}} T \chi_T)^{-1}, \tag{2.23a}$$

which is useful in normalizing the measured $c(r)$ data.

A further discussion of these distribution and correlation functions may be found in Chapter 4.

III. Dynamic Structure of Liquids

To discuss the properties of liquids by considering the pair correlation function alone is incomplete in the sense that it is, in principle, incapable of describing dynamic processes, such as the self-diffusion of an atom (or a molecule) and the collective motions of many atoms in liquids, which are of course of great interest when one deals with transport properties. The extension of $g(r)$ to $G(r, t)$, which describes an instantaneous structure of liquids, was originally done by Van Hove (1954), and since this work a great deal of effort has been made to understand its general properties and to calculate it for some model systems. The extension is meaningful, especially in view of the fact that the inelastic neutron-scattering technique has opened a way to measure this quantity experimentally.

A. General Properties of the Space–Time Correlation Function

It was shown by Van Hove that the inelastic neutron cross section for an N-particle system at temperature T is expressible as a space–time Fourier transform of the following two-particle correlation function:

$$G(\mathbf{r}, t) = (1/N) \sum_{l,m} \int d\mathbf{r}' \,\langle \delta(\mathbf{r} + \mathbf{r}_l(0) - \mathbf{r}')\, \delta(\mathbf{r}' - \mathbf{r}_m(t)) \rangle, \tag{3.1}$$

where $\mathbf{r}_l(0)$ and $\mathbf{r}_m(t)$ are position vectors, considered as Heisenberg observables, of the lth atom at time zero and the mth atom at time t. The summation over l and m extends over all atoms in the system. The thermal average of an operator A is defined as

$$\langle A \rangle = \mathrm{Tr}\{A e^{-\beta H}\} / \mathrm{Tr}\{e^{-\beta H}\}, \tag{3.2}$$

where $\beta = 1/k_{\mathrm{B}} T$ and H is the Hamiltonian of the system.

As defined in (3.1), $G(\mathbf{r}, t)$ is called the space–time correlation function and is in general a complex function, i.e.,

$$[G(\mathbf{r}, t)]^* = G(-\mathbf{r}, -t) \neq G(\mathbf{r}, t). \tag{3.3}$$

This complex nature of the function arises from the fact that, regarded as Heisenberg operators, $\mathbf{r}_l(0)$ and $\mathbf{r}_m(t)$ do not in general commute, and therefore the order of the two delta-functions in (3.1) has to be maintained in the calculation. However, most liquids, except helium and hydrogen, are classical, in the sense that the thermal de Broglie wavelength $\Lambda = h/(2\pi M k_B T)^{1/2}$ is smaller than the interparticle distance R, and for these classical liquids it can be shown that for time t larger than the Debye relaxation time T, which is the time needed by an average particle of the system to travel over a distance R (of the order of 10^{-11} sec as estimated in Section I) the function $G(\mathbf{r}, t)$ is even in t. Furthermore the isotropy of liquids guarantees that $G(\mathbf{r}, t)$ is also even in $\mathbf{r}$ and thus, from (3.3), $G(\mathbf{r}, t)$ is then a real function. In other words, for most liquids, except for the time scale much less than 10^{-11} sec, the quantum nature of $G(\mathbf{r}, t)$ does not have to be considered. In this classical limit, the position vectors in (3.1) can be taken as commuting variables and thc integration over $\mathbf{r}'$ can be carried out to give

$$G_{\text{cl}}(\mathbf{r}, t) = (1/N)\Big\langle \sum_l \sum_m \delta(\mathbf{r} + \mathbf{r}_l(0) - \mathbf{r}_{\text{m}}(t))\Big\rangle_{\text{cl}}, \tag{3.4}$$

where the average $\langle \cdots \rangle_{\text{cl}}$ is to be performed over the classical canonical or grand canonical ensemble as is in (2.2) and (2.12). The expression (3.4) suggests the following interpretation of $G_{\text{cl}}(\mathbf{r}, t)$: the term $G_{\text{cl}}(\mathbf{r}, t)\, d\mathbf{r}$ is the number of particles at $\mathbf{r}$, within $d\mathbf{r}$, at time t, knowing that there is a particle at $\mathbf{r} = 0$ at $t = 0$.

The physical interpretation of $G(\mathbf{r}, t)$ can be made more plausible if we note that the instantaneous local density of a liquid can be written as

$$\varrho(\mathbf{r}, t) = \sum_l^N \delta(\mathbf{r} - \mathbf{r}_l(t)). \tag{3.5}$$

Using this representation one can then rewrite (3.1) as

$$\begin{aligned} G(\mathbf{r}, t) &= (1/N) \int d\mathbf{r}' \,\langle \varrho(\mathbf{r}' - \mathbf{r}, 0)\varrho(\mathbf{r}', t)\rangle \\ &= (1/N) \int d\mathbf{r}'' \,\langle \varrho(\mathbf{r}'', 0)\, \varrho(\mathbf{r}'' + \mathbf{r}, t)\rangle. \end{aligned}$$

If we now take explicit account of the homogeneity of the liquid, the quantity within the brackets is independent of $\mathbf{r}''$ and therefore integration can be carried out to give

$$G(\mathbf{r}, t) = (1/\varrho)\,\langle \varrho(0, 0)\varrho(\mathbf{r}, t)\rangle. \tag{3.6}$$

We thus see that $G(r, t)$ is a quantum-mechanical density–density correlation function. In the classical limit, $G(\mathbf{r}, t)/\varrho$ is then the conditional probability of finding a particle at $\mathbf{r}$ at time t knowing that there was a particle (which may or may not be the same particle) at the origin at $t = 0$.

For distances larger than the correlation length or after time long compared to the correlation time of density fluctuations in liquids,

$$G(\mathbf{r}, t) \underset{r\to\infty, t\to\infty}{\approx} (1/\varrho)\langle \varrho(0, 0)\rangle\langle \varrho(\mathbf{r}, t)\rangle = \varrho. \tag{3.7}$$

At $t = 0$, $\mathbf{r}_l(0)$ and $\mathbf{r}_m(0)$ commute and (3.1) reduces to

$$G(\mathbf{r}, 0) = \delta(\mathbf{r}) + (1/N) \sum_{l \neq m} \langle \delta(\mathbf{r} + \mathbf{r}_l - \mathbf{r}_m)\rangle.$$

For a classical liquid, the average in the second term can be computed in the canonical ensemble and it is not hard to envince oneself that the result is $\varrho g(r)$. Therefore†

$$G(r, 0) = \delta(r) + \varrho g(r). \tag{3.8}$$

Thus the time evolution of $G(r, t)$ in the classical limit can be pictured as follows: At $t = 0$, there is a delta-function representating a particle situated at the origin and a term representing the static pair correlation between the particle at the origin and all other particles. As time goes on, the particle at the origin begins to move out and the pair correlation begins to smear out due to the relaxation of the local structure. For time much longer than the correlation time T, the localization of the particle initially at the origin is completely lost due to the self-diffusion in liquids and the pair correlation term also approaches the asymptotic value ϱ. In view of these two types of correlations in liquids, one can naturally split $G(\mathbf{r}, t)$ into two parts,

$$G(r, t) = G_s(r, t) + G_d(r, t), \tag{3.9}$$

† Owing to isotropy of the liquid, $G(\mathbf{r}, t)$ should depend on the scalar r only. Henceforth, we denote it by $G(r, t)$.

where

$$G_s(r, t) = (1/N)\left\langle \sum_{l=1}^{N} \int d\mathbf{r}' \, \delta(\mathbf{r} + \mathbf{r}_l(0) - \mathbf{r}') \, \delta(\mathbf{r}' - \mathbf{r}_l(t)) \right\rangle \quad (3.10)$$

$$G_d(r, t) = (1/N)\left\langle \sum_{l \neq m} \int d\mathbf{r}' \, \delta(\mathbf{r} + \mathbf{r}_l(0) - \mathbf{r}') \, \delta(\mathbf{r}' - \mathbf{r}_m(t)) \right\rangle. \quad (3.11)$$

From these definitions, we can interpret G_s as the correlation function which expresses the probability that a particle which was at the origin

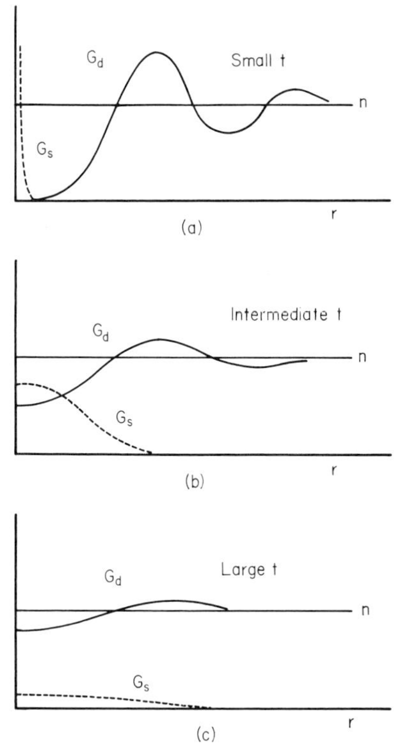

FIG. 6. Qualitative behavior of Van Hove correlation functions (Van Hove, 1954).

at time $t = 0$ will be at $\mathbf{r}$ at time t; and G_d the analogous conditional probability of finding a different particle at $\mathbf{r}$ at time t. Therefore, $G_d(r, t)$ can be regarded as a direct extension of $\varrho g(r)$ to the time-dependent case. We sketch in Fig. 6 the qualitative behavior of G_s and G_d at different times.

It is instructive to look at a few simple systems for which the space–time correlation function can be written down explicitly. Referring for proofs to Sjölander (1965), we only discuss here their general properties.

1. *Perfect Gas*

There is no correlation between the particles in this case. Therefore

$$G_{\rm d} = 0$$

$$G_{\rm s}(r, t) = [4\pi W(t)]^{-3/2} \exp[-r^2/4W(t)] \tag{3.12}$$

$$W(t) = (k_{\rm B}T/2M)t(t - i\hbar/k_{\rm B}T), \qquad t > 0 \tag{3.13}$$

$$\langle r^2(t)\rangle = 6W(t). \tag{3.14}$$

For $t \ll \hbar/k_{\rm B}T$, $W(t) \simeq -(i\hbar/2M)t$, and the quantum effect dominates, while for $t \gg \hbar/k_{\rm B}T$, $W(t) = (k_{\rm B}T/2M)t^2 = \frac{1}{2}\langle(v_x t)^2\rangle$, i.e., $\langle r^2\rangle = \langle(vt)^2\rangle$, which is a purely classical expression that the particle is diffusing with speed v.

2. *Continuous Diffusion*

This is a single-particle motion and therefore $G_{\rm d} = 0$. Denoting the diffusion constant by D, we have

$$\begin{aligned} G_{\rm s}(r, t) &= [4\pi W(t)]^{-3/2} \exp[-r^2/4W(t)] \\ W(t) &= Dt = \tfrac{1}{6}\langle r^2\rangle, \qquad t > 0. \end{aligned} \tag{3.15}$$

3. *Harmonic Solid*

Consider only a case of a Bravais lattice with cubic symmetry. Then $G(r, t)$ has the same Gaussian form as (3.12) with a width function $W(t)$ which at small time agrees with the perfect-gas expression (3.13). However, at large time, it oscillates around a constant value because in a harmonic lattice an atom does not diffuse away from the lattice site but oscillates in time. In this case, $G_{\rm d}(r, t)$ is not zero, because there are strong correlations between atoms in the lattice which propagate in the form of phonons. We shall not bother to write down the complicated expression here.

3. *Real Liquid*

No rigorous expression either for G_s or G is known. We can, however, discuss the qualitative behavior of G_s here. It can be shown (see Sjölander, 1965) that in a real liquid G_s is also approximately Gaussian in form with a width function which varies in time as follows: At short times ($t < \hbar/k_B T$ when the potential acting on a particle is practically constant, the width, has the perfect-gas behavior (3.13), which, neglecting the imaginary part, goes essentially like t^2, whereas at large time, the particle is diffusing and therefore the width function goes like t, as in (3.15). The intermediate time range, corresponding to the transition from t^2 to t bahavior, should reflected the vibratory motion of the particle. Sketches of the time dependences of the width function for the different cases are shown in Fig. 7. The $G(r, t)$ for a real liquid is a much more complicated function

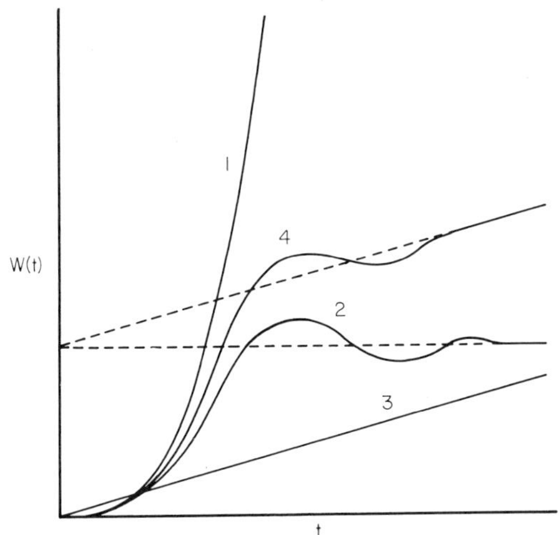

FIG. 7. Schematic form of the width function as function of time for different systems. (1) Perfect gas; (2) crystal; (3) diffusion model; (4) liquid.

to describe since it contains not only the single-particle motion but also the collective motions of all frequencies (the low-frequency collective motions being sound waves). Many different approximations and models have been proposed in recent years for this function. These are mostly geared toward explaining inelastic neutron scattering data and we refer interested readers to the literature (Sjölander, 1965); Chung and Yip 1969).

B. Space–Time Correlation Functions and Their Relation to the Transport Properties of Liquids

It useful to define the following two Fourier transforms of the space–time correlation function:

$$I(\mathbf{q}, t) = \int (\exp i\mathbf{q} \cdot \mathbf{r}) G(r, t)\, d\mathbf{r}$$

$$= \frac{1}{N} \sum_{l,m}^{N} \langle \exp[-i\mathbf{q} \cdot \mathbf{r}_l(0)] \exp[i\mathbf{q} \cdot \mathbf{r}_m(t)] \rangle \tag{3.16}$$

$$S(\mathbf{q}, \omega) = (1/2\pi) \int_{-\infty}^{\infty} e^{-i\omega t} I(\mathbf{q}, t)\, dt. \tag{3.17}$$

$I(\mathbf{q}, t)$ is called the "intermediate scattering function," a name derived from its appearance in the inelastic neutron cross section, as we shall see in Section IV. The second line of (3.16) follows from using definition (3.1) for $G(\mathbf{r}, t)$. The $S(\mathbf{q}, \omega)$ is usually called the "dynamic structure factor" because it is an extension of the static structure factor $S(\mathbf{q})$ familiar in X-ray scattering work. We can likewise define $I_s(\mathbf{q}, t)$ and $S_s(\mathbf{q}, \omega)$ from the corresponding $G_s(\mathbf{r}, t)$ defined in (3.10). The physical meaning of $I(\mathbf{q}, t)$ is that it describes the time behavior of the collective density oscillation of wave vector $\mathbf{q}$ in an N-particle system. Then $S(\mathbf{q}, \omega)$ the spectral density of this collective density oscillation. Many important properties such as the ground-state energy and the excitation spectrum of a neutral N-particle system are related to the $S(\mathbf{q}, \omega)$ function. The most important property of $S(\mathbf{q}, \omega)$ is the so-called detailed-balance condition

$$S(-\mathbf{q}, -\omega) = \exp(-\hbar\omega/k_B T) S(\mathbf{q}, \omega), \tag{3.18}$$

which can be shown from its definition (see Sjölander, 1965), and is the consequence of the fact that the system is in thermal equilibrium. We recall that for liquids $G(r, t)$ depends only on the scalar r and so does $S(q, \omega)$ on q. Schofield (1960) was the first to realize that the detailed-balanced condition (3.18) is intimately connected with the quantum nature of $G(r, t)$ i.e., it is a complex function. In the classical limit when $G(r, t)$ is an even function of t, $S(q, \omega)$ becomes even in ω. We therefore see that the classical limit corresponds to the high-temperature limit $k_B T \gg \hbar\omega$ and the exponential factor on the right-hand side of (3.18) approaches unity. It is easy to see from definitions (3.16) and (3.17) that the nth frequency moment of $S(\mathbf{q}, \omega)$ is related to the nth time derivative of $I(\mathbf{q}, t)$ by

$$\langle\omega^n\rangle \equiv \int_{-\infty}^{\infty} \omega^n S(\mathbf{q}, \omega)\, d\omega = (-i)^n I^{(n)}(\mathbf{q}, 0)$$
$$\langle\omega^n\rangle_s \equiv \int_{-\infty}^{\infty} \omega^n S_s(\mathbf{q}, \omega)\, d\omega = (-i)^n I_s^{(n)}(\mathbf{q}, 0). \tag{3.19}$$

It follows that the frequency moments are given in terms of the static property of the liquid, and their explicit expressions were first given by Placzek (1952). We here quote only the classical result given by De Gennes (1959).

Zeroth moments:

$$\int_{-\infty}^{\infty} S(\mathbf{q}, \omega)\, d\omega = S(\mathbf{q}) = I(\mathbf{q}, 0). \tag{3.20}$$

$S(\mathbf{q})$ defined here is called the structure factor. $I(\mathbf{q}, 0)$ can be evaluated using (3.8) and (3.16) to get

$$S(q) = 1 + \varrho \int (\exp i\mathbf{q}\cdot\mathbf{r}) g(r)\, d\mathbf{r}. \tag{3.21}$$

We note here an important limiting value of $S(q)$,

$$\lim_{q\to 0} S(q) = \varrho k_B T \chi_T, \tag{3.22}$$

which follows from (3.21) and (2.23). For the "self" part, we have

$$\int_{-\infty}^{\infty} S_s(\mathbf{q}, \omega)\, d\omega = I_s(\mathbf{q}, 0) = 1. \tag{3.23}$$

Second moments:

$$\langle\omega^2\rangle = (k_B T/M) q^2 = \langle\omega^2\rangle_s. \tag{3.24}$$

Note that both moments are independent of atomic interactions. This means that in the small-time expansion of the coefficients of I up to the t^2 term are independent of the interaction; hence, this substantiates the the statement we made earlier that at short times the motion of an atom is like that in a perfect gas.

Fourth moments:

$$\langle\omega^4\rangle = (k_B T q^4/M^2)[3k_B T + \int d\mathbf{r}\, g(r)\{(1 - \cos qx)/q^2\} \partial^2 u/\partial x^2] \tag{3.25}$$

$$\langle\omega^4\rangle_s = (k_B T q^4/M^2)[3k_B T + \int d\mathbf{r}\, g(r)(1/q^2)(\partial^2 u/\partial x^2)]. \tag{3.26}$$

All the odd moments are zero because in the classical limit both S and S_s are even functions of ω. These frequency moments are sum rules which restrict the general forms of S and S_s whenever one tries to calculatethem from a model. They are also very helpful in normalizing experimental data on them.

We next discuss relations of $S(q, \omega)$ and $S_s(q, \omega)$ to transport coefficients in liquids.

1. *Self-Diffusion Coefficient*

We define the wave-vector and frequency-dependent diffusion coefficient $D(q, \omega)$ by the generalized Fick's law

$$\mathbf{j}(q, \omega) = -D(q, \omega)(\nabla \varrho)_{q,\omega}. \tag{3.27}$$

Here, $\mathbf{j}(q, \omega)$ is the (q, ω)th Fourier component of the local current density of the diffusing particles as a response to the (q, ω)th Fourier component of the gradient $\nabla \varrho$ of the local particle density. Zwanzig (1964) showed that there is a very simple rigorous relation existing between $S_s(q, \omega)$ and $D(q, \omega)$, namely

$$S_s(q, \omega) = [\hbar\beta\omega q^2/\pi\omega^2(1 - e^{-\beta\hbar\omega})] \operatorname{Re}[D(q, \omega)]. \tag{3.28}$$

In the classical limit where $\beta\hbar\omega \ll 1$, (3.28) reduces to

$$\operatorname{Re}[D(q, \omega)] = \pi\omega^2[S_s(q, \omega)/q^2]. \tag{3.29}$$

If we define a frequency-dependent function $Z(\omega)$ by

$$Z(\omega) = \lim_{q\to 0} \omega^2\left[\frac{S_s(q, \omega)}{q^2}\right]. \tag{3.30}$$

It can be shown (see Egelstaff, 1967) that $Z(\omega)$ is the spectral density of the velocity autocorrelation function,

$$Z(\omega) = (1/2\pi) \int_{-\infty}^{\infty} \langle v_x(0)v_x(t)\rangle e^{-i\omega t}\, dt, \tag{3.31}$$

which, in liquids, plays a similar role as the phonon frequency distribution function in solids, except that it is normalized to

$$\int_{-\infty}^{\infty} Z(\omega)\, d\omega = \langle v_x{}^2\rangle = k_B T/M. \tag{3.32}$$

From (3.29) and (3.30), we see that $Z(\omega)$ has a limiting value which is proportional to the diffusion coefficient D,

$$[Z(\omega)]_{\omega\to 0} = (1/\pi)[\operatorname{Re} D(q, \omega)]_{\substack{q\to 0\\ \omega\to 0}} = (1/\pi)D. \tag{3.33}$$

This relation was noted earlier by Egelstaff (1961) and it shows that incoherent neutron scattering can be used to determine the frequency- and wave-vector-dependent transport coefficient. Since, as in (3.31), $Z(\omega)$ can be determined if one can trace the time dependence of the velocity of a particle (averaged over many particles) in a liquid, the molecular-dynamics experiment using a computer is ideal for its calculation. Figure 8 shows such a calculation carried out by Rahman (1964)

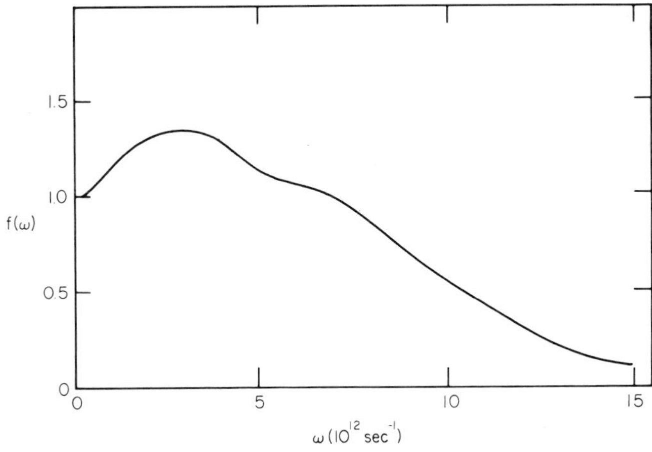

FIG. 8. The frequency distribution function for liquid argon at 94.4°K calculated by Rahman (1964) using molecular dynamics.

for argon at 94.4°K. The function plotted is $f(\omega) = (2M/k_B T)Z(\omega)$ defined only for $\omega > 0$. The nonzero value of $f(\omega)$ at $\omega = 0$ shows that the low-frequency mode of motion for an atom in a liquid is the diffusive motion and the fairly broad peak at finite ω shows that the particle undergoes a damped oscillation between diffusive steps.

2. *Viscosity Coefficient*

It can be shown rigorously (Rice and Gray, 1965) that the viscosity is a time integral of the fluctuating part of the stress correlation function in liquids, namely

$$\eta = (1/Vk_{\mathrm{B}}T) \int_{-\infty}^{\infty} \langle \sigma'_{zx}(0)\sigma'_{zx}(t)\rangle \, dt \tag{3.34}$$

$$\tfrac{4}{3}\eta + \zeta = (1/Vk_{\mathrm{B}}T) \int_{-\infty}^{\infty} \langle \sigma'_{zz}(0)\sigma'_{zz}(t)\rangle \, dt, \tag{3.35}$$

where η and ζ are the shear and the bulk viscosity, respectively. The combination $(\frac{4}{3}\eta + \zeta)$ is called the longitudinal viscosity because it is related to the longitudinal stress correlation function. Since $S(q, \omega)$ is essentially a longitudinal density fluctuation spectrum, only the longitudinal viscosity can be related to it. Define again the spectral density of the longitudinal stress correlation function by

$$Z_v(\omega) = (1/2\pi) \int_{-\infty}^{\infty} \langle \sigma'_{zz}(0)\sigma'_{zz}(t)\rangle e^{-i\omega t} \, dt. \tag{3.36}$$

Egelstaff (1967) then showed that

$$Z_v(\omega) = \omega^4 [S(q, \omega)/q^4]_{q\to 0} \tag{3.37}$$

Using (3.35) and (3.36), we then obtain

$$\tfrac{4}{3}\eta + \zeta = (\pi/Vk_{\mathrm{B}}T) Z_v(0). \tag{3.38}$$

In principle, $Z_v(\omega)$ be measured by the neutron scattering. But in view of the small-q limit to be taken in (3.37), it should be easier by light scattering. There are so far no data on $Z_v(\omega)$ available.

3. *Thermal Conductivity*

In a simple liquid in the small-q limit, or the so-called hydrodynamic limit, $S(q, \omega)$ consists of three distinct peaks, one centered around $\omega = 0$, and the other two symmetrically displaced by $\pm\omega_{\mathrm{B}}$. The side peaks are called Brillouin components and the central peak Landau–Placzek peak, all named after their discoverers. Using the hydrodynamic equations (see Mountain, 1966) one can show that the LP peak is given by

$$S_{\mathrm{LP}}(q, \omega) = \varrho k_{\mathrm{B}} T \left(\frac{C_p - C_v}{C_p} \right) \frac{1}{\pi} \frac{D_T q^2}{\omega^2 + (D_T q^2)^2}, \tag{3.39}$$

where $D_{\mathrm{T}} = \lambda/\varrho C_p$ is the thermal diffusivity and λ the thermal conductivity. This is a Lorentzian peak with a width

$$(\Delta\omega)_{1/2} = \frac{\lambda}{\varrho C_p} q^2. \tag{3.40}$$

Therefore measurement of the half-width as a function of q^2 gives directly the thermal conductivity.

4. *Electrical Conductivity*

In a liquid metal, the electrical resistivity is mostly determined by the scattering of electrons from the moving ions. Since $S(q, \omega)$ describes motions of the ions, the conductivity σ (inverse of the resistivity) should be related to $S(q, \omega)$ and the pseudopotential of the electron–ion interaction $W(q)$. Baym (1964), indeed, showed that

$$\sigma = \frac{\varrho e^2}{m} \tau, \tag{3.41}$$

where the relaxation time τ is given by

$$\frac{1}{\tau} = \frac{m}{12\pi^3 Z} \int_0^{2k_F} dq\, q^3 \mid W(q) \mid^2 \int_{-\infty}^{\infty} \frac{d\omega}{2\pi} \frac{S(q, \omega)\beta\hbar\omega}{e^{\beta\hbar\omega} - 1} \tag{3.42}$$

Here, $\hbar k_F$ is the Fermi momentum, Z the valence of the ion, and m the mass of the electron.

IV. Investigation of Liquid Structure by Radiation Scattering

As mentioned in Section I, radiation scattering offers the most direct determination of both the static and the dynamic structure of liquids. The radiation, considered as a probe, should be able to couple to motions of atoms (or molecules) in liquids directly or indirectly. Neutrons interact primarily with the nucleus of an atom and thus are able to couple to the atomic motion directly. Both X-rays and light, on the other hand, interact with electrons. In an insulating liquid, where electrons can be regarded as firmly attached to individual atoms, the electromagnetic wave couples to motions of atoms indirectly via atomic electrons. In the case of X-ray scattering from a liquid metal, where there are free electrons, one can show that X-rays are mainly scattered by the core electrons, which are, to a good approximation, tightly bound to an atom.

In a radiation scattering process, there are two basic probe parameters: the momentum transfer $\hbar\mathbf{Q}$ and the energy transfer $\hbar\Omega$ to the system. Denoting the vave vector and the energy of the incident and the scattered waves by $(\mathbf{k}_i, E_i)$ and $(\mathbf{k}_f, E_f)$, respectively, we have

$$\mathbf{Q} = \mathbf{k}_i - \mathbf{k}_f \tag{4.1}$$

$$\hbar\Omega = E_i - E_f. \tag{4.2}$$

Figure 9 shows the vector diagram of (4.1), where ϕ is the scattering angle.

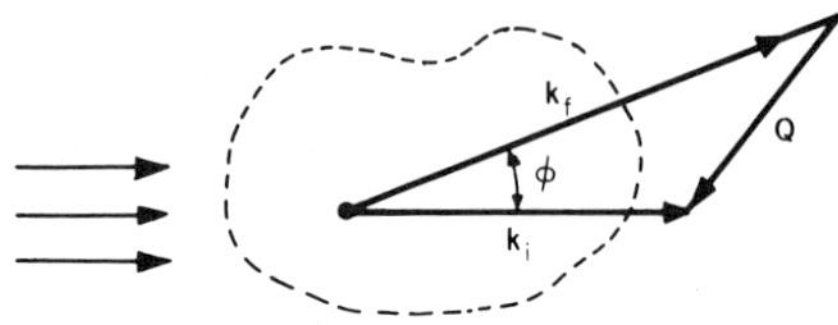

FIG. 9. Vector diagram of $\mathbf{Q} = \mathbf{k}_i - \mathbf{k}_f$.

Briefly, the angular distribution of the scattered intensity, or more specifically, the scattered intensity at constant momentum transfer $\mathbf{Q}$, gives the $\mathbf{Q}$th Fourier component of the spatial correlation between atoms in the system. Such measurements are useful in obtaining the structure factor or the static structure of a liquid. On the other hand, if one were able to analyze the energy distribution at a constant $\mathbf{Q}$, the intensity at a particular energy transfer $\hbar\Omega$ gives the Ωth Fourier component of the time correlation for the fluctuation with wavevector $\mathbf{Q}$. This latter quantity is measured through the so-called double differential cross section and is useful for the dynamic structure factor determination.

For neutrons, the energy and the wave vector are related by

$$E = \hbar^2 k^2/2m, \tag{4.3}$$

and for electromagnetic waves,

$$E = c\hbar k. \tag{4.4}$$

Because of these energy–wave vector dispersion relations, the momentum transfer $\hbar\mathbf{Q}$ and the energy transfer $\hbar\Omega$ in (4.1) and (4.2) are related. Therefore in an experiment where both $\mathbf{Q}$ and Ω are varied, for a given initial energy and scattering angle, $\Omega(\mathbf{Q})$ forms a track in the $\mathbf{Q}$–Ω space. Since liquids are isotropic, their scattering properties depend on the

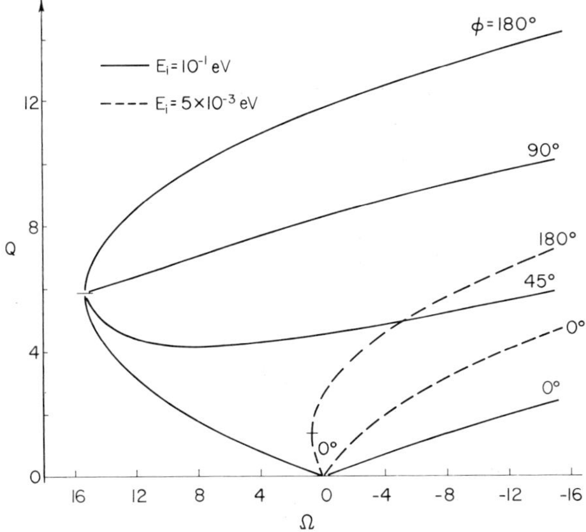

FIG. 10. Experimental Q–Ω tracks in neutron-scattering experiments.

scalar **Q** only. Figure 10 shows some of the $\Omega(Q)$ tracks for different initial neutron energies. It shows, for a given initial energy, what part of Q–Ω space it can cover and it is essential for understanding what types of fluctuations a given probe can effectively measure. Neutrons are massive particles and are able to transfer a large amount of momentum. Equation (4.1) can be rewritten as

$$\hbar^2 Q^2 = 2m[2E_i - \hbar\Omega - 2\sqrt{E_i(E_i - \hbar\Omega)}\cos\phi), \tag{4.5}$$

which shows that if one fixes the scattering angle ϕ and measures the energy distribution, then a different energy transfer $\hbar\Omega$ would correspond to a different momentum transfer $\hbar\mathbf{Q}$. This is slightly inconvenient since one really likes to obtain intensity distributions at different $\hbar\Omega$ for a fixed **Q**. However, in practice, there is a way to circumvent this difficulty by a clever programming of the neutron spectrometer known as the constant-Q method (Brockhouse, 1961). Since for thermal neutrons E_{i} and E_{f} are of the order of 10^{-2} eV, the probe variables are $Q \sim 10^8$ cm^{-1} and $\Omega \sim 10^{13}$ sec^{-1}. This implies that neutrons are capable of probing the liquid structure within a distance of $Q^{-1} \sim 10^{-8}$ cm and within a time interval of $\Omega^{-1} \sim 10^{-13}$ sec. On the other hand, for the electromagnetic wave, $k_{\mathrm{i}} = k_{\mathrm{f}}$ to good accuracy for energy transfers less than 1 eV, or so, and in this case (4.1) reduces to

$$Q = 2k\sin(\phi/2), \tag{4.6}$$

where k is the wave vector in the medium. Since $k = nk_i$, where n is the index of refraction of the medium, we see that measurement at a constant angle does give a constant-Q measurement. This is an advantage of X-ray and light scattering over neutron scattering. For X-rays, Q is again of the order of 10^8 cm^{-1}, but since the X-ray photon has an energy of about 50 keV, the energy change of interest, which may be of the order of a fraction of 1 eV, cannot be measured with present-day instruments. X-rays are therefore only useful for the structure-factor measurement where one measures the integrated intensity at constant Q. As for light, taking the red light from a helium–neon gas laser as an example, in the backward scattering geometry, $Q \sim 2 \times 10^5$ cm^{-1} and Q^{-1} is then about a couple of hundred times the interatomic distance in liquids. Light scattering is thus only suited for the study of long-range fluctuations in liquids. Since, for visible light, the energy analysis can be carried out with great accuracy, the light-scattering technique has a unique advantage in the study of the dynamic structure of dielectric liquids in the hydrodynamic regime. In recent years, this advantage has also been exploited in the study of the dynamic structure near the critical points of single-component and two-component liquids (see Section V).

A. Neutron Scattering

Consider a beam of monoenergetic neutrons traveling in the Z direction represented by a plane wave

$$\psi_i(z, t) = \exp[i(k_i z - \omega_i t)]. \tag{4.7}$$

The neutron flux can be written in terms of the speed of the neutrons v_i as

$$J_i = v_i \, | \psi_i |^2 = v_i \quad \text{sec}^{-1}\,\text{cm}^{-2}. \tag{4.8}$$

Let the neutrons be scattered by a stationary target nucleus situated at $\mathbf{R}_l$. Referring to Fig. 11, the scattered wave amplitude far from the target can be represented by a spherical outgoing wave of the form (S-wave potential scattering)

$$\begin{aligned}\psi_f^l(\mathbf{r}, l) &= -a_l \exp[i\mathbf{k}_i \cdot \mathbf{R}_l] \exp[i(k_i \, | \mathbf{r} - \mathbf{R}_l \, | - \omega_i t)] / \, |\mathbf{r} - \mathbf{R}_l \, | \\ &\simeq -a_l \exp[i\mathbf{Q} \cdot \mathbf{R}_l] \exp[i(k_i r - \omega_i t)]/r\end{aligned} \tag{4.9}$$

a_l is the bound-atom scattering length of the nucleus and the negative

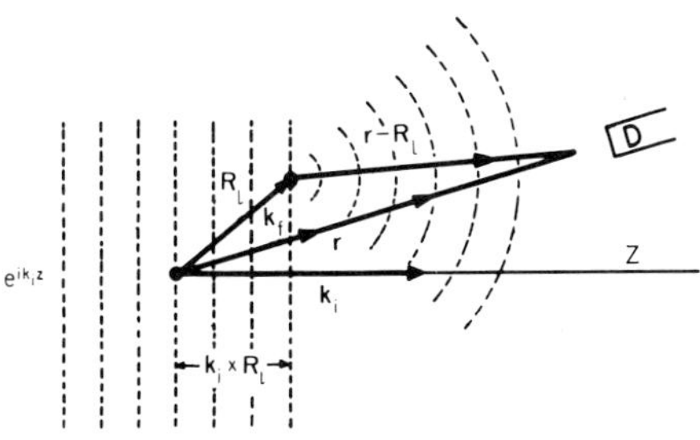

FIG. 11. Scattering geometry.

sign is a convention such as to make a_l positive for most nuclei. The first exponential takes into account the additional phase factor introduced because the nucleus is not at the origin and where $\mathbf{Q} = \mathbf{k}_i - \mathbf{k}_f\hat{r} = \mathbf{k}_i - \mathbf{k}_f$. If we now consider a target system of N nuclei, we have, instead of (4.9), the expression

$$\psi_f(\mathbf{r}, t) = f(Q) \exp[i(k_i r - \omega_i t)]/r, \tag{4.10}$$

where

$$\mathrm{f}(Q) = -\sum_{l=1}^{N} a_l \exp[i\mathbf{Q} \cdot \mathbf{R}_l]. \tag{4.11}$$

We obtain the differential cross section by evaluating $d\sigma/d\Omega$, which is equal to the number of neutrons scattered into a unit solid angle in the direction k_f, divided by the incident neutron flux:

$$\begin{aligned} d\sigma/d\Omega &= \langle\langle v_f \,|\, \psi_f \,|^2 r^2/v_i \,|\, \psi_i \,|^2\rangle_{\text{av}}\rangle \\ d\sigma/d\Omega &= (k_f/k_i)\langle\langle f^*(\mathbf{Q}) f(\mathbf{Q})\rangle_{\text{av}}\rangle \\ &= \Big\langle \sum_{l,l'} \langle a_l a_{l'}\rangle_{\text{av}} \exp[i\mathbf{Q} \cdot (\mathbf{R}_l - \mathbf{R}_{l'}]\Big\rangle. \end{aligned} \tag{4.12}$$

We have used the fact that the scattering from stationary nuclei is elastic and therefore $\omega_i = \omega_f$, $k_i = k_f$. The double average is necessary because first we have to average over the distribution of nuclei over various spin and isotopic states and then over the distribution of the system in various members of the canonical ensemble. The first average in the case of a liquid is easily done because the spin and the isotopic states are randomly distributed among atoms at various positions, and thus

$$\begin{aligned} \langle a_l a_{l'}\rangle_{\text{av}} &= \bar{a}_l^2 \;\; = \bar{a}^2, \qquad l = l' \\ &= \bar{a}_l \bar{a}_{l'} = \bar{a}^2, \qquad l \neq l', \end{aligned}$$

or combined together,

$$\begin{aligned}\langle a_l a_{l'} \rangle_{\text{av}} &= [\overline{a^2} - \bar{a}^2]\, \delta_{ll'} + \bar{a}^2 \\ &= a_{\text{inc}}^2\, \delta_{ll'} + a_{\text{coh}}^2 .\end{aligned} \tag{4.13}$$

Therefore, from (4.12) and (4.13), we obtain

$$\begin{aligned} d\sigma/d\Omega &= Na_{\text{inc}}^2 + a_{\text{coh}}^2 \Big\langle \Big|\sum_l \exp(i\mathbf{Q}\cdot\mathbf{R}_l)\Big|^2 \Big\rangle \\ &= Na_{\text{inc}}^2 + Na_{\text{coh}}^2 I(\mathbf{Q}, 0) \\ &= Na_{\text{inc}}^2 + Na_{\text{coh}}^2 S(\mathbf{Q}). \end{aligned} \tag{4.14}$$

The second and third steps follow from (3.16) and (3.20). We see that in general the differential cross section in liquids consists of an incoherent part and a coherent part, where the latter is proportional to the liquid structure factor $S(Q)$. The incoherent contribution comes from the mean-square fluctuation of the scattering length, and since in this case the phases of the scattered waves from each nuclei are random, they do not interfere with each other at the detector position. The scattered intensity is thus proportional to N, the number of nuclei present. On the other hand, the mean value of the scattering length gives rise to the coherent scattering and in this case the wevelets from each nuclei interfere with each other at the detector. In fact, $S(\mathbf{Q})$ is just the interference factor, and the degree of interference depends on the correlation of atomic positions in the liquid.

We next consider a more realistic situation in which atoms in the liquid move, and hence $\mathbf{R}_l(t)$ is time-dependent. Our previous scattering formulation can easily be extended to this time-dependent case, as was noted earlier by Waller (1966). We should write, instead of (4.11), a time-dependent scattering amplitude

$$f(\mathbf{Q}, t) = -\sum_{l=1}^{N} a_l \exp[i\mathbf{Q}\cdot\mathbf{R}_l(t)]. \tag{4.11a}$$

However, we should also take into account the fact that now the scattering can be inelastic, i.e., $\omega_{\text{i}} \neq \omega_{\text{f}}$, $k_{\text{i}} \neq k_{\text{f}}$. The scattered wave at the detector is thus

$$\psi_{\text{f}}(\mathbf{r}, t) = f(\mathbf{Q}, t) \exp[i(k_{\text{f}} r - \omega_{\text{i}} t)]/r. \tag{4.15}$$

If we now perform an energy analysis of the scattered wave, we should be extracting one of the Fourier components

$$\psi_{\text{f}}(\mathbf{r}, t)\,|_{\omega_{\text{f}}} = f(\mathbf{Q}, \Omega_p) \exp[i(k_{\text{f}} r - \omega_{\text{f}} t)]/r, \tag{4.16}$$

where we first decompose

$$f(\mathbf{Q}, t) = \sum_p f(\mathbf{Q}, \Omega_p) \exp i\Omega_p t \tag{4.17}$$

$$\Omega_p = 2\pi p/\tau, \qquad p = \text{integer}, \tag{4.18}$$

and obtain

$$f(\mathbf{Q}, \Omega_p) = \lim_{\tau\to\infty} (1/\tau) \int_{-\tau/2}^{\tau/2} f(\mathbf{Q}, t) \exp[-i\Omega_p t]\, dt. \tag{4.19}$$

Using (4.17) in (4.15) and comparing with (4.16), we obtain

$$\Omega_p = \omega_{\mathrm{i}} - \omega_{\mathrm{f}},$$

which, in the limit $\tau \to \infty$, tends to the Ω defined in (4.2).

When one performs the energy analysis, it is more natural to define the double differential cross section per unit solid angle, per unit outgoing energy (i.e., divided by $\Delta E = \hbar\, \Delta\Omega = \hbar 2\pi/\tau$), namely

$$\frac{d^2\sigma}{dE\, d\Omega} = \frac{k_{\mathrm{i}}}{k_{\mathrm{f}}} \left[\frac{\langle\langle f^*(\mathbf{Q}, \Omega_p) f(\mathbf{Q}, \Omega_p)\rangle_{\mathrm{av}}\rangle}{\hbar(2\pi/\tau)} \right]_{\tau\to\infty}. \tag{4.20}$$

It should be noted that we have used $d\Omega$ for the differential solid angle, which sould not be confused with the energy transfer variable Ω. Using (4.19) and the stationarity of the time correlation function, we can easily show that

$$\langle f^*(\mathbf{Q}, \Omega_p) f(\mathbf{Q}, \Omega_p)\rangle_{\tau\to\infty} = \frac{1}{\tau} \int_{-\tau/2}^{\tau/2} \langle f^*(\mathbf{Q}, t) f(\mathbf{Q}, t)\rangle \exp[-i\Omega_p t]\, dt \tag{4.21}$$

and therefore the double scattering cross section becomes

$$\begin{aligned} \frac{d^2\sigma}{dE\, d\Omega} &= \frac{k_{\mathrm{f}}}{k_{\mathrm{i}}} \frac{1}{2\pi\hbar} \int_{-\infty}^{\infty} \langle\langle f^*(\mathbf{Q}, t) f(\mathbf{Q}, t)\rangle_{\mathrm{av}}\rangle \exp[-i\Omega t]\, dt \\ &= \frac{k_{\mathrm{f}}}{k_{\mathrm{i}}} \frac{1}{2\pi\hbar} \\ &\times \int_{-\infty}^{\infty} \left\langle \sum_{l,l'} \langle a_l a_{l'}\rangle_{\mathrm{av}} \exp[-i\mathbf{Q}\cdot\mathbf{R}_l(0)] \exp[i\mathbf{Q}\cdot\mathbf{R}_{l'}(t)] \right\rangle \mathrm{e}^{-i\Omega t}\, dt \\ &= \frac{d^2\sigma_{\mathrm{inc}}}{dE\, d\Omega} + \frac{d^2\sigma_{\mathrm{coh}}}{dE\, d\Omega}, \end{aligned} \tag{4.22}$$

where

$$\frac{d^2\sigma_{\text{inc}}}{dE\,d\Omega} = \frac{k_{\text{f}}}{k_{\text{i}}} Na_{\text{inc}}^2 \frac{1}{2\pi\hbar} \int_{-\infty}^{\infty} \frac{1}{N} \left\langle \sum_l \exp[-i\mathbf{Q}\cdot\mathbf{R}_l(0)] \exp[i\mathbf{Q}\cdot\mathbf{R}_l(t)] \right\rangle \times \mathrm{e}^{-i\Omega t}\,dt = Na_{\text{inc}}^2 \frac{k_{\text{f}}}{k_{\text{i}}\hbar} S_{\text{s}}(\mathbf{Q}, \Omega), \tag{4.23}$$

$$\frac{d^2\sigma_{\text{coh}}}{dE\,d\Omega} = \frac{k_{\text{f}}}{k_{\text{i}}} Na_{\text{coh}}^2 \frac{1}{2\pi\hbar} \int_{-\infty}^{\infty} \frac{1}{N} \left\langle \sum_{l\,l'} \exp[-i\mathbf{Q}\cdot\mathbf{R}_l(0)] \exp[i\mathbf{Q}\cdot\mathbf{R}_{l'}(t)] \right\rangle \times \mathrm{e}^{-i\Omega t}\,dt = Na_{\text{coh}}^2 \frac{k_{\text{f}}}{k_{\text{i}}\hbar} S(\mathbf{Q}, \Omega), \tag{4.24}$$

and definition (3.17) has been used in the second lines.[†]

We thus see that the incoherent and the coherent double differential cross sections can be simply related to $S_{\text{s}}(\mathbf{Q}, \Omega)$ and $S(\mathbf{Q}, \Omega)$, respectively. In other words, the incoherent scattering probes the single-particle motion and the coherent scattering the collective motion of atoms in a liquid. If one performes the enegy analysis by the constant-Q method, then it can be shown that the extra factor $k_{\text{f}}/k_{\text{i}}$ is constant during the measurement. In this case, the area under $S(\mathbf{Q}, \Omega)$ would give the structure factor $S(Q)$. In general, one measures $S(\mathbf{Q}, \Omega)$ and $S_{\text{s}}(\mathbf{Q}, \Omega)$ by either the triple-axis spectrometer or the time-of-flight spectrometer. The constant-Q method is applicable only when one uses the former. For details of the instruments and method, we refer readers to an authoritative review article by Brockhouse (1961).

As seen from (4.22), in practice, measurement of the double differential cross section includes contributions from both coherent and incoherent scatterings. For example, for natural argon, the ratio $a_{\text{inc}}^2/a_{\text{coh}}^2$ is 0.34/0.66; therefore experimental separation of the two is farily difficult. However, from (4.23) and (4.24), we notice that, except for a constant factor, the double differential cross section is equal to the combination

$$(a_{\text{inc}}^2/a_{\text{coh}}^2) S_{\text{s}}(\mathbf{Q}, \Omega) + S(\mathbf{Q}, \Omega). \tag{4.25}$$

Or, using (3.9), the above expression can be written as the space–time

[†] We denote the dynamic structure factor in Section III by $S(\mathbf{q}, \omega)$ where $\mathbf{q}$ and ω are merely space and time Fourier transform variables. In this section we purposely switch the notation to $S(\mathbf{Q}, \Omega)$ to emphasize that in a scattering experiment these Fourier transform variables mean the momentum and energy transfers defined by (4.1) and (4.2).

Fourier transform of

$$\left(\frac{a_{\text{inc}}^2}{a_{\text{coh}}^2} + 1\right) \mathbf{G}_s(\mathbf{r}, t) + \mathbf{G}_d(\mathbf{r}, t). \tag{4.26}$$

Dasannacharya and Rao (1965) performed neutron scattering measurements with liquid argon by the constant-Q method, and Fig. 12 shows the expression (4.25) plotted as a function of Q and outgoing wavelength λ' [since the incident wavelength is fixed, the presentation is related uniquely to $S(\mathbf{Q}, \Omega)$]. The area under the peak gives the structure factor $S(Q)$ and we see that it is peaked at $Q = 2.0 \times 10^8$ cm^{-1}, corresponding to the first diffraction peak of $S(Q)$. The above authors showed that for the time range from zero up to 10×10^{-13} sec, $G_s(\mathbf{r}, t)$ is well localized and $G_s(\mathbf{r}, t)$ and $G_d(\mathbf{r}, t)$ are well separated from each other so that presentation of G_s and G_d alone is possible. This is shown in Fig. 13. We see that the curves agree with the general behavior of G_s and G_d mentioned in Section III.

B. X-Ray and Light Scattering

The scattering of an electromagnetic (EM) wave by an atomic system occurs as a result of interaction between the EM wave and the atomic electrons. Classifically, one imagines that the incident wave induces vibrations of the electrons and the vibrating electrons in turn radiate EM waves in all directions, which is observed as the scattered radiation. If the electron were free, the frequency of the induced vibration would be the same as the incident wave frequency ω_i and the scattering is elastic. A common example of this is the Thompson scattering of X-rays. On the other hand, if the electron were bound with a characteristic frequency ω_0, then the induced frequency would have components at $\omega_0 \pm \omega_i$ (imagine a forced harmonic oscillation) and the scattering is inelastic. An example of this type is the Brillouin scattering of light. The main difference between X-ray and light scattering from liquids lies in the fact the their wavelengths are very different. In the former case, the EM wavelength is comparable to the atomic size and therefore one must treat the scattering from individual electrons. In the latter case, on the other hand, the EM wevelength is so much larger compared to the interatomic distance that atoms in a macroscopic volume element in the medium see the same electric field. In this case, it is meaningful to talk about the macroscopic dielectric property of the medium and light can be thought of as being scattered by inhomogeneities of the dielectric constant of the medium.

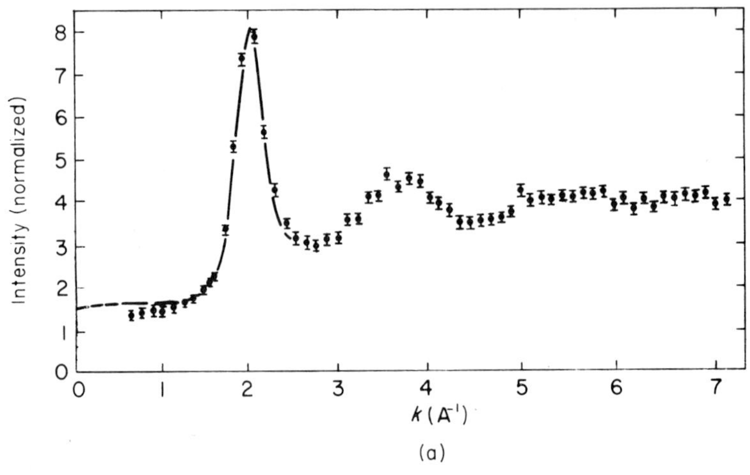

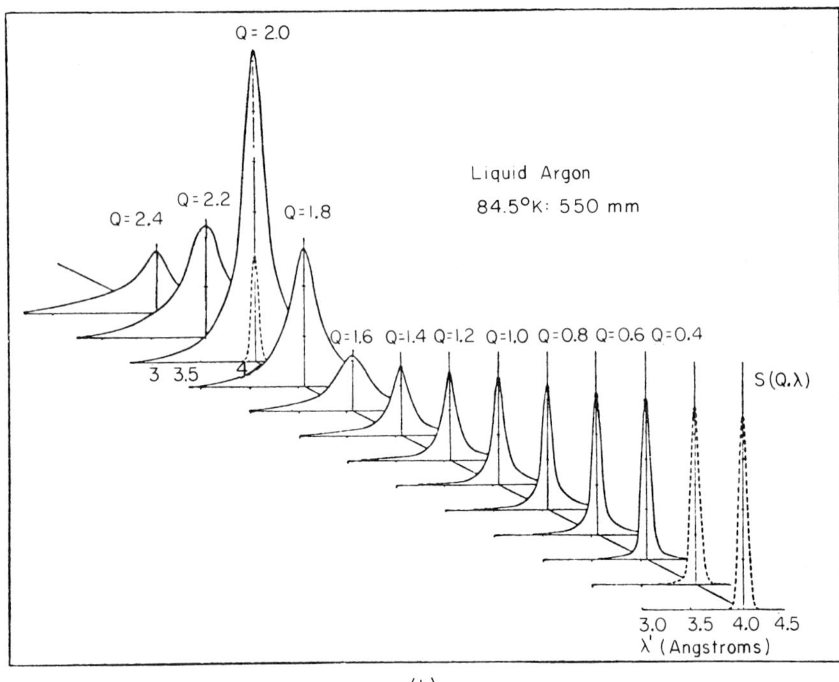

FIG. 12. (a) The integrated intensity of neutrons scattered from liquid argon. The ordinate is proportional to the structure factor $S(Q)$. Dots are points obtained by Henshaw (1957) at $T = 84°K$ and the solid line by Dasannacharya and Rao (1965) at $T = 85.5°K$ based on data shown in (b): the dynamic structure factor $S(Q, \omega)$ for liquid argon plotted as a function of Q and wavelength of outgoing neutrons λ' (Dasannacharya and Rao, 1965).

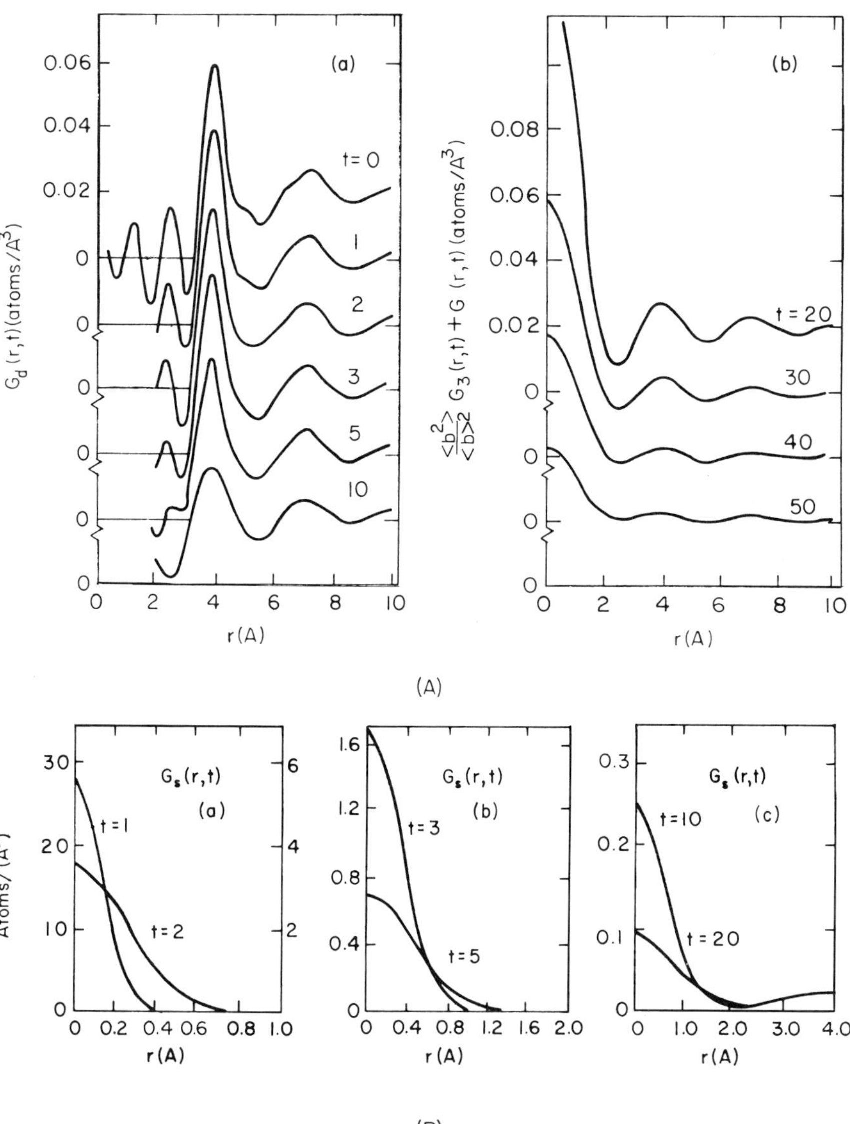

(B)

Fig. 13. Space–time correlation functions for liquid argon reduced from neutron-scattering measurements such as in Fig. 12 (Dasannacharya and Rao, 1965). (A) The pair correlation function $G_d(r, t)$ and a weighted combination of the self- and pair-correlation function at different times. The unit of time is 10^{-13} sec. (B) Self-correlation function for different times (in units of 10^{-13} sec).

1. *X-Ray Scattering Cross Section*

Let the incident EM wave be described by a plane wave

$$\mathbf{E}_i(z, t) = \mathbf{E}_0 \exp i(k_i z - \omega_i t). \tag{4.27}$$

The scattered wave in the wave zone can be written in a form analogous to the neutron case (4.15), as was shown by Landau and Lifshitz (1960),

$$\mathbf{E}_s(\mathbf{r}, t) = (e^2/mc^2)\hat{\mathbf{k}} \times (\hat{\mathbf{k}}_f \times \mathbf{E}_0)\varrho_e(\mathbf{Q}, t) \exp i(k_f r - \omega_i t)/r. \tag{4.28}$$

$\hat{\mathbf{k}}_f$ here denotes the unit vector in the scattered direction and $\varrho_e(\mathbf{Q}, t)$ is the Fourier transform of the electron number density,

$$\varrho_e(\mathbf{Q}, t) = \int \varrho_e(\mathbf{r}, t) \exp[i\mathbf{Q} \cdot \mathbf{r}]\, d\mathbf{r}. \tag{4.29}$$

Since the incident frequency ω_i is much larger than the typical frequency of the motion of electrons in atoms, one can disregard the time dependence in $\varrho_e(\mathbf{Q}, t)$ and calculate the differential cross section according to

$$\frac{d\sigma}{d\Omega} = \frac{\langle\langle(c/4\pi) \mid E_s \mid^2 r^2\rangle_{av}\rangle}{(c/4\pi) \mid E_i \mid^2}, \tag{4.30}$$

where $\langle\ \rangle_{av}$ denotes an average over the incident polarization directions. For an unpolarized X-ray, the incident polarization vector is distributed equally between the two perpendicular directions (both perpendicular to $\hat{\mathbf{k}}_i$), and one can easily show

$$\langle \mid \hat{\mathbf{k}}_f \times (\hat{\mathbf{k}}_f \times \mathbf{E}_0) \mid^2\rangle_{av} = \tfrac{1}{2}(1 + \cos^2 \phi) \mid E_0 \mid^2, \tag{4.31}$$

where ϕ is the scattering angle. Using this expression in (4.30), we then have

$$d\sigma/d\Omega = (e^2/mc^2)\tfrac{1}{2}(1 + \cos^2 \phi)\langle\varrho_e(-\mathbf{Q}, t)\varrho_e(\mathbf{Q}, t)\rangle. \tag{4.32}$$

Now, if electrons can be regarded as tightly bound to atoms of coordinates $\mathbf{R}_l(t)$, we can write

$$\begin{aligned}\varrho_e(\mathbf{r}, t) &= \sum_i \delta(\mathbf{r} - \mathbf{r}_i(t)) \\ &= \sum_l \varrho_l(\mathbf{r} - \mathbf{R}_l(t), t),\end{aligned} \tag{4.33}$$

where ϱ_l is the charge density at the atom l. The Fourier transform is then

$$\begin{aligned}\varrho_e(\mathbf{Q}, t) &= \sum_l \int (\exp i\mathbf{Q} \cdot \mathbf{r})\varrho_l(\mathbf{r} - \mathbf{R}_l(t), t)\, d\mathbf{r} \\ &= \sum_l [\exp i\mathbf{Q} \cdot \mathbf{R}_l(t)] \int (\exp i\mathbf{Q} \cdot \mathbf{r}')\varrho_l(\mathbf{r}', t)\, d\mathbf{r}' \\ &= \sum_l f_l(\mathbf{Q}, t) \exp[i\mathbf{Q} \cdot \mathbf{R}_l(t)]. \end{aligned} \tag{4.34}$$

We next use the adiabatic approximation that electrons in an atom move much faster than the speed of the atom in a liquid; therefore it is legitimate to replace $f_l(\mathbf{Q}, t)$ by an average form factor $f(Q)$, so

$$\begin{aligned}\langle \varrho_e(-\mathbf{Q}, t)\varrho_e(\mathbf{Q}, t)\rangle &= N\,|\,f(\mathbf{Q})\,|^2 \\ &\quad \times \left\langle \frac{1}{N} \sum_{l,l'} \exp[-i\mathbf{Q} \cdot \mathbf{R}_l(t)] \exp[i\mathbf{Q} \cdot \mathbf{R}_{l'}(t)]\right\rangle \\ &= N\,|\,f(\mathbf{Q})\,|^2 I(\mathbf{Q}, 0) \\ &= N\,|\,f(\mathbf{Q})\,|^2\, S(\mathbf{Q}) \end{aligned} \tag{4.35}$$

where we have used (3.20). The differential cross section can finally be written as

$$d\sigma/d\Omega = N r_e^2\,|\,f(\mathbf{Q})\,|^2\, \tfrac{1}{2}(1 + \cos^2\phi) S(\mathbf{Q}), \tag{4.36}$$

from which we can see that the effective scattering length is $r_e f(\mathbf{Q})$, where r_e denotes the classical electron radius e^2/mc^2. We also note that $f(\mathbf{Q})$ approaches Z, the atomic number, as $\mathbf{Q} \to 0$. Pings (1968) has given a comprehensive summary of the experimental measurement of X-ray scattering from liquids.

2. *Light-Scattering Cross Section*

As we remarked earlier, light scattering is caused by the fluctuation in the dielectric constant due to thermal fluctuations in a liquid. Let

$$\varepsilon(\mathbf{r}, t) = \varepsilon + \Delta\varepsilon(\mathbf{r}, t), \tag{4.37}$$

where ε is the average dielectric constant and $\Delta\varepsilon(\mathbf{r}, t)$ is the local fluctuation of the dielectric constant, which we take to be a scalar in a simple liquid. We can write down an analogous formula to (4.28) (see Landau and

Lifshitz):

$$\mathbf{E}_s(\mathbf{r}, t) = -(\omega_f^2/4\pi c^2)(\hat{\mathbf{k}}_f \times \hat{\mathbf{k}}_f \times \mathbf{E}_0)\, \Delta\varepsilon(\mathbf{Q}, t) \exp i(k_f r - \omega_i t)/r \quad (4.38)$$

where $\Delta\varepsilon(\mathbf{Q}, t)$ is the Fourier transform of $\Delta\varepsilon(\mathbf{r}, t)$. Using the same argument as in the case of neutron scattering, one easily arrives at

$$\frac{d^2\sigma}{dE\, d\Omega} = \left(\frac{\omega_f}{c}\right)^4 \frac{|\hat{\mathbf{k}}_f \times \hat{\mathbf{E}}_0|^2}{16\pi^2} \frac{1}{2\pi\hbar} \int_{-\infty}^{\infty} dt\, e^{-i\Omega t} \langle \Delta\varepsilon(-\mathbf{Q}, 0)\, \Delta\varepsilon(\mathbf{Q}, t)\rangle. \quad (4.39)$$

Now, it can be shown (Fabelinskii, 1968) that in a simple liquid the fluctuation in dielectric constant is mainly caused by the isothermal fluctuation in density; therefore one can write

$$\Delta\varepsilon(\mathbf{Q}, t) = (\partial\varepsilon/\partial\varrho)_T \varrho(\mathbf{Q}, t). \quad (4.40)$$

Using this relation, one can rewrite (4.39) as

$$\begin{aligned} \frac{d^2\sigma}{dE\, d\Omega} &= \left(\frac{\omega_f}{c}\right)^4 \frac{|\hat{\mathbf{k}}_f \times \hat{\mathbf{E}}_0|}{16\pi^2} \left(\frac{\partial\varepsilon}{\partial\varrho}\right)_T^2 \frac{1}{2\pi\hbar} \\ &\quad \times \int_{-\infty}^{\infty} dt\, e^{-i\Omega t} \langle \varrho(-\mathbf{Q}, 0)\varrho(\mathbf{Q}, t)\rangle \\ &= \left(\frac{\omega_f}{c}\right)^4 \frac{|\hat{\mathbf{k}}_f \times \hat{\mathbf{E}}_0|}{16\pi^2\hbar} \left(\frac{\partial\varepsilon}{\partial\varrho}\right)_T^2 NS(\mathbf{Q}, \Omega). \end{aligned} \quad (4.41)$$

In the usual experimental configuration, one sends in a linearly polarized light perpendicular to the scattering plane and in this case $|\hat{\mathbf{k}}_f \times \hat{\mathbf{E}}_0|^2 = 1$. If the intensity only is measured, then one integrates over the energy in (4.41) for a constant $\mathbf{Q}$ and obtains

$$\begin{aligned} \frac{d\sigma}{d\Omega} &= \left(\frac{\omega_f}{c}\right)^4 \frac{1}{16\pi^2} \left(\frac{\partial\varepsilon}{\partial\varrho}\right)_T^2 NS(\mathbf{Q}) \\ &\simeq \frac{N}{16\pi^2} \left(\frac{\omega_f}{c}\right)^4 \left(\frac{\partial\varepsilon}{\partial\varrho}\right)_T^2 \varrho k_B T \chi_T. \end{aligned} \quad (4.42)$$

In deriving the second line, we use the fact that $\mathbf{Q}$ is very small in light scattering and therefore (3.22) can be used. This formula was first obtained by Einstein in 1910.

For details of modern light-scattering measurements, we refer the reader elsewhere (Chen and Yip, 1972). As we can see from (4.36) and (4.41), both X-ray and light scattering from a simple liquid contain only the coherent scattering and therefore do not give information on the single-particle motions of a liquid.

C. Experimental Results on the Structure of Simple Dielectric Liquids

Among the many dielectric liquids which have been studied by neutron and X-ray scattering (see review articles by Larson *et al.*, 1968); Kruh, 1962); Furukawa, 1962), we shall single out liquid argon for our discussion because it is one of the simplest liquid we know of and also there are extensive data available both from neutron and X-ray as well as computer studies.

1. *Radial Distribution Function near the Triple Point*

Measurements of the liquid structure factor for argon near the triple point ($T = 84°\text{K}$, $p = 0.675$ atm) have been reported by Henshaw (1957) and Dasannacharya and Rao (1965) using neutrons. Henshaw used the conventional diffraction technique in which the total scattering intensity is measured as a function of scattering angle. The wavelength of the incident neutrons was 1.04 Å and it is assumed that the change in wavelength due to the inelastic scattering is so small that the integrated intensity is obtained effectively at a constant Q. This approximation (called the static approximation) is reasonable for wavelengths of neutrons around or less than 1 Å. On the other hand, Dasannacharya and Rao used 4.06-Å neutrons to measure $S(Q, \Omega)$ (see Section A) and then numerically integrated over the energy transfer $\hbar\Omega$ to obtain $S(Q)$. Figure 12(a) shows the result of Henshaw's data as black dots corrected for such things as the incoherent scattering, background, resolution, double scattering, and also the change of number of scattering atoms with angle. The data of Dasannacharya and Rao are shown by a solid line. Considering the different measurement technique used and a slightly different temperature of the sample used (Dasannacharya and Rao used a temperature $T = 85.5°\text{K}$), the agreement is quite reasonable. Gingrich and Thompson (1962) used X-ray scattering where the static approximation is excellent and obtained a result which differs from the data of Henshaw and Dasannacharya and Rao by more than the experimental error. Rahman (1965) pointed out that this might be due to the difficulty in normalizing the intensity data in the X-ray case and that the neutron data are more reliable. We shall therefore not show the X-ray result.

Recall Eq. (3.21), which relates the structure factor $S(Q)$ to the radial distribution function $g(r)$. Since $g(r)$ has an asymptotic value of unity as $r \to \infty$, $S(Q)$ has a $\delta(\mathbf{Q})$ component in the forward direction. One can drop this term on the grounds that one can never distinguish between the directly transmitted beam and the forward scattered beam. We therefore write

$$S(Q) = 1 + \varrho \int (\exp i\mathbf{Q} \cdot \mathbf{r})[g(r) - 1] \, d\mathbf{r}, \qquad Q \neq 0. \quad (4.43)$$

On taking the inverse Fourier transform and rearranging terms, we arrive at a more practical expression,

$$4\pi r^2 \varrho[g(r) - 1] = (2r/\pi) \int_0^\infty Q[S(Q) - 1] \sin Qr \, dQ, \quad (4.44)$$

which connects rather directly the scattered intensity to the "atomic density distribution function" $4\pi r^2 \varrho g(r)$. Figure 14 shows this function derived from data of Fig. 12(a) together with a dotted line representing $4\pi r^2 \varrho$ for $\varrho = 2.13 \times 10^{-2}$ atoms Å^{-3}, which is the atomic density at

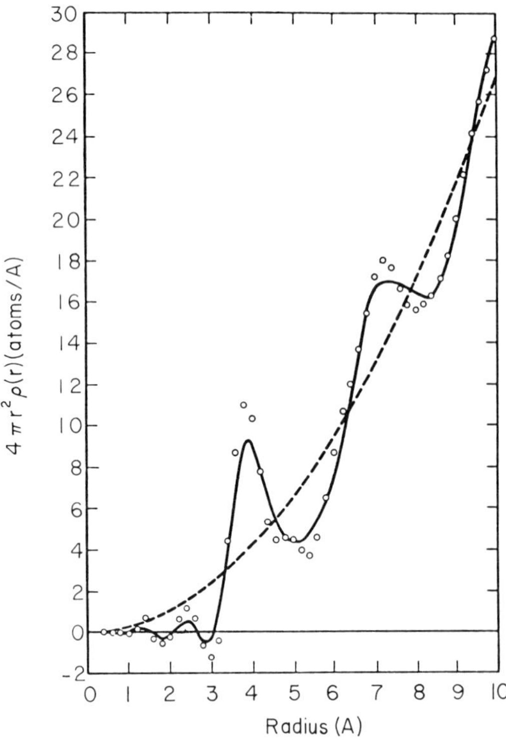

FIG. 14. (——) The atomic density distribution function $4\pi r^2 \varrho g(r)$ measured by Henshaw (1957) ($T = 84°$K), (○); the result obtained by Dasannacharya and Rao (1965) ($T = 84.5°$K); (- - -) the average atomic density $4\pi r^2 \varrho$ with $\varrho = 2.13 \times 10^{-2}$ atoms Å^{-3}.

$T = 84°K$. The small oscillations for $r < 2$ Å arise because the integration on the right-hand side of (4.44) is terminated at $Q = 7$ Å^{-1} before $S(Q)$ has reached its limiting value of unity. The data of Dasannacharya and Rao are shown as open circles in the figure.

The atomic density distribution shows two distinct peaks at $r = 4$ Å and $r = 7$ Å and perhaps a third weak peak at 10 Å. These correspond to the first, the second, and perhaps the third neighbor shells surrounding an arbitrary atom in the liquid. At distances greater than 7 Å, the shell structure becomes fuzzy and thus one obtains a quantitative idea of the short-time order in a liquid. Besides showing this overall short-range order, the curve contains two more pieces of information. First, one can obtain the distance of closest approach of two atoms from the point of intercept of the curve with the abscissa, which is about 3.05 Å. Secondly, Henshaw calculated the number of nearest neighbors by integrating the area under the first peak and obtained a value 8.0–8.5 depending on the shape assumed for the peak when extrapolating it to the right. This so-called "coordination number" is not a well-defined quantity because of different possible ways of estimating the area under the peak. Pings (1968) has investigated different ways of computing this number and showed examples from his own measurements (see the following).

2. *Density and Temperature Dependence of the RDF*

Using X-rays, Mikolaj and Pings (1967) have made a systematic measurement of $S(Q)$ for argon both along the critical isochore and on isotherms near the critical temperature. The data are significant in that one is able to make a quantitative assessment of the temperature and density dependence from them. Figure 15 shows the pvT diagram of argon from measurements of Levelt (1958). The dark circles and triangles are some of the thermodynamic states for which the X-ray measurements were made. Figure 16 shows the net correlation function $h(r) = g(r) - 1$ for different densities along the isotherm $T = 148°K$ computed from Eq. (4.44). Figure 17 shows $h(r)$ so obtained along the critical isochore for different temperatures. We see that, while the position of the first peak is relatively independent of the states, the second and further peaks wash out fairly rapidly as the density is lowered. The dependence on temperature is seen to be weak. One can understand the strong dependence of the structure of a liquid on density because the majority of the short-range order is brought about from the mutual exclusion effect of the hard core of atoms. In fact, Ashcroft and Lekner (1966) computed the structure factor of liquid metals from the Percus–Yevick equation (2.43) using just a hard-sphere potential with an effective packing density of 0.45 (packing density $= \frac{1}{6}\pi\sigma^3\varrho$) and obtained a fair agreement with that obtained by neutron measurements of Gingrich and Heaton (1961).

3. *The Direct Correlation Function*

As we have noted in Section II, B, the Ornstein–Zernike direct correlation function $c(r)$ is a theoretically important quantity since it is conjectured to be a short-ranged function and simpler in structure than the net correlation function $h(r)$. As we saw, the three approximate equations (2.39), (2.41), and (2.43) are based on this physical intuition. Now, $c(r)$ is also an experimentally accessible quantity based on the following equation. From (4.43) and the Fourier transform of (2.38), we easily obtain

$$C(Q) = [S(Q) - 1]/S(Q), \tag{4.45}$$

where

$$C(Q) = \varrho \int c(r) \exp[i\mathbf{Q} \cdot \mathbf{r}]\, d\mathbf{r}. \tag{4.46}$$

Thus, from (4.45) one can obtain $c(Q)$ experimentally and then, by com-

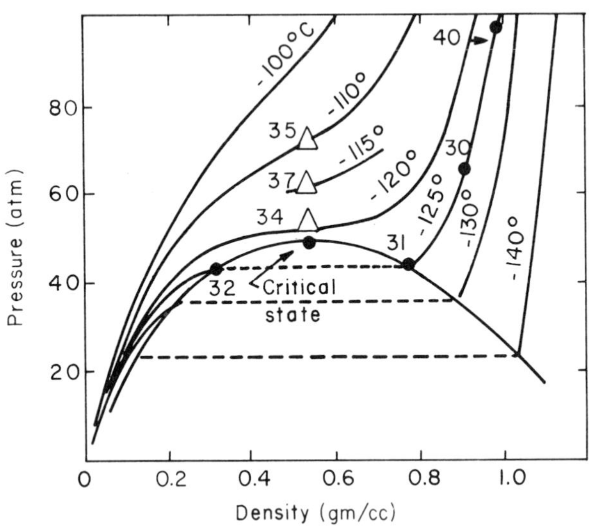

FIG. 15. Argon pvT data from Levelt (1958). The circles and triangles are some of the thermodynamic states for which the X-ray measurements of Mikolaj and Pings (1967) were made.

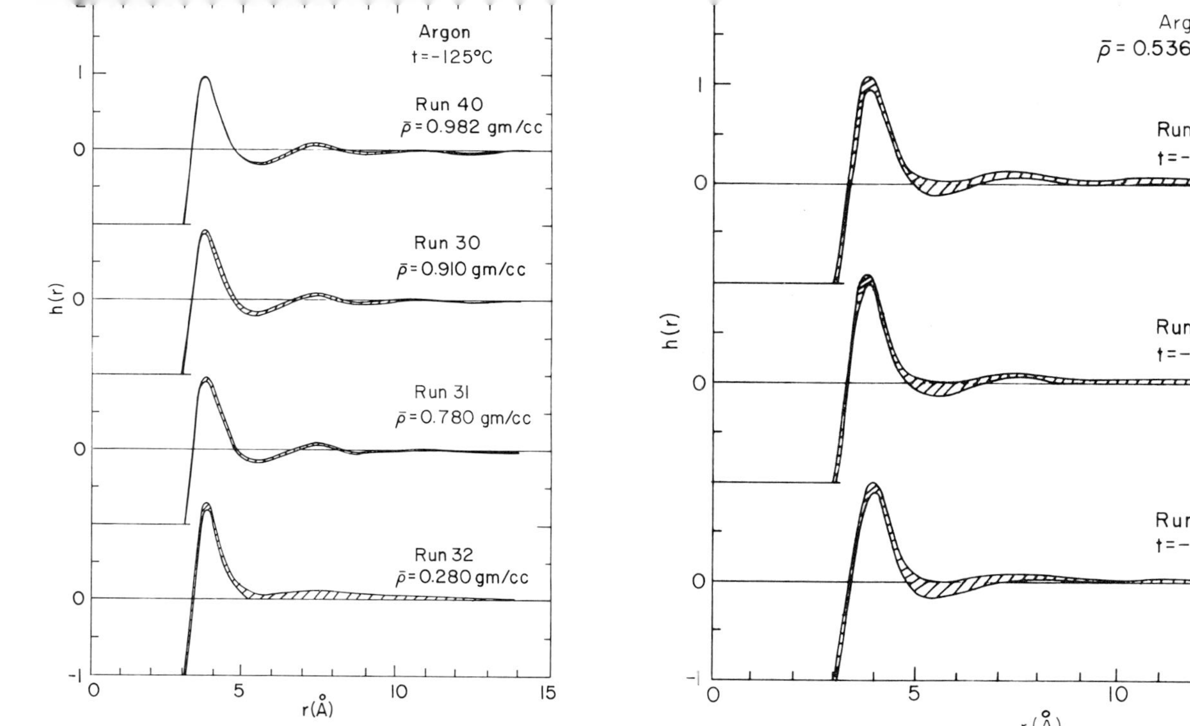

FIG. 16. The net correlation function $h(r)$ of argon for different densities along the isotherm $T = 148°K$ computed from Eq. (4.44). The thermodynamic states correspond to circles in Fig. 15 (Mikolaj and Pings, 1967).

FIG. 17. The net correlation function $h(r)$ of argon for different temperatures along the critical isochore $\varrho_c = 0.536$ g cm^{-3}. The thermodynamic states correspond to triangles in Fig. 15 (Mikolaj and Pings, 1967).

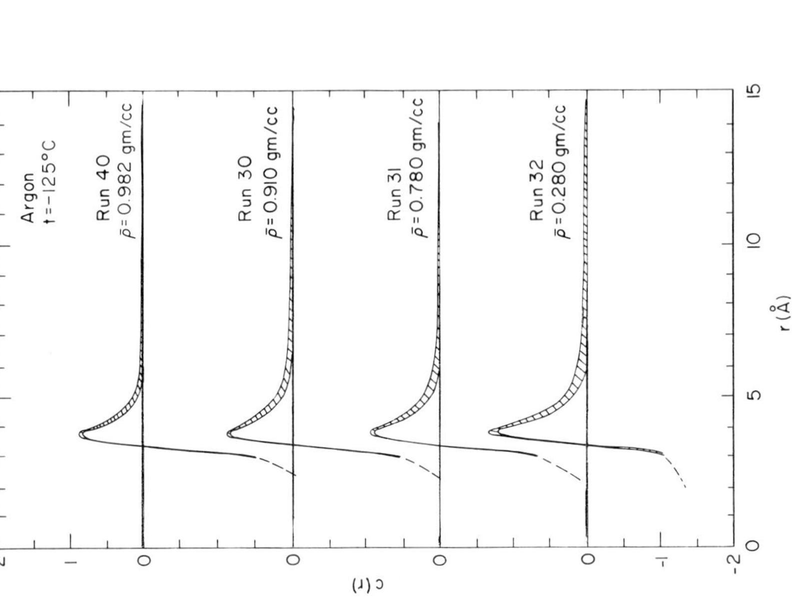

FIG. 18. The direct correlation function $c(r)$ for argon reduced from data in Fig. 16 (Mikolaj and Pings, 1967).

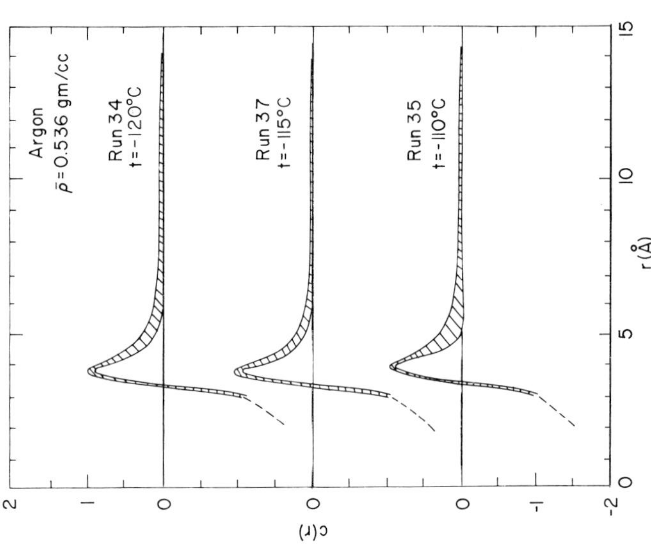

FIG. 19. The direct correlation function $c(r)$ for argon reduced from data in Fig. 17 (Mikolaj and Pings, 1967).

puting the inverse Fourier transform from (4.41), one gets $c(r)$. Mikolaj and Pings (1967; see also Pings, 1968) have carried out his computation and obtain the results shown in Figs. 18 and 19. It is indeed pleasing to see, at least qualitatively, that the direct correlation function in argon exhibits the following features: (1) $c(r)$ is shorter-ranged and is basically a simpler function than $h(r)$. (2) $c(r)$ is negative at distances less than the subatomic diameter, rising steeply through zero toward a single positive maximum and then decaying fairly rapidly toward zero with a negative slope.

4. *Density Dependence of the First Coordination Number*

One direct way of seeing the packing effect of atomic spheres around an atom in a liquid is perhaps through the density dependence of the first coordination number. Pings (1968) defines this number N_1 by

$$N_1 = 2 \int_0^{r_{\max}} 4\pi r^2 \varrho g(r) \, dr, \tag{4.47}$$

where $r_{\max}$ is the position of the first maximum of the function $r^2 g(r)$. The density dependence of N_1 in argon is shown in Fig. 20, where the solid curve represents a quadratic equation fitted to these points. We sse that N_1 is a rapidly rising function of the density with the coordination number of about 6 at $\varrho_{\mathrm{m}} = 1$ g cm^{-3}. An ideal close-packed structure would have a coordination number of 12.

5. *Computer Experiment: The Dynamic Structure of Liquids*

The method of studying a system of interacting particles by numerical computation of their trajectories is now commonly referred to as "molecular dynamics." In this method, one takes a finite number of particles, puts them in a box, and computes all their trajectories in time using Newton's equation and the cyclic boundary conditions. If one uses a realistic interatomic potential and adjusts the size of the box so that one obtains a proper density, one can simulate the actual condition in a liquid fairly well. The use of Newton's equation is quite appropriate in classical liquids such as argon and liquid metals.

Both Rahman (1964) and Verlet (1967) have performed the molecular-dynamic calculation with 864 argon atoms. The trajectory of the atoms was computed in steps of 10^{-14} sec from time zero to about 10^{-11} sec. Four quantities can immediately be obtained by analyzing the recorded

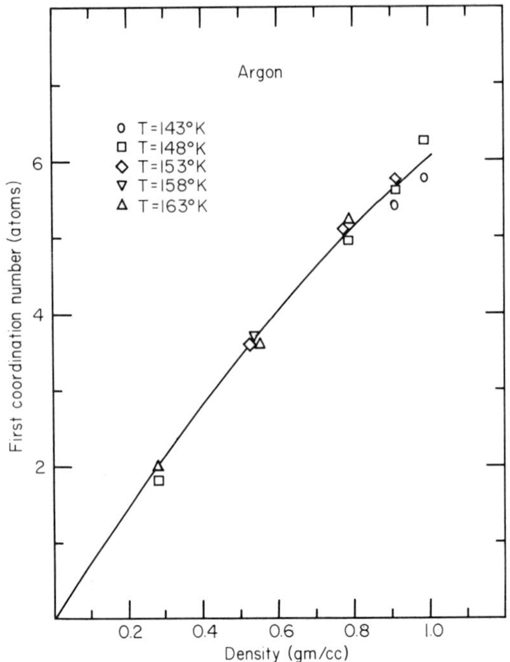

FIG. 20. The first coordination number of argon as a function of its density (Pings, 1968).

trajectories. If $n(r)$ denotes the number of particles situated at a distance between r and $r + \Delta r$ from a given particle, one has for the pair correlation functions:

$$g(r) = \frac{V}{N} \frac{n(r)}{4\pi r^2 \, \Delta r}, \tag{4.48}$$

$$\mathbf{G}_{\mathrm{d}}(r, t) = \frac{V}{N} \frac{n(r, t)}{4\pi r^2 \, \Delta r}, \tag{4.49}$$

where $n(r, t)$ is the time-dependent extension of $n(r)$. For the quantities related to the single-particle motion, one has

$$\langle r^{2n} \rangle = (1/N) \sum_{i=1}^{N} [\mathbf{r}_i(t) - \mathbf{r}_i(0)]^2, \tag{4.50}$$

which is the even moment of $G_s(r, t)$, i.e.,

$$\langle r^{2n} \rangle = \int r^{2n} G_s(r, t) \, d\mathbf{r}, \tag{4.51}$$

and the velocity autocorrelation function

$$\langle \mathbf{v}(0) \cdot \mathbf{v}(t) \rangle = (1/N) \sum_{i=1}^{N} \mathbf{v}_i(0) \cdot \mathbf{v}_i(t). \tag{4.52}$$

Having obtained $g(r)$ for different temperatures and densities, one can then use the formula derived in Section II to calculate the equilibrium properties of the fluid. This is the study that Verlet (1967) undertook and he found good agreement with experimentally measured values in all quantities he calculated. This is a strong indication that the Lennard-Jones potential is a fairly good representation of interatomic interaction in liquid argon and that a 864-atom system can simulate the bulk properties of a liquid fairly well. Verlet (1968) has an extensive tabulation of $g(r)$ in the critical region of argon ($T_c = 157°\text{K}$, $\varrho_c = 0.279$ g cm^{-3}) and an example was already shown in Fig. 4. Rahman, on the other hand, concentrated his effort in obtaining the dynamic properties of argon and he obtained the following two important quantities. Figure 21 shows $G_d(r, t)$ at three times obtained from molecular dynamics. We see that, starting from $g(r)$ at $t = 0$ with well-defined peaks corresponding to the well-defined neighbor shells, as time goes on, $G_d(r, t)$ gradually damps out. Combining Fig. 21 and Fig. 13(b), we can estimate that the time it takes for the local structure of argon near the triple point to relax is about 5×10^{-12} sec. It is difficult theoretically to calculate $G_d(r, t)$ in this time range rigorously, and further, the neutron-scattering data available at present are rather limited in accuracy [see the Dasannacharya–Rao data in Fig. 13(b)]. Therefore molecular dynamics is very valuable in providing information in this area. Figures 22 and 23 show the time dependence of the mean-square displacement and the velocity autocorrelation function defined in (4.50) and (4.52). The $\langle r^2 \rangle$ vs. t curve shows that, at small times, $\langle r^2 \rangle$ depends quadratically on t, as in a gas, and at long times, $\langle r^2 \rangle$ depends linearly on t, like a continuously diffusing particle. The diffusion constant $D = 2.43 \times 10^{-5}$ cm^2 sec^{-1} thus obtained ($\langle r^2 \rangle = 6Dt + C$) agrees well with the experimental value of Naghizadeh and Rice (1962). The velocity autocorrelation function shows a negative value around 0.5×10^{-12} sec, meaning that an argon particle undergoes a kind of vibratory motion on this time scale. This is consistent with a picture that in liquid an atom is surrounded by the near neighbors in a cagelike configuration and as soon as the atom moves it is likely to hit the wall of the cage and will temporarily be pushed back. Thus, in this sense, one can say that, within a time scale of 0.5×10^{-12} sec or so, an atom in a

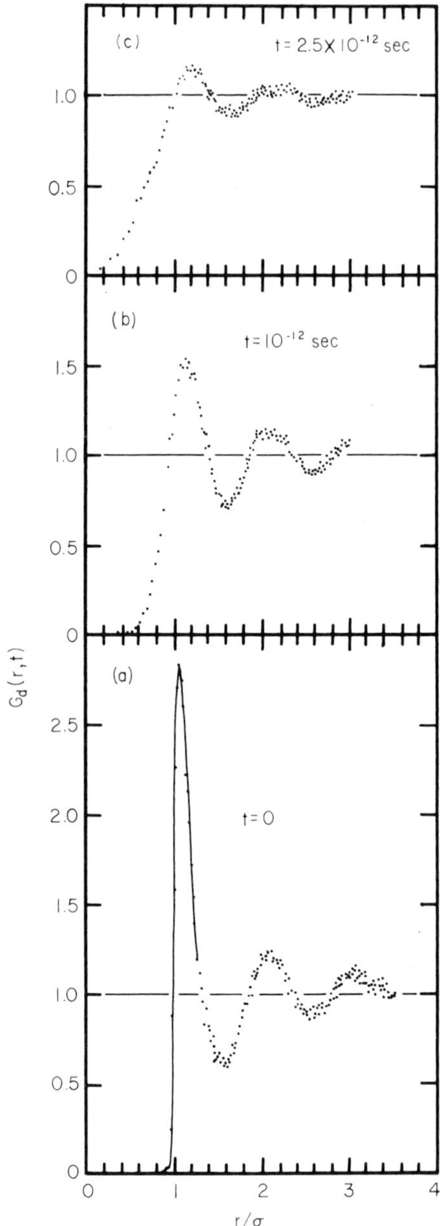

FIG. 21. Time-dependent pair-correlation function $G_d(r, t)$ for liquid argon ($T = 94.4°K$) at different times from molecular dynamics (Rahman, 1964).

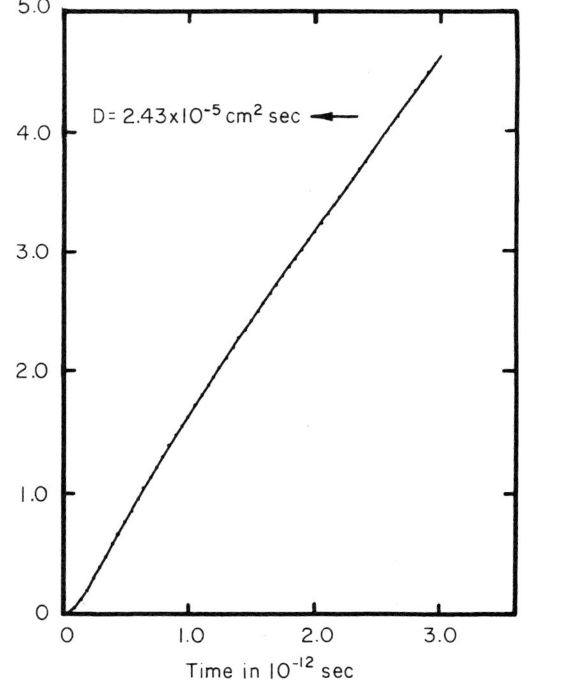

FIG. 22. Mean-square displacement of an atom in liquid argon ($T = 94.4°K$) from molecular dynamics (Rahman, 1964).

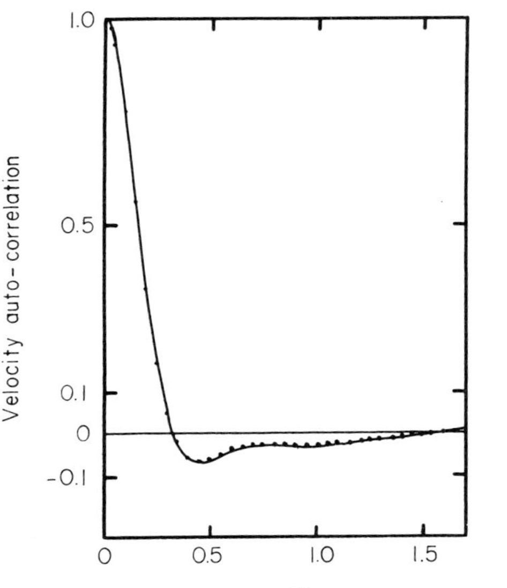

FIG. 23. The velocity autocorrelation function in liquid argon ($T = 94.4°K$) from molecular dynamics (Rahman, 1964).

liquid performs an oscillatory motion and this is solidlike in its behavior. Combining Fig. 22 and 23, we can also draw a conclusion that if we observe the single-particle motion in a liquid on a time scale longer than 10^{-12} sec, we should see that each atom is merely diffusing continuously around.

The molecular-dynamics method is also discussed in Chapter 3.

D. Experimental Results on the Structure of Simple Liquid Metals

There have been a large number of structure studies of liquid metals by X-rays and neutrons (see, for example, review articles by Kruh, 1967 and Enderby, 1968). Most of the liquids studied are low-melting-point metals near the triple point. As we have noted in Section I,C, the effective pair potential in even a simple metal is a subtle one and the theoretical considerations leads one to believe that the potential is long-range and oscillatory. It is therefore natural to ask whether the long-range oscillatory behavior of the potential would manifest itself in the measured structure factor of the liquid; or conversely, can one use the structure factor data to deduce the behavior of the pair potential at large distances? The first attempt to extract the pair potential from experimental structure was made by Johnson *et al.* (1963). They used the Born–Green approximate equation (2.39) for his purpose and found that for sodium the potential thus obtained has a marked oscillation, whereas for argon the potential is of the expected van der Waals type. This exciting possibility stimulated others to further study in this direction and we shall discuss some of the results below.

1. *Structure Factor and the Oscillatory Potential*

For a simple liquid near its triple point, the local structure comes mostly from geometric excluded volume effects and should be unaffected by the details of the interaction. As we mentioned in Section IV,C, Ashcroft and Lekner's (1966) work demonstrates this point very well. Figure 24 shows a comparison of the theoretical hard-sphere structure factor and the reduced structure factor for potassium and rubidium near the triple point. The circles and triangles are neutron-diffraction data of Gingrich and Heaton (1961). The agreement in general is seen to be fair. This suggests that perhaps with the present experimental accuracy of neutron and X-ray data it is very difficult to detect the small effect on the struc-

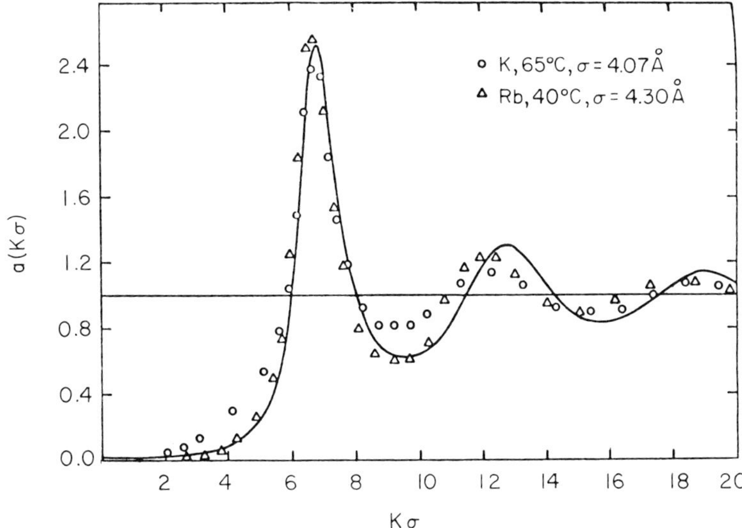

FIG. 24. Structure factor (denoted by $a(K\sigma)$) for two alkali metals. Comparison between PY hard-sphere calculation (Ashcroft and Lekner, 1966) and neutron-diffraction data (Gingrich and Heaton, 1961). Momentum transfer is denoted by K and σ is the assumed hard-sphere radius.

ture factor coming from the oscillatory part of the interaction potential. This point is further confirmed by the recent molecular-dynamics studies of Paskin and Rahman (1966, 1967) and Schiff (1969), One can use, for example, a Lennard-Jones potential and an oscillatory potential, respectively, to calculate the radial distribution function of a liquid metal near its melting point using molecular dynamics. Figure 25 shows one of such calculations made by Paskin and Rahman (1966). One can see the radial density distribution function thus obtained does not differ greatly for such a large difference in the potential function used. In fact, if the structures of liquids near the melting point are largely determined by the size of the hard core of the potential, then the radial distribution function ought to be similar for different liquids if one simply scales the distance r by appropriate core parameter α. Figure 26 is such a plot taken from Paskin (1967), which shows that one can almost superpose the RDF of argon on top of that of liquid alkalis if one chooses the scale parameter as shown in the figure. It is possible that the structure factor near the small-Q region is sensitive to the long-range nature of the oscillatory potential because at large distances the behavior of the RDF is simply related to the pair potential (see Ascarelli *et al.*, 1967). But the

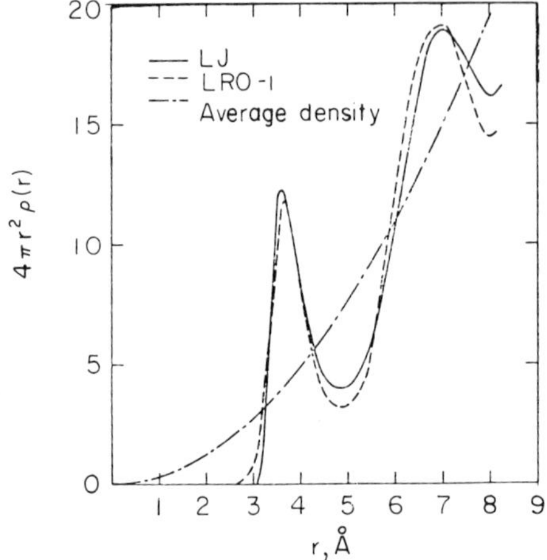

FIG. 25. A comparison of the atomic density distribution function obtained by Paskin and Rahman (1966) for a Lennard-Jones potential (solid line) and for a long-range oscillatory potential (dashed line) for liquid Na near the melting point.

accurate measurement of the structure factor in the small-Q region is expecially difficult due to the multiple-scattering effect. It appears therefore that from the static structure of liquid metals it is not possible to deduce the oscillatory behavior of the pair potential. However, this is not so for the dynamic structure, as has also been shown by the above authors. As we have mentioned in Section III,B one of the quantities which can

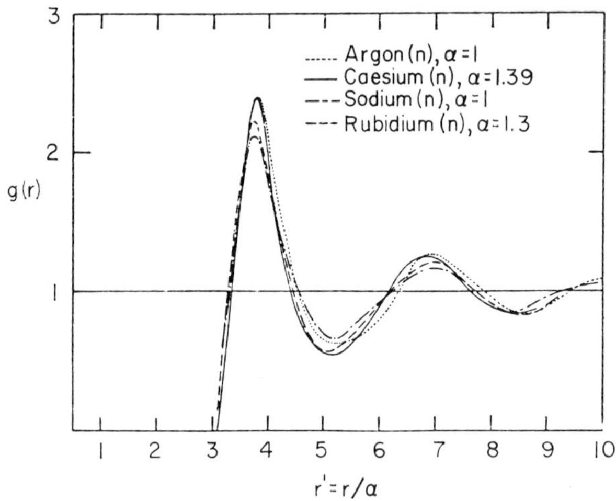

FIG. 26. The pair correlation functions $g(r)$ of liquid alkali metals and liquid argon near the melting point scaled so that the position of the first maximum coincides (Paskin, 1967).

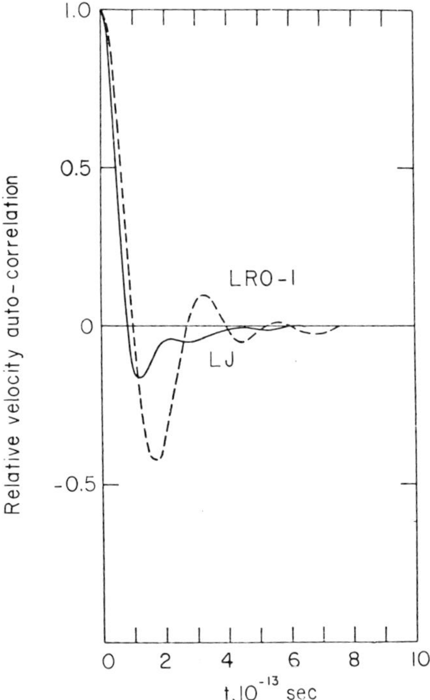

FIG. 27. The velocity autocorrelation functions for a Lennard-Jones and a long-range oscillatory potential obtained from a molecular dynamics of liquid Na (Paskin and Rahman, 1966). Both potentials give reasonable fits to the X-ray diffraction data.

be deduced from $S_s(q, \omega)$ is the frequency distribution function $Z(\omega)$ which is the Fourier transform of the velocity autocorrelation function. Paskin and Rahman (1966) have demonstrated that, while the oscillatory potential would produce an oscillatory tail in the velocity autocorrelation function, the Lennard-Jones potential would not. Figure 27 shows these two velocity autocorrelation functions produced by two potentials which are used to calculate the radial density distributions in Fig. 25. Thus it is hopeful that when more accurate experimental dynamic structure factors become available one might be able to subject the proposed oscillatory potentials to a stringent test.

2. *Partial Structure Factors for Binary Liquid Alloys*

So far, we have limited ourselves to discussions of simple one-component systems. However, in liquid metals one can find binary systems which are sufficiently simple, in the sense that each constituent is a simple atomic species, and yet they show some interesting structural properties. In a binary system, the second-order structure of a liquid cannot be completely described by a single pair correlation function. Denoting the individual atomic species by 1 and 2, one can then define the partial structure factor, in analogy to Eq. (4.43) for the one-component system, by

$$\begin{aligned} S_{\alpha\beta}(Q) &= 1 + \varrho \int [g_{\alpha\beta}(r) - 1] \exp i\mathbf{Q} \cdot \mathbf{r}\, d\mathbf{r} \\ &= \Big\langle (1/N) \sum_{l,m} \exp[i\mathbf{Q} \cdot (\mathbf{r}_{l\alpha} - \mathbf{r}_{m\beta})] \Big\rangle, \end{aligned} \tag{4.53}$$

where $g_{\alpha\beta}(r)$ has as its physical meaning the probability of finding an α atom in unit volume at radius r from the center of a β atom, normalized in such a way that at large distances r it tends to unity. The $\mathbf{r}_{l\alpha}$ denotes the position vector of lth atom and α, $\beta = 1, 2$ the species. The intensity of the radiation scattered coherently from the system is proportional to [cf. (4.12)].

$$I(Q) = \Big\langle \sum_{l} \sum_{m} f_l f_m \exp i\mathbf{Q} \cdot (\mathbf{r}_l - \mathbf{r}_m) \Big\rangle, \tag{4.54}$$

where f_l, f_m are the coherent scattering amplitudes given in (4.34) for the X-ray case and in (4.14) for the neutron case. In taking the ensemble average in (4.54), one must also consider the fact that, at a given position r_l, the chance of having the atomic species α is $(N_\alpha/N) = c_\alpha$, i.e., the atomic fractional concentration. If we recall from (4.35) that for a one-component system we have

$$\begin{aligned} I(Q) &= Nf^2 S(Q) \\ &= N[f^2 + f^2(S(Q) - 1)], \end{aligned} \tag{4.55}$$

then the direct extension of this formula to the two-component system is

$$\begin{aligned} I(Q) = N[&c_1 f_1^2 + c_2 f_2^2 + c_1^2 f_1^2 (S_{11}(Q) - 1) \\ &+ c_2 f_2^2 (S_{22}(Q) - 1) + 2c_1 c_2 f_1 f_2 (S_{12}(Q) - 1)], \end{aligned} \tag{4.56}$$

where the partial structure factors $S_{\alpha\beta}(Q)$ are related to the partial radial distribution function $g_{\alpha\beta}(r)$ by (4.53).

In neutron-diffraction measurement, there is an important advantage that the scattering lengths f_1 and f_2 can be varied by using different isotopes of the same element. Since the measured quantity can be expressed as

$$S(Q) = \gamma[F(Q) + \Delta], \tag{4.57}$$

with

$$F(Q) = c_1^2 f_1^2 (S_{11}(Q) - 1) + c_2^2 f_2^2 (S_{22}(Q) - 1) + 2c_1 c_2 f_1 f_2 (S_{12}(Q) - 1), \tag{4.58}$$

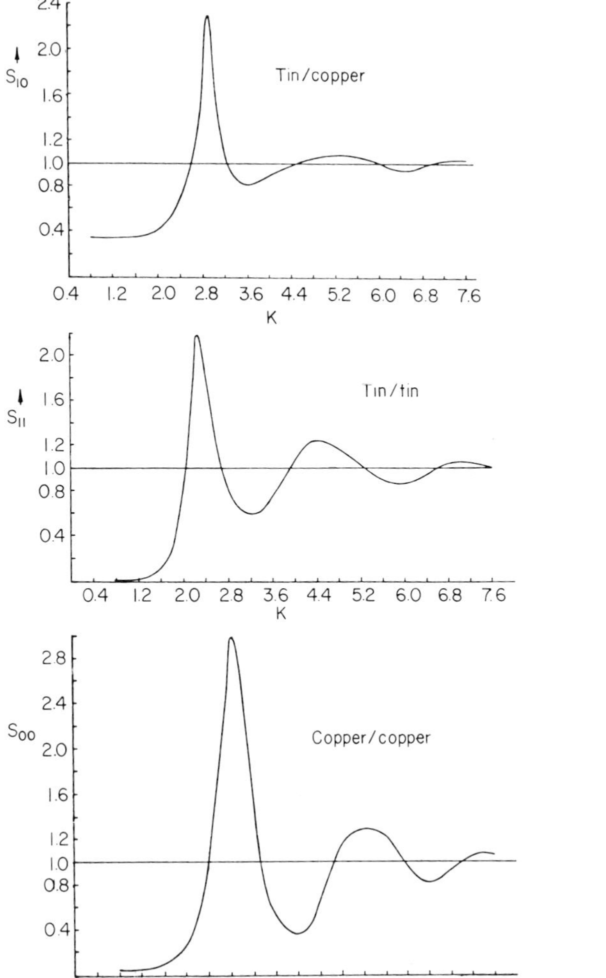

FIG. 28. Partial structure factors for liquid Cu–Sn system measured by Enderby *et al.* (1966). The wave-vector transfer is denoted by K.

then, if one can experimentally determine constants γ, Δ, and $F(Q)$ for three different values of f_1 of f_2, one can then deduce S_{11}, S_{22}, and S_{12}, respectively. Enderby *et al.* (1966) applied this technique to a liquid Cu–Sn alloy and were able to obtain all three partial structure factors as shown in Fig. 28. The scattering length adjusted was that of Cu, for which two isotopes ^{63}Cu and ^{65}Cu are available. Three values of f_{cu} used where relatively 1, 0.73, and 1.96.

Two interesting features are immediately apparent on inspection of Fig. 28. First, Enderby *et al.* noticed that the positions of the first few extrema in S_{11} and S_{22} correspond rather closely to those of pure copper and pure tin as shown in Table III. This feature is to be expected if the form of the structure factor is dominated by the random mixture of hard spheres of two different diameters R_1 and R_2 ($R_2 > R_1$). Enderby *et al.* (1967) therefore solved the Percus–Yevick equation for a mixture of hard spheres to obtain the partial structure factors, and one can see from the table that the theoretical positions of extrema agree reasonably well with the experimental values. However, the hard-sphere model fails to explain the experimental observation that extrema of S_{12} are quite close to those of S_{11}. This observation clearly points to the physical fact that there is departure from the simple additivity of hard-sphere diameters. The second interesting feature is that S_{12} remains substantial as $Q \to 0$. This presumably reflects the importance of long-range fluctuations in composition and should therefore be more pronounced near the critical mixing point of a binary system.

E. Structure of Liquids near the Critical Point

So far, we have focused our attention on the short-range part of the pair correlation functions. This is natural in terms of the available neutron and X-ray diffraction data because it is difficult to have accurate data for $g(r)$ for $r > 10$ Å. As far as the $r < 10$ Å region is concerned, we have shown in Section IV,C that $g(r)$ depends strongly on density but only weakly on temperature. In fact, one sees in Fig. 17 that, as one approaches along the critical isochore of argon toward the critical point, the first and second peaks in $h(r)$ are still clearly visible. Therefore one can safely conclude that the short-range part of the structure is relatively unchanged even near the critical point. This is not so for the long-range part of the structure, however, as can easily been shown from the small-angle scattering data. Recall Eq. (3.22), which relates the limiting be-

TABLE III

Positions of Extrema in the Partial Structure Factor of Liquid Cu–Sn System[a]

Extrema ($Å^{-1}$)	Pure Cu, Exp.	Cu–Cu		Pure Sn, Exp.	Sn–Sn		Cu–Sn	
		Exp.	Theory		Exp.	Theory	Exp.	Theory
First maximum	3.0	2.9	2.9	2.3	2.25	2.4	2.9	2.65
First minimum	3.95	4.0	4.4	3.4	3.3	3.3	3.6	3.8
Second maximum	5.4	5.3	5.8	4.3	4.4	4.3	5.2	4.9

[a] From Enderby (1968).

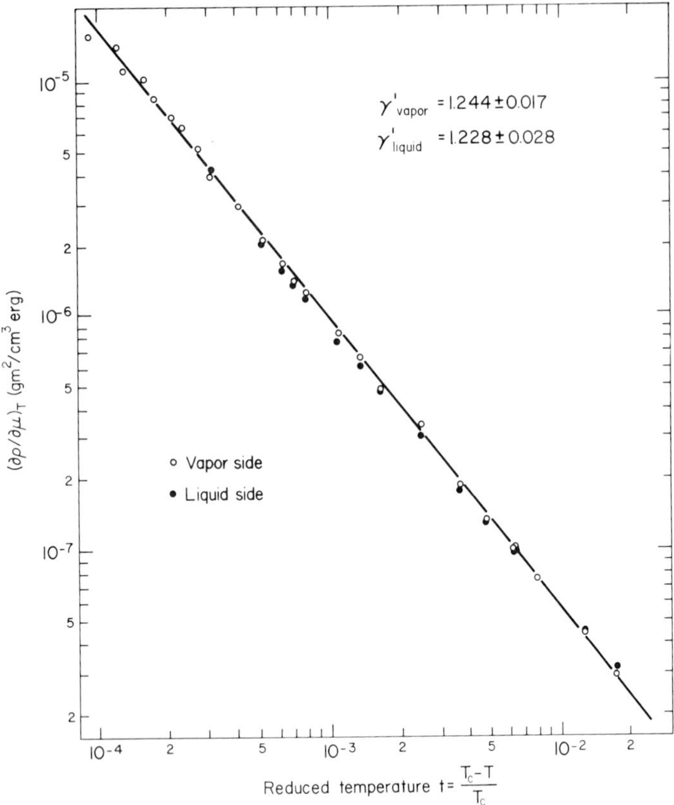

FIG. 29. Plot of $(\partial \varrho / \partial \mu)_T$, which is proportional to the compressibility, along the vapor and liquid sides of the coexistence curve of xenon as a function of the reduced temperature. Open circles are points on the vapor side and solid circles are those on the liquid side (Giglio and Benedek, 1969).

havior of $S(Q)$ as $Q \simeq 0$ to the isothermal compressibility. Figure 29 shows the experimental data on the compressibility of xenon near the critical point as a function of $T - T_c$. One sees that as $T \to T_c$ the compressibility diverges like $(T - T_c)^{-\gamma}$ with $\gamma = 1.23$. The graph is taken from a recent light-scattering experiment of Giglio and Benedek (1969). The fact that the compressibility and hence the forward scattering intensity should grow very large near the critical point is well known since the turn of the century and is often called the critical opalescence. Using (4.43), one can then infer the behavior of $g(r)$ as $T \to T_c$. Specifically, one can write

$$S(Q)_{Q \to 0} = 1 + \varrho \int h(r)\, d\mathbf{r} \tag{4.59}$$

and note that under normal conditions $h(r)$ is short-ranged and therefore the integral is finite. However, as $T \to T_c$, $S(Q)$ diverges, while the small-r part of $h(r)$ is relatively unchanged, which leads one to the conclusion that $h(r)$ has to become long-ranged in such a way that the integral in (4.59) diverges. Ornstein and Zernike [see Zernike (1916)] were the first to show that the asymptotic form of $h(r)$ is given by

$$h(r) \simeq (1/4\pi\varrho R^2)(1/\gamma) \exp[-r/\xi], \qquad r \to \infty \tag{4.60}$$

where R^2 is the second moment of the direct correlation function,

$$R^2 = \tfrac{1}{6}\varrho \int r^2 c(r)\, d\mathbf{r}, \tag{4.61}$$

and the correlation length ξ is a temperature-dependent quantity given by

$$\xi^2 = \varrho k_B T \chi_T R^2. \tag{4.62}$$

The form (4.60) was derived by using a heuristic argument (see Fisher, 1967) based on the fact that the direct correlation function $c(r)$ is a short-range function even near the critical point and therefore both the first moment [cf. (2.23a) at the end of Section II] and the second moment R^2 exist. The consequence of the assumed form (4.60) is that the structure factor at small Q is now a Lorentzian function in Q, i.e.,

$$S(Q) = (1/R^2)\xi^2/\{1 + \xi^2 Q^2\}, \qquad Q \to 0. \tag{4.63}$$

Figure 30 shows $1/S(Q)$ plotted against Q^2 for argon at the critical pressure ($P_c = 48.35$ atm) and at various temperatures near T_c. The data are due to Thomas and Schmidt (1963) and the measurement was done using small-angle X-ray scattering. The fact that the data points fall into straight lines reasonably well indicates that the asymptotic form of the net correlation function (4.60) is essentially a correct one.

Physically, one can picture the growth of the long-range order near the critical point as follows. As $T \to T_c$, say from above, the gaseous phase is trying to condense into the liquid phase. We may say that/some macroscopic regions of the fluid are trying to form droplets of the liquid density, with the consequence that the density fluctuation is growing larger and larger (this is connected to the fact that the compressibility is also growing larger). At the same time, the range of the fluctuation has to become longer in order to accomplish the ultimate goal of forming the droplet of sufficiently large size so that the phase separation is possible. We can therefore take the correlation length ξ to be the measure of the size of the droplet.

Questions regarding the form of the dynamic structure factors near the critical point are also very interesting and challenging, expecially in view of the fact that good experimental data are beginning to become available

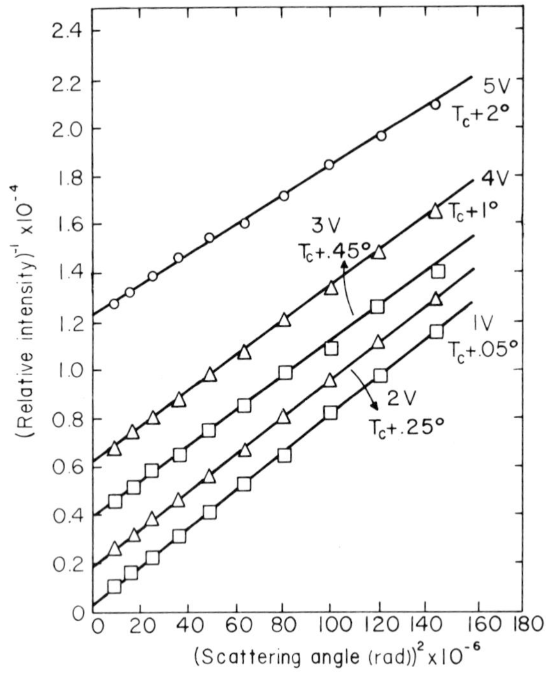

FIG. 30. Ornstein–Zernike plots of $S^{-1}(q)$ as a function of Q^2 (proportional to scattering angle squared for small scattering angles) for argon vapor at 48.3 atm and five different temperatures above the critical temperature (Thomas and Schmidt, 1963).

due to advances in light-scattering techniques in recent years. However, even a brief discussion of this fascinating topic would take us too far in this chapter. We therefore merely refer the interested reader to the literatures (see, for example, the review article by Heller, 1967).

ACKNOWLEDGMENT

This work was sponsored by the U.S. Atomic Energy Commission.

General References

Chen, S. H., and Yip, S. (1972). "Inelastic Neutron and Photon Scattering." Academic Press, New York (to be published).

Egelstaff, P. A. ed. (1965). "Thermal Neutron Scattering." Academic Press, New York.

Egelstaff, P. A. (1967). "An Introduction to the Liquid State." Academic Press, New York.

Special References

Ascarelli, P., Harrison, R. J., Paskin A. (1967). *Advan. Phys.* **17**, 717.

Ashcroft, N. W., and Lekner, J. (1966). *Phys. Rev.* **145**, 83.

Barker, J. A. (1963). "Lattice Theories of Liquids." Pergamon Press, Oxford.

Baym, G. (1964). *Phys. Rev.* **135**, A1691.

Brockhouse, B. N. (1958). *Nuovo Cimento* **9** (suppl.), 45.

Brockhouse, B. N. (1961). *In* "Inelastic Scattering of Neutrons in Solids and Liquids," p. 113. IAEA, Vienna.

Chung, C. H., and Yip, S. (1969). *Phys. Rev.* **182**, 323.

Dasannacharya, B. A., and Rao, K. R. (1965). *Phys. Rev.* **137**, A417.

Debye, P., and Menke, H. (1931). Ergebnisse der Techn. Rongtgenk. II.

De Gennes, P. G. (1959). *Physica* **25**, 825.

Egelstaff, P. A. (1961). *In* "Inelastic Scattering of Neutrons in Solids and Liquids," p. 25. IAEA, Vienna.

Enderby, J. E. (1968). *In* "Physics of Simple Liquids" (Temperley *et al.*, eds.). Wiley, New York.

Enderby, J. E., North, D. M., and Egelstaff, P. A. (1966). *Phil. Mag.* **14**, 131.

Enderby, J. E., North, D. M., and Egelstaff, P. A. (1967). *Advan. Phys.* **16**, 171.

Fabelinskii, I. L. (1968). "Molecular Scattering of Light." Plenum Press, New York.

Fisher, M. E. (1967). *Rep. Progr. Phys.* **39**, II, 615.

Furukawa, K. (1962). *Rep. Progr. Phys.* **25**, 395.

Giglio, M., and Benedek, G. (1969). *Phys. Rev. Lett.* **23**, 1145.

Gingrich, N. S., and Heaton, L. (1961). *J. Chem. Phys.* **34**, 873.

Gingrich, N. S., and Thompson, C. W. (1962). *J. Chem. Phys.* **36**, 2398.

Harrison, W. A. (1966). "Pseudo-Potentials in the Theory of Metals." Benjamin, New York.

Heller, P. (1967). *Rep. Progr. Phys.* **30**, II, 731.

Henshaw, D. G. (1957). *Phys. Rev.* **105**, 976.

Johnson, M. D., Hutchinson, P., March N. H. (1964). *Proc. Roy. Soc.* **A282**, 283.

Kruh, R. F. (1967). *Chem. Rev.* **62**, 319.

Landau, L. D. and Lifshitz, E. M. (1960). "Electrodynamics of Continuous Media." Pergamon Press, New York.

Larson, K. E., Dahlborg, V., Skold, K.,(1968). *In* "Simple Dense Fluids" (H. Frisch and Z. W. Salsburg, eds.). Academic Press, New York.

Levelt, J. M. H. (1958). *Physica* **24**, 769 (1958).

Mikolaj, P. G., and Pings, C. J. (1967), *J. Chem. Phys,* **46**, 1401.

Mikolaj, P. G., and Pings, C. J. (1967). *J. Chem. Phys.* **46**, 1412.

Mountain, R. D. (1966). *Rev. Mod. Phys.* **38**, 205.

Naghizadeh, J., and Rice, S. A. (1962). *J. Chem. Phys.* **36**, 2710.

Ornstein, L. S., and Zernike, F. (1914). *Proc. Akad. Sci.* (*Amsterdam*) **17**, 793.

Paskin, A. (1967). *Advan. Phys.* **16**, 223.

Paskin, A., and Rahman, A. (1966). *Phys. Rev. Lett.* **16**, 300.

Pings, C. J. (1968). *In* "Physics of Simple Liquids" (H. N. Temperley *et al.*, eds.). Wiley, New York.

Placzek, G. (1952). *Phys. Rev.* **86**, 377.

Rahman, A. (1964). *Phys Rev.* **136**, A405.

Rahman, A. (1965). *J. Chem. Phys.* **42**, 3540.

Rice, S. A., and Gray, P. (1965). "The Statistical Mechanics of Simple Liquids." Wiley (Interscience), New York.

Schiff, D. (1969). Doctor Thesis submitted to Faculté des Sciences d'Orsay, Université de Paris.

Schofield, P. (1960). *Phys. Rev. Lett.* **4**, 239.

Sjolander, A. (1965). *In* "Thermal Neutron Scattering" (P. A. Egelstaff, ed.). Academic Press, New York.

Thomas, J. E., and Schmidt, P. W. (1963). *J. Chem. Phys.* **39**, 2506.

Van Hove, L. (1954). *Phys. Rev.* **95**, 249.

Verlet, L. (1967). *Phys. Rev.* **159**, 98.

Verlet, L. (1968). *Phys. Rev.* **165**, 201.

Waller, I. (1966). *In* "Advanced Methods of Crystallography" (G. N. Ramanchandran, ed.), p. 157. Academic Press, New York.

Zernike, F. (1916). *Proc. Akad. Sci.* (*Amsterdam*) **18**, 1520.

Zernike, F., and Prins, J. A. (1927). *Z. Phys.* **41**, 184.

Zwanzig, R. (1964). *Phys. Rev.* **133**, A50.

Zwanzig, R., and Mountain, R. D. (1965). *J. Chem. Phys.* **43**, 4464.

Chapter 3

Computer Calculations for Model Systems[†]

FRANCIS H. REE

I. Introduction

Aside from a few nontrivial cases, most problems in classical equilibrium statistical mechanics have no analytic solution. Mathematical difficulty rather than lack of understanding of the physical phenomena has been the chief deterrent to progress in this field. In the past fifteen years or so, however, this difficulty has been gradually disappearing, and the chief reason has been the increasing availability and sophistication

[†] Work performed under the auspices of the U.S. Atomic Energy Commission.

of computers. From earlier days, when a computer was regarded as a simple labor-saving device, both its speed and capacity have grown tremendously, accordingly enabling us to handle much larger and more complex problems. The material presented in this chapter is intended to demonstrate this. For this purpose, the contents of this chapter are divided into two categories according to the nature of the problem. Included in the first category are problems dealing with "computer experiments." All other problems, requiring equally heavy use of computer time, are considered, for convenience, to be in the second category.

The "computer experiment" uses a computer to simulate a finite system whose particles interact under an *a priori* force law and which must satisfy certain constraints on particle number N, volume V, pressure P, temperature T, and boundary condition. The "experiment" can be performed either by following the time history of the particles in the system or by generating configurations with frequencies proportional to an appropriate weight prescribed by a statistical-mechanical equilibrium ensemble of interest. The first method is commonly known as the *molecular-dynamic method*, and the second method as the *Monte Carlo method* (a term adopted primarily for convenience, but, in a strict sense, a misnomer because of its wider connotation besides the one used in the present context). The chief advantage of a "computer experiment" lies in the fact that we can *control* the shape of the intermolecular potential energy to be used as well as, within the significant figures retained in the computer, independent external variables (N, V, T, etc.) as accurately as desired. This brings numerous benefits. Some of these are the ability (a) to check existing approximate theories in statistical mechanics and possibly to suggest new ones, (b) to extract knowledge on the importance of various microscopic variables (for example, shapes of intermolecular potential, three-body forces, etc.) affecting macroscopic quantities which are measured by performing a real and a corresponding computer experiment, (c) to gain physical information at extended range, otherwise unavailable by a real experiment, and (d) to check on the validity of the ergodic hypothesis. A main shortcoming in the "computer experiment" is that only a *finite* system can be dealt with and only for a *finite* computing time. While the number dependence of a finite system can be theoretically studied in simple situations (Oppenheim and Mazur, 1957; Lebowitz and Percus, 1961), similar studies in the "experimental" situations can only be done empirically by using a system with different sizes. So far, this sort of study has yielded the information that differences between the equation-of-state data of the finite system and the corresponding thermodynamically infinite system (N, $V \to \infty$ and the number density $\varrho = N/V =$ finite) are not larger than $1/N$, if the system under observation is in the one-phase region. Within the transition region, however, this is not the case and the equilibrium properties of a finite system depend strongly on its size. The largest three-dimensional system we can study on a computer is not yet large enough to properly sample the equilibrium configurations within the transition region, which contains locally large density fluctuations. It will be shown in Sections III,A and III,D that a "computer experiment" that eliminates this undesirable feature can still be set up. However, no currently available "computer experiment" can be used (due to their finite size) to investigate the long-range behavior of a correlation function near the critical point. Furthermore, the finite computation time available may be too short to study certain nonequilibrium phenomena which take place at relatively longer times. A typical machine experiment takes about 1 hr on a CDC 6600 computer, which, in the case of the Lennard-Jones system of 864 particles simulating argon molecules, amounts to a real-time progression of only 10^{-11} sec (Verlet, 1967).

In Section II, various techniques available for the "computer experiments" will be described. In Section III, we choose several examples of classical equilibrium statistical mechanics and apply the different techniques developed in Section II. We give further detailed discussions on the resulting solutions for each problem. These problems are chosen mainly from those which have not been reviewed elsewhere, because of their relatively recent development. In Sections III,E and III,G, we pick two examples (radial distribution functions and virial expansions) to show the numerical complexities associated with the second class of problem and to illustrate the techniques available to solve them. Since rigorous bounds on both the virial and fugacity expansions have been established recently, inclusion of these results into the section on expansion methods is felt to be timely.

There is a large amount of literature dealing with the computational aspects of problems in classical statistical mechanics. For further information, the reader should consult Neece and Widom's review article (1969), which lists the literature through 1968. For the Monte Carlo method, Wood's review article (1968a) is very useful. In Section III,H, we classify the literature according to subject. The list supplements earlier publications, including Neece and Widom's article, as well as the articles quoted in the earlier sections of the present chapter.

II. Computer Experiments in Classical Statistical Mechanics

A. Monte Carlo Methods

The term "Monte Carlo method" as used in equilibrium statistical mechanics refers to a probabilistic method of sampling configurations specified in various ensembles. Using configurations generated in this manner, estimates can be made for most equilibrium averages of interest in statistical mechanics and thermodynamics.

Ensembles that have been used in Monte Carlo calculations are the canonical (NVT) ensemble, the isothermal–isobaric (NPT) ensemble, and the grand canonical (zVT) ensemble. The configurational probability densities for finding N particles at positions at $\mathbf{r}^N = (\mathbf{r}_1, \mathbf{r}_2, \mathbf{r}_3, \ldots, \mathbf{r}_N)$ in these ensembles are as follows:

Canonical ensemble:

$$P_{NVT}(\mathbf{r}^N, V, T) = [\exp(-\beta\Phi_N)]/(N!\, Q_{NVT}). \tag{2.1}$$

Isobaric-isothermal ensemble:

$$P_{NPT}(\mathbf{r}^N, P, T) = [\exp(-\beta\Phi_N - \beta PV)]/(N!\, Q_{NPT}). \tag{2.2}$$

Grand canonical ensemble:

$$P_{zVT}(\mathbf{r}^N, z, T) = [\exp(-\beta\Phi_N)]z^N/(N!\, \Xi) \tag{2.3}$$

Here, z and β are, respectively, fugacity and $1/(kT)$, and Φ_N denotes the total potential energy of the N particles. The normalizing constants Q_{NVT}, Q_{NPT}, and Ξ are the (configurational NVT and NPT) partition functions of the respective ensembles and are related to the configurational Helmholtz (A) and Gibbs (G) free energies and PV by the following relationships:

$$\exp(-\beta A) = Q_{NVT} = (1/N!) \int_V d\mathbf{r}_N \exp(-\beta\Phi_N), \tag{2.4}$$

$$c \exp(-\beta G) = Q_{NPT} = (1/N!) \int_0^\infty dV \int_V d\mathbf{r}^N \exp(-\beta\Phi_N - \beta PV), \tag{2.5}$$

$$\exp(\beta PV) = \Xi = \sum_{N=0}^{\infty} (z^N/N!) \int_V d\mathbf{r}^N \exp(-\beta\Phi_N), \tag{2.6}$$

where the factor c in Eq. (2.5) has the dimension of volume and was taken to be either constant (Hill, 1956) or equal to $(\beta P)^{-1}$ (Sack, 1959).

Calculation of thermodynamic volume ($\equiv \partial G/\partial P$) using the expression (2.5) for G leads to a value of V which is smaller by $(\beta P)^{-1} \approx O(1/N)$ if the latter interpretation is used [see Eq. (2.26) and also Wood (1968a,b)]. Note that the right-hand sides of expressions (2.4)–(2.5) are the configurational part of the partition function expressed in an arbitrary unit of length. Choice of the unit of length is immaterial, since the entire partition function is a dimensionless quantity, a product of a configurational part and a kinetic-energy part with contribution λ^{-Nd}, where λ is the thermal de Broglie length and is equal to $[\beta h^2/(2\pi m)]^{1/2}$ (d is the number of dimensions, h is Planck's constant, and m is the particle mass).

Denoting by $\boldsymbol{\xi}$ a vector representing a configuration specified in each ensemble [i.e., $\boldsymbol{\xi} = (\mathbf{r}^N)$ for the NVT-ensemble, $\boldsymbol{\xi} = (\mathbf{r}^N, V)$ for the NPT-ensemble, and $\boldsymbol{\xi} = (\mathbf{r}^N, z^N)$ for the zVT-ensemble], an averaged value of the function $F(\boldsymbol{\xi})$ is

$$\langle F \rangle \equiv \int d\boldsymbol{\xi}[F(\boldsymbol{\xi})P(\boldsymbol{\xi})], \tag{2.7}$$

where the integral is understood (in the case of the zVT-ensemble) to be the summation over N, followed by integrations over $\mathbf{r}^N$, and $P(\boldsymbol{\xi})$ represents any one of the probability densities (2.4)–(2.6).

In general, the total potential energy Φ_N can be expressed as sums of external effects (gravitational, boundary, etc.) and contributions by interparticle forces (pairwise additive force, three- or higher-body forces). At present, except for very extreme situations in density or temperature, there is no convincing evidence suggesting appreciable contributions by higher than two-body forces to any equilibrium properties for inert gases. The experimental data have not been accurately determined over the wide range of temperatures and densities necessary to estimate quantities such as the third virial coefficients, which can in turn yield information on the three-body force. An additional difficulty is that the higher virial coefficients depend sensitively, even in the absence of the many-body force, on the shape of the *pair* potential, which has not yet been accurately determined either theoretically or experimentally. From now on, we will restrict discussions to a pairwise additive potential. The neglected higher-body force may influence the equilibrium properties of a real system at an extremely condensed state or affect solid-state transformations with small free-energy differences. These influences might even extend to the normal densities and temperatures.

For a finite system with periodic boundary, the associated Φ_N can be calculated using two essentially different methods. The first method is the more conventional one, in which the interparticle distance required for calculation of a pair potential energy is chosen to be the smaller of two possible distances between two particles in a periodic box. This is the so-called "minimum-image distance" convention. The second method is the so-called Ewald procedure. In this method, the net potential energy of a particle is calculated as the sum of contributions by the other particles in the box as well as contributions by their "images" (including its own) belonging to other identical cells which span to make an infinite system. For a short-range potential (for example, hard-sphere potential), the two conventions obviously lead to identical Φ_N, provided the dimensions of the system are larger than the range of the potential. For a long-range potential, such as a Coulomb potential (discussed in Section III,F), the use of the second convention leads to thermodynamic quantities that are appreciably less sensitive to the particle number.

The minimum-image distance convention, which has a fixed cutoff distance (i.e., half the linear dimension of the box), can be slightly generalized by introducing an arbitrary cutoff distance r_c (Wood and Parker, 1957). The r_c can be chosen to be the radius of a spherical region from a given particle so that particles outside this boundary contribute "residually" to equilibrium properties. In this convention as well as in the minimum-image distance convention, if a potential has a long-range tail ($>r_c$), residual long-range corrections due to particles lying outside of r_c must be made to the average potential energy and other thermodynamic quantities so that the corrected quantities are less sensitive to N and V. This can be done by using the "lattice correction" or the "fluid correction" (Wood and Parker, 1957) for the potential energy per particle. The former correction is simply the lattice energy sum for particles outside r_c, which are assumed to be fixed at lattice sites, while the "fluid" correction calculates potential-energy contributions for $r > r_c$ by assuming that particles are uniformly distributed. The use of one type of correction or the other is dictated, as their names indicate, by the kind of phase in which the system is being examined.

The most direct way of performing a Monte Carlo computer "experiment" is by a random sampling of configurations. In the case of a system of N hard spheres in a box V, for example, this is done by placing a particle successively into random positions in the box. The particle is left in the box if it does not overlap with any other particles already in the box. Otherwise, a new random position is selected and this position is checked for overlap. The probability P_{n+1} of a successful placement of a hard sphere at a random position in a box containing n hard spheres is related exactly to the configurational canonical partition function Q_N ($\equiv Q_{NVT}$) (Byckling, 1961) by

$$Q_N = (V^N/N!) \prod_{n=1}^{N} P_n. \tag{2.8}$$

In practice, P_n decreases exponentially as the pressure increases. Thus, only at low densities (up to 30% of the close-packed density ϱ_0) was this procedure proved to be of any value (Alder *et al.*, 1955; Byckling, 1961).

An analogous but improved "unweighted" procedure which is applicable for a more general class of potentials has been proposed recently (Singer, 1966; McDonald and Singer, 1967a). For this purpose, the configurational canonical partition function (2.4) is rewritten as

$$Q_N = (V^N/N!) \int_E dE[\gamma(E) \exp(-\beta E)], \tag{2.9}$$

where $\gamma(E)\, dE$ denotes the fraction of configurations ($\mathbf{r}^N$) with potential energy between E and $E + dE$. The object of McDonald and Singer is to tabulate a term proportional to $\gamma(E)$ by means of a Monte Carlo method. This is done by selecting a particle randomly and giving a random displacement. If the potential energy E is less than a preset upper bound, the new configuration replaces the old; otherwise, the old configuration remains as the next configuration. The weighting factor $\gamma(E)$ calculated in this manner rises very rapidly with increasing E [therefore, with decreasing $\exp(-\beta E)$]. As was the case in the random sampling procedure for hard spheres, this method is unsuitable for a system that is at the condensed phase or that contains a large number of particles.

The above methods break down at high density and large N, since the majority of configurations attempted in this manner correspond to states of low probabilities, whose expressions are given by Eqs. (2.1)–(2.3). This difficulty was overcome by Metropolis *et al.* (1953). They invented a method of generating a nonterminating Markov chain whose individual steps sample the configurations proportional to the probability $P(\boldsymbol{\xi})$ given by Eqs. (2.4)–(2.6). Since the configuration $\boldsymbol{\xi}(i)$ generated in the ith step in a realization of the Markov chain has the probability $P(\boldsymbol{\xi}(i))$, the average $\langle F \rangle$ can be replaced by the average of $F(\boldsymbol{\xi}(i))$ over the total number of steps taken in a realization of the Markov chain, i.e.,

$$\langle F \rangle \approx n^{-1} \sum_{i=1}^{n} F(\boldsymbol{\xi}(i)). \tag{2.10}$$

From the computational point of view, the above formula is an approximation. For a given amount of computing time, the Markov chain length n should be terminated at some finite n, and the space variables $\mathbf{r}^N$ and V in $\xi(i)$ are no longer continuous variables and can be specified up to an accuracy determined by the number of bits carried by a computer to represent them. Therefore, strictly speaking, the present Markov chain forms a discrete set composed of a finite number of states ξ_k ($k = 1, \ldots, \Gamma$), with a stationary probability $P_k \equiv P(\xi_k)$. The above two points are directly related to the speed and size of the computer. In view of the high-speed computation ability of present-generation computers, the time limitation is not a formidable problem, unless for some reason we need very accurate data ($<0.1\%$ for the pressure) on equilibrium quantities in the one-phase region. In the two-phase region, where a large density fluctuation occurs, longer calculation time may help, but may not be justified by the result. Indirect procedures described in Sections III A and D are better suited in this case. The second limitation (of discretizing the originally continuous variables) is not likely to cause any noticeable error either. This observation comes from quantitative agreement (ignoring small discrepancies from the other sources of error) of an equilibrium quantity obtained by using computers with different numbers of bits to represent the variables. One must note, however, that a continuous system is treated by a computer as an equivalent lattice-gas system in which particles interact across many lattice sites. Therefore, an extremely coarse-grained lattice distorts the available phase space so drastically that the melting density as well as the order of transition are seriously altered (Hoover *et al.*, 1964; Runnels, 1965; Ree and Chesnut, 1966, 1967; Bellemans and Nigam, 1966).

Besides these computational limitations, there are theoretical conditions under which the procedure of Metropolis *et al.* is strictly equivalent to the corresponding ensemble average. The conditions for the equivalence (2.10) (Feller, 1950; Wood, 1968a) are:

(a) The states ξ_i and ξ_j ($i, j = 1, \ldots, \Gamma$) must be mutually accessible. That is, if the stochastic matrix element describing the transition probability between two successive states ($\xi_i \to \xi_l$) of the Markov chain is represented as p_{il}, then, after a finite number of transitions ($p_{il} \neq 0$), the state ξ_j must always be reached from the state ξ_i.

(b) The transition probability p_{ij} should satisfy the relation:

$$\sum_{i=1}^{\Gamma} p_{ij} P_i = P_j, \quad j = 1, \ldots, \Gamma; \qquad \sum_{i=1}^{\Gamma} p_{ij} = 1. \tag{2.11}$$

Condition (2.11) can be satisfied, for example, for the set $\{p_{ij}\}$ whose elements obey the following reciprocity condition:

$$p_{ij}P_i = p_{ji}P_j. \tag{2.12}$$

Condition (2.12) has been used for the presently available Monte Carlo calculations. This is obviously not a necessary condition for Eq. (2.11). In this respect, it will be worthwhile to devise and examine a new set $\{p_{ij}\}$ that violates Eq. (2.12) but not Eq. (2.11). Elements p_{ij} that satisfy the condition (a) are referred to as belonging to the same *ergodic class* (or simply, same *class*). For a finite system of hard-core particles, if the density ($\varrho = N/V$) is sufficiently close to the close-packed density ϱ_0, the configurational phase space is broken up into $(N-1)!$ pieces of allowed configurations. The transition probability p_{ij} that has been used so far cannot connect between any two of these pieces. However, this difficulty does not affect calculating the average $\langle F\rangle$ by Eq. (2.10) for a symmetric function $F(\xi)$, since the average of $F(\xi)$ over configurations belonging to one of these pieces is identical to similar averages over configurations belonging to a different piece. For further discussion on whether or not a system satisfies a "quasiergodic" condition within a computationally feasible time, refer to Section 5 in Wood's review article (1968a). Assuming that the above problem has been resolved, there still needs to be a way of choosing the transition probability p_{ij}.

The following transition probability p_{ij} has been most frequently used in Monte Carlo calculations (Wood and Parker, 1957; Wood, 1968a). Let $\eta(i)$ denote a specified set containing $\Gamma^{\neq}$ ($\in \Gamma$) neighboring states of the state i (itself counted in $\Gamma^{\neq}$). Furthermore, $\Gamma^{\neq}$ is chosen to be independent of the state i, so that $i \in \eta(j)$ if and only if $j \in \eta(i)$. Then, p_{ij} is chosen as

$$p_{ij} = 0 \qquad \text{if} \quad j \notin \eta(i), \tag{2.13a}$$

$$= 1/\Gamma^{\neq} \qquad \text{if} \quad j \in \eta(i), \quad j \neq i, \quad \text{and} \quad P_j \geq P_i, \tag{2.13b}$$

$$= P_j/(P_i\Gamma^{\neq}) \qquad \text{if} \quad j \in \eta(i) \quad \text{and} \quad P_j < P_i, \tag{2.13c}$$

$$p_{ii} = 1 - \sum_{j\neq i} p_{ij} = (1/\Gamma^{\neq}) + (1/\Gamma^{\neq}) \sum_{j,P_j<P_i} [1 - (P_j/P_i)]. \tag{2.13d}$$

Note that Eq. (2.13) satisfies Eq. (2.12) and, consequently, Eq. (2.11). Note also that only the ratio P_j/P_i (therefore, no normalizing constant) enters in Eq. (2.12). The quantity p_{ii} represents the probability of the next configuration in the Markov chain being an identical configuration i of its predecessor. As can be seen from the second equality in Eq. (2.13d), this is composed of a probability $(1/\Gamma^{\neq})$ that a random selection out of the $\Gamma^{\neq}$ configurations happens to be an identical configuration (that is, the configuration i) plus all other possibilities for which a selected configuration fails the test prescribed by Eq. (2.13c).

The initial configuration usually chosen for the Monte Carlo calculations or the molecular-dynamic calculations corresponds in a three-dimensional system to the face-centered-cubic lattice with particles at its lattice points. A cubic periodic box with equal side lengths is used to enclose the particles. In this way, the number of particles used can vary as $4n^3 = 4$, 32, 108, 256, and 500.

In the following, more specific descriptions will be given of the Monte Carlo procedure of generating a Markov chain with the transition probability of Eq. (2.13) for the NVT- and the NPT-ensembles for a continuum system. The Monte Carlo procedure for the zVT-ensemble applicable to a lattice-gas system will also be described.

1. *Markov Chain for the Canonical Ensemble*

For the NVT-ensemble, if $\xi_k(i) = (\mathbf{r}_1, \mathbf{r}_2, \ldots, \mathbf{r}_N)$ is a configuration at the ith step in a Markov chain at the state k, select randomly a particle (say, particle j at $\mathbf{r}_j$) and give a random displacement $\delta\mathbf{r}_j$ from $\mathbf{r}_j$, provided that $\delta\mathbf{r}_j$ is confined within a box (with its center at $\mathbf{r}_j$) having equal side lengths 2δ. Let the new configuration generated in this manner be $\xi_{k'}$, which is identical to $\xi_k(i)$ except for translation $(\mathbf{r}_j \rightarrow \mathbf{r}_j + \delta\mathbf{r}_j)$ of the position of particle j. Let the stationary probability corresponding to the configuration $\xi_{k'}$ be $P_{i'}$. Then, if $P_{i'}/P_i \geq 1$, we take $\xi(i+1) = \xi_{k'}$. Otherwise, a uniform random number R $(0 < R < 1)$ is chosen, and a test is carried out to check if $P_{i'}/P_i \geq R$ or $P_{i'}/P_i < R$. If the former is the case, $\xi(i+1) = \xi_{k'}$; otherwise, $\xi(i+1) = \xi_k(i)$. In this procedure, the total number of neighboring states $\Gamma^{\neq}$ is equal to $N(2\delta)^d$, i.e., the N possible ways of picking a random particle times the available volume $[(2\delta)^d$, with d the number of dimensions$]$ in which it can be moved. In this scheme, the magnitude of the "jump" parameter δ is left unspecified. The maximum jump length, however, governs the efficiency with which the Markov chain can sample the phase space. In the past, δ was empirically determined so that about half of the total number of the Monte Carlo moves generate successful new configurations. The configurational internal energy U and the pressure P can be obtained from Eq. (2.4) as follows:

$$U = \beta(\partial A/\partial \beta)_{N,V} = \langle \Phi_N \rangle, \tag{2.14a}$$

$$\beta P/\varrho = 1 - (\beta/dN)\langle \Psi_N \rangle, \tag{2.14b}$$

where the average $\langle \ldots \rangle$ over the NVT-probability density (2.1) is replaced by the average over the total number of Monte Carlo steps taken in a realization of the NVT Markov process. The quantity is the average virial, defined as

$$\langle \Psi_N \rangle = \langle \sum_i [\mathbf{r}_i \cdot \partial \Phi_N/\partial \mathbf{r}_i] \rangle. \tag{2.15}$$

The above expression for P is not directly applicable for hard-core particles, because of their shapes. For a pairwise additive potential $\phi(r)$ such as the hard-core potential, alternate expressions for U and P in terms of the *radial distribution function* $g(r)$ can be sometimes more useful:

$$U/N = (\varrho/2) \int d\mathbf{r}[\phi(r)g(r)], \tag{2.16}$$

$$\beta P/\varrho = 1 - (\beta\varrho/2d) \int d\mathbf{r}\,\{r[\partial\phi(r)/\partial r]g(r)\}. \tag{2.17}$$

In Section III,E, we describe the Monte Carlo determination of $g(r)$ using the NVT-ensemble Markov chain.

The data available from the NVT-ensemble Markov chain at a single point of ϱ, T can be shown to give the thermodynamic data at neighboring points ϱ', T', as well as at different sets of potential parameters ε and σ such as those occurring in the Lennard-Jones potential [Eq. (3.33)]. This was done by McDonald and Singer (1967a; 1969) by scaling the potential parameters or temperatures. The class of applicable $\phi(r)$ does not include hard-core potentials. This is achieved by accumulating histograms of $\tilde{\Phi}(n)$ and $\tilde{\Psi}(n)$ at each step of the Markov chain. These histograms show, respectively, the number of configurations in the Markov chain lying between $\Phi(n) - (\delta\Phi/2)$ and $\Phi(n) + (\delta\Phi/2)$ ($\delta\Phi = \text{const}$), and the virial Ψ_N averaged over these configurations belonging to this energy interval. As long as $\delta\Phi$ is chosen small, the average $\langle \Psi_N \rangle$ is then equal to

$$\langle \Psi_N \rangle = \Big[\sum_n \tilde{\Phi}(n)\tilde{\Psi}(n)\Big] \Big/ \sum_n \tilde{\Phi}(n). \tag{2.18}$$

The corresponding average $\langle \Psi_N \rangle'$ at a neighboring temperature T' is evaluated from Eq. (2.18) by substituting the following expression for $\tilde{\Phi}'(n)$:

$$\tilde{\Phi}'(n) = \tilde{\Phi}(n) \exp[-(\beta' - \beta)\Phi(n)]. \tag{2.19}$$

If a pair potential is given by a Lennard-Jones-like potential with potential parameters (ε, σ), both Φ_N and Ψ_N are expressed as linear combinations of r^{-12} and r^{-6} (or r^{-m} and r^{-n}, respectively, for a slightly more general class of potentials). Similarly, Φ_N' and Ψ_N' for the same type of potential energy with different parameters (ε', σ') can be expressed as linear combinations of r^{-12} and r^{-6}. This implies that both Φ_N' and Ψ_N' are linear combinations of Φ_N and Ψ_N. The averages $\langle \Phi_N' \rangle$ and $\langle \Psi_N' \rangle$ calculated from the Markov chain corresponding to the parameters (ε', σ') can be shown to be obtained from the histograms of $\tilde{\Phi}(n)$ and $\tilde{\Psi}(n)$ for the parameters (ε, σ) (McDonald and Singer, 1967a; 1969). Since the parameters ε and σ scale T and distance (thus, $V^{1/3}$), respectively, changing ε to ε' and σ to σ' is equivalent to changing T to $T' = (\varepsilon'/\varepsilon)T$ and V to $V' = (\sigma'/\sigma)^3 V$. Therefore, P and U at T' and V' can be evaluated from the data on $\tilde{\Phi}(n)$ and $\tilde{\Psi}(n)$ at T and V. This type of reweighting procedure has a practical limitation on the applicable range of ε', σ', V' and T'. For example, the accuracy of the "reweighting" function $\tilde{\Phi}'(n)$ [Eq. (2.19)] is determined by the accuracy of $\tilde{\Phi}(n)$. Therefore, β' should not be much different from β so that $\tilde{\Phi}(n)$ and $\tilde{\Phi}'(n)$ can overlap sufficiently over the statistically significant range of n. The applicable ranges of the potential parameters ε' and σ' are estimated by McDonald and Singer to lie approximately within $\pm 15\%$ from ε and less than $\pm 1\%$ from σ.

2. *Markov Chain for the Isothermal–Isobaric Ensemble*

The Markov chain for the NPT-ensemble can be generated in a computationally more convenient way (Wood, 1968a,b, 1970) if the position vector $\mathbf{r}^N$ of the N particles in a box $(V \equiv L^d)$ is reduced by L and expressed by a dimensionless displacement vector $\boldsymbol{\eta}^N$ $(\equiv \boldsymbol{\eta}_1, \boldsymbol{\eta}_2, \ldots, \boldsymbol{\eta}_N)$, with $\boldsymbol{\eta}_j = (x_j/L, y_j/L, z_j/L)$. With these variables, the stationary probability (2.2) of the NPT-ensemble being at the ith configuration $\boldsymbol{\xi}_i = (L, \boldsymbol{\eta}^N)$ is

$$P(\boldsymbol{\xi}) = C_{NPT} \exp\{-\beta[\Phi(L\boldsymbol{\eta}^N) + PL^d - dN \ln L]\}, \tag{2.20}$$

where the normalizing constant C_{NPT} is $(N!\,Q_{NPT})^{-1}$. The NPT-

ensemble average of a function $F(L\boldsymbol{\eta}^N)$ is then

$$\langle F\rangle = \int_0^\infty dL \int d\boldsymbol{\eta}^N [F(L\boldsymbol{\eta}^N)P(\boldsymbol{\xi})], \tag{2.21}$$

where integrations over $\boldsymbol{\eta}^N$ are carried over a Nd-dimensional unit cube.

A Markov chain with a stationary probability (2.20) can be generated as follows: let the ith step in a realization of the Markov chain happen to be in the kth configuration, i.e., $\boldsymbol{\xi}_k(i) = (L, \boldsymbol{\eta}_1, \boldsymbol{\eta}_2, \ldots, \boldsymbol{\eta}_N)$. In order to generate a configuration $\boldsymbol{\xi}(i+1)$ for the $(i+1)$th step, we pick a random number δL lying between $-\delta_L$ and δ_L, where δ_L is a preset constant. Next, as was done for a realization of the NVT-Markov chain, we select a random particle (say j) and give a random displacement $\delta\boldsymbol{\eta}_j$ within a cube $(2\delta)^d$, where δ is another preset constant. Let the configuration generated in this manner be denoted as $\boldsymbol{\xi}_{k'}$ $\{\equiv(L+\delta L, \boldsymbol{\eta}_1, \ldots, \boldsymbol{\eta}_j + \delta\boldsymbol{\eta}_j, \ldots, \boldsymbol{\eta}_N)\}$. If the corresponding $P(\boldsymbol{\xi}_{k'})$ obtained by using Eq. (2.20) is greater than or equal to P_i $\{\equiv P(\boldsymbol{\xi}_k(i))\}$, we set $\boldsymbol{\xi}(i+1) = \boldsymbol{\xi}_{k'}$. If $P(\boldsymbol{\xi}_{k'}) < P_i$, a random number R $(0 < R < 1)$ is chosen. If $P(\boldsymbol{\xi}_{k'})/P_i > R$, then $\boldsymbol{\xi}_k(i+1) = \boldsymbol{\xi}_{k'}$; otherwise, $\boldsymbol{\xi}(i+1) = \boldsymbol{\xi}_k(i)$. The total number of states $\Gamma^{\neq}$ in Eq. (2.12) is $L_{\max}N(2\delta)^d$, where the maximum linear dimension $L_{\max}$ of the box can be truncated at a reasonably large value whose probability of occurrence can be practically regarded as nil. Again, suitable values of the parameters δ_L and δ must be determined prior to the Monte Carlo calculations. The above method has been used by McDonald (1969) to calculate the excess enthalpy and volume of a mixture of Lennard-Jones particles having two different sets of potential parameters to simulate argon and krypton atoms.

An alternative method of carrying out the average $\langle F\rangle$ [Eq. (2.21)] applicable for any hard-core particle was proposed by Wood (1968a,b). If the order of integration in Eq. (2.21) for L and $\boldsymbol{\eta}^N$ is reversed and the step-function property of the Boltzmann factor for a hard-core potential is used, the average $\langle F\rangle$ can be expressed as an average of a related function $F(\boldsymbol{\eta}^N)$ over a new probability density $P(\boldsymbol{\eta}^N)$:

$$\langle F\rangle = \int d\boldsymbol{\eta}^N [\tilde{F}(\boldsymbol{\eta}^N)P(\boldsymbol{\eta}^N)], \tag{2.22}$$

$$\tilde{F}(\boldsymbol{\eta}^N) \equiv [N^{N+1}\phi^{N+1}/\Gamma(N+1, N\phi\tau_{\mathrm{m}})] \int_{\tau_{\mathrm{m}}}^\infty d\tau \{[\exp(-N\phi\tau)]\tau^N F(L\boldsymbol{\eta}^N)\}, \tag{2.23}$$

$$P(\boldsymbol{\eta}^N) = \Gamma(N+1, N\phi\tau_{\mathrm{m}}) \Big/ \int d\boldsymbol{\eta}^N \Gamma(N+1, N\phi\tau_{\mathrm{m}}), \tag{2.24}$$

where τ is the reduced volume V/V_0 (the close-packed volume, $\phi = PV_0/NkT$, and τ_m is the smallest allowable reduced volume of the box with particles at $\boldsymbol{\eta}^N$ and $F(L\boldsymbol{\eta}^N) \neq 0$. The quantity $\Gamma(N+1, N\phi\tau_m)$ is defined as follows:

$$\Gamma(N+1, N\phi\tau_m) = (N\phi)^{N+1} \int_{\tau_m}^{\infty} d\tau [\tau^N \exp(-N\phi\tau)]. \tag{2.25}$$

Calculations for $\langle V \rangle$ and $g(r)$ using the Markov chains with the probability given by Eq. (2.25) for hard disks at different values of ϕ have been reported by Wood (1968b, 1970).

There are several distinguishing features present in the *NPT* calculations of the equilibrium properties. First of all, in the *NPT* calculations, P is given, and the thermodynamic volume V is related to the average volume through G [Eq. (2.5)],

$$V = \partial G/\partial P = \langle V \rangle - (1/\beta P), \tag{2.26}$$

in which the factor c in Eq. (2.5) is taken as $(\beta P)^{-1}$. In the case of hard-core potentials, the *NPT*-ensemble equation-of-state data can be generated more conveniently than in the corresponding *NVT*-ensemble procedure, where P is related to ϱ through the virial theorem (2.17), in which the value of $g(r)$ at the contact distance of two particles is required. In practice, this has to be numerically extrapolated from available data at slightly larger distances. Furthermore, in contrast to the *NPT*-ensemble, the *NVT*-ensemble $g(r)$ is obtained by numerical differentiation of the Monte Carlo accumulated averages of pair distances with respect to r.

The second point of interest in the *NPT*-ensemble can be best described by using the following relationship (Wood, 1968a) between the *NVT*-ensemble pressure and the *NPT*-ensemble pressure:

$$[PV_0/(NkT)]_{NVT} = (PV_0/NkT) + N^{-1}\, \partial[\ln P_{NPT}(V)]/\partial V, \tag{2.27}$$

where $P_{NPT}(V)\, dV$ is the *NPT*-ensemble probability of finding volume between V and $V + dV$. Using Eq. (2.2), this is defined as

$$P_{NPT}(V) = [Q_{NVT}/Q_{NPT}] \exp(-\beta PV). \tag{2.28}$$

Equation (2.27) results from differentiating $\ln P_{NPT}(V)$, given by the above expression, with respect to V. Since the *NVT*-ensemble P is likely to yield a van der Waals loop within the phase-transition region [see, for example, Alder and Wainwright, 1962; Mayer and Wood, 1965) and since the NPT-ensemble P is a monotonic decreasing function in V (Wood, 1968a), Eq. (2.27) implies that $P_{NPT}(V)$ becomes bimodal within the transition region. Since $P_{NPT}(V)$ can be "experimentally" tabulated, the NVT-ensemble P over all volumes can, in principle, be evaluated from data on $P_{NPT}(V)$ at a single value of P together with Eq. (2.27). This reweighting procedure has been carried out for hard disks at several PV_0/NkT by Wood (1970). If the difference between V and the NPT-ensemble $\langle V \rangle$ is larger than $O(1/N)$, the corresponding $P_{NPT}(V)$ has a large statistical uncertainty. Thus, a good pressure estimate at some other volume using this procedure is limited to the close proximity of P itself.

It is of practical interest to compare the relative efficiencies of the NVT- and NPT-Monte Carlo procedures. This has been carried out for hard disks by Wood (1970), who compared the NVT-ensemble $g(r)$ at $\varrho = 0.6\varrho_0$ [$PV_0/NkT = 2.975$] obtained by Chae *et al.* (1969) with his NPT-ensemble $g(r)$ at $PV_0/NkT = 3$. Wood notes that, for a given computation time, the NVT-procedure gives smaller standard deviations in $g(r)$ and, therefore, is probably slightly more efficient than the NPT-procedure. Further theoretical and computational studies to account for this will be worthwhile.

3. *Markov Chain for the Grand Canonical Ensemble*

No Monte Carlo calculation for equilibrium properties has been reported for a continuum (hard-sphere, Lennard-Jones, etc.) system using the grand canonical zVT-ensemble at densities of practical interest. When using the Metropolis *et al.* (1953) procedure, difficulties are encountered in sampling configurations with the weight of the zVT-ensemble probability density $P_{zVT}(\mathbf{r}^N, z, T)$ [Eq. (2.3)]. According to this scheme, one would have to either delete or add a particle in the box. As was described earlier, this is not easy, especially when a box already contains a relatively large number of particles. The probability of a successful addition of a particle diminishes exponentially with P and N. Due to a smaller number of available states, this difficulty, however, is less of a concern for a small lattice-gas system with a point hard-core and an attraction covering several neighboring sites (Salsburg *et al.*, 1959). Chesnut and Salsburg (1963) and Chesnut (1963) carried out the zVT-ensemble Monte Carlo calculations of statistical-mechanical quantities for a two-dimensional triangular lattice gas. Since a lattice gas has no dynamic analog, pressure cannot be calculated from the usual virial theorem [Eq. (2.17)]. Instead, the pressure calculation makes use of the relationship $\beta PV = \ln \Xi$. Therefore, it is equivalent to the troublesome evaluation of the grand partition function itself. Nevertheless, in case of a lattice gas with a point core and nearest-neighbor interaction, this procedure is feasible, as was demonstrated by Chesnut and Salsburg, who evaluated the zVT-ensemble average $\langle[\exp(\beta\Phi_N)]z^{-N}N!\rangle$. Using the zVT-ensemble probability density (2.3), and $\mathbf{k}$ to represent one of 2^V possible discrete configurations for V lattice sites, a relation linking this average with Ξ can be established:

$$\begin{aligned}\langle[\exp(\beta\Phi_N)]z^{-N}N!\rangle &= \sum_{\mathbf{k}} \left(P_{zVT}(\mathbf{k})\{\exp[\beta\Phi_N(\mathbf{k})]\}z^{-N(\mathbf{k})}N!\right) \\ &= \sum_{\mathbf{k}} \left(\Xi^{-1}\{\exp[-\beta\Phi_N(\mathbf{k})]\}z^{N(\mathbf{k})} \right. \\ &\qquad \left. \times \{\exp[\beta\Phi_N(\mathbf{k})]\}z^{-N(\mathbf{k})}\right) \\ &= \Xi^{-1}\sum_{\mathbf{k}} 1 = \Xi^{-1}2^V, \end{aligned} \tag{2.29}$$

where $\Phi_N(\mathbf{k})$ and $N(\mathbf{k})$ represent total potential energy and number of particles for the configuration $\mathbf{k}$. Note that, at high z (and accordingly at high density), the average $\langle[\exp(\beta\Phi_N)]z^{-N}N!\rangle$ cannot be efficiently evaluated by this procedure, for precisely the same reason as encountered in the case of the earlier random-sampling methods. Note also that at present there is no comparable formula to Eq. (2.29) that is applicable to a system with an extended hard-core covering neighboring sites. This is so because, in contrast to the known number of states (2^V) for the Ising-like lattice gas, a similar number for the extended lattice gas requires a solution of a notoriously difficult combinatorial problem of placing particles with a finite size in the lattices. This is still an unsolved problem. A realistic way to evaluate P for such systems can still be done by way of numerical integration of $\varrho(z)$ [$= \langle N\rangle/V$] evaluated at several different values of z; i.e.,

$$\begin{aligned}\beta P &= \beta P_0 + \int_{z_0}^{z} dz'[\varrho(z')/z'], \\ &= \beta P_0 + V^{-1}\int_{z_0}^{z} dz'[\langle N(z')\rangle/z'], \end{aligned} \tag{2.30}$$

where z_0 can be taken sufficiently small so that P_0 evaluated at z_0 can be successfully approximated by its truncated fugacity series. This procedure, when used in conjunction with the Chesnut and Salsburg sampling method, should be able to give a satisfactory P for values of z that are not too large.

In this respect, we mention the recent work and the earlier work quoted therein of Ree and Chesnut (1967), Runnels (1965, 1967), Bellemans (1966, 1967), and Orban (1968, 1969), who evaluated Ξ for a semiinfinite strip by the Kramers and Wannier (1941) matrix method. Provided that the interaction potential energy of a lattice gas does not extend too far, the matrix method of evaluating statistical-mechanical quantities is superior to the Monte Carlo calculations. Incidentally, the matrix method presents another representative example belonging to the second class of problems discussed in Section I which require the heavy use of a computer.

B. Molecular-Dynamic Methods

An alternate method of obtaining data on the equations of state and statistical-mechanical quantities can be achieved by numerically solving a set of $N \times d$ Newton equations of motion for a system of N individual particles with mass m, fixed total energy, and momentum:

$$m\, d^2\mathbf{r}_i/dt^2 = \mathbf{f}_i(\mathbf{r}_1, \mathbf{r}_2, \ldots, \mathbf{r}_N), \qquad i = 1, \ldots, N, \tag{2.31}$$

where $\mathbf{f}_i(\mathbf{r}_1, \mathbf{r}_2, \ldots, \mathbf{r}_N)$ represents the force acting on particle i. This method, commonly known as the molecular-dynamic method, provides information on the time evolution of the position $\mathbf{r}_i$ and velocity $\mathbf{v}_i$ of the N particles. Hence, using it rather than the Monte Carlo method enables one to gain information on nonequilibrium properties as well.

In setting up the molecular-dynamic calculations, the initial configuration of particles and the long-range corrections to P (as discussed in the previous section on the Monte Carlo method) can be applied equally well. In the following discussion, we give two separate descriptions of the molecular-dynamic method, each suitable for an interparticle potential energy belonging to either one of the two classes (a) a hard-core or square-well potential, or (b) a soft-core potential such as the Lennard-Jones potential.

1. *Hard-Core Potentials*

Only the molecular-dynamic calculations for a system of N identical hard-core molecules with mass m will be described. Generalization to the square-well potential follows in a straightforward manner, although in this case the corresponding computer program becomes a little more complicated because of additional logic steps required to consider separately the states of particles (bound or unbound) as well as the nature of collisions (core or attractive collisions). For details, refer to the original paper by Alder and Wainwright (1959).

Except at the time of collision, the hard-core or square-well particles do not experience any force and hence they move in a straight-line path with a constant speed. Upon making an elastic collision, transfer of energy and momentum takes place between the two colliding particles so that net energy and momentum are unchanged. No change in energy or momentum occurs for the other particles except this pair. Therefore, the time evolution of a system of N hard spheres of diameter σ can be treated as a sequence of successive binary collisions which take place at uneven time intervals. At any instant, if the relative position and velocity of two particles i and j are represented as $\mathbf{r}_{ij}$ $(\equiv \mathbf{r}_i - \mathbf{r}_j)$ and $\mathbf{v}_{ij}$ $(\equiv \mathbf{v}_i - \mathbf{v}_j)$, respectively, and if these two are to collide (i.e., $r_{ij} = \sigma$) after elapsed time t_{ij}, the following identity must be satisfied:

$$|\, \mathbf{r}_{ij} + t_{ij}\mathbf{v}_{ij} \,| = \sigma. \tag{2.32}$$

this is a quadratic equation in t_{ij} and can be immediately solved for t_{ij}. Discarding one of the two possible roots for the failure to satisfy a condition that $t_{ij} = 0$ if $r_{ij} = \sigma$, the remaining root is

$$t_{ij} = v_{ij}^{-2}\{-b_{ij} + [b_{ij}^2 - v_{ij}^2(r_{ij}^2 - \sigma^2)]^{1/2}\}, \tag{2.33}$$

where

$$b_{ij} = \mathbf{r}_{ij} \cdot \mathbf{v}_{ij}. \tag{2.34}$$

If $b_{ij} < 0$, then the centers of the two particles are receding and consequently no collision occurs. However, if $b_{ij} > 0$, the two centers are approaching; thus, depending on the quantity $b_{ij}^2 - v_{ij}^2(r_{ij}^2 - \sigma^2)$ being either greater or less than zero, the two particles do or do not collide. The first part of the molecular-dynamic code evaluates t_{ij} according to Eq. (2.33) and stores it in a computer memory allocated for the pair i and j for all $N(N - 1)/2$ pairs of particles. Interparticle distances required in Eq. (2.33) are calculated using the minimum-image distance convention discussed previously. Those pairs that have no solution (hence, no collision) for t_{ij} are also recorded accordingly. If it happens that no t_{ij} is less than a preset constant t_{m}, the program advances particles from the current positions to the positions corresponding to a future time t_{m}, and then repeats the test for a possible collision. This procedure is repeated, if necessary, until at least one t_{ij} eventually becomes less than

t_m. The maximum time t_m is chosen short enough so as to prevent any pair from separating as far away as $(L/2) - \sigma$ in a single time step. The pair of particles (say, k and l) with the smallest t_{ij} calculated in this manner are participants in the immediately following collision process. Changes in directions of motion and speeds after the collision for particles k and l can be calculated from the momentum and energy conservation before and after (denoted by a prime) the collision for these particles,

$$\mathbf{v}_k' - \mathbf{v}_k = -(\mathbf{v}_l' - \mathbf{v}_l) = -(b_{kl}/\sigma^2)\mathbf{r}_{kl}. \qquad (2.35)$$

In the second part of the program, particles not involved in the collision are moved with constant velocities for a time t_{kl} to new positions and the collision time for each pair of these particles is updated. This is done simply by subtracting t_{kl} from the old value t_{ij} to obtain the new collision time for the pair (i, j). The remaining $2N - 3$ pairs include at least either particle k or l of the collided pair. Hence, the collision times involving these pairs have to be recalculated from Eq. (2.33) with new positions and velocities after the collision. If at least one pair in the table of $N(N-1)/2$ values of t_{ij} is found to be less than t_m, the program selects the pair with the smallest t_{ij} as the next colliding pair and repeats the second part. If no t_{ij} is found, the program returns to the first step and repeats the above procedure. Since the second part of the program involves calculation of $2N - 3$ collision times, in contrast to the first part with $N(N-1)/2$ values of t_{ij}, use of the two-stage scheme results in a considerable reduction in the computing time. This is especially true for the high-density calculations, where most of the computing time is spent in the second part. For example, even in a not-so-dilute fluid region, Alder and Wainwright (1959) report that the ratio of the computing times required in the first and second parts is approximately 1 : 36, the net computing time going up proportional to N.

The pressure for the hard-core particles can be calculated from the virial theorem (2.14) in which β is replaced by $d/m\langle v^2\rangle$ when the average (2.15) of the virial is replaced by its time-average over a sufficiently long time. Because of an impulsive force felt only at time of each collision, the integration over time can be replaced by a summation over all collision events; this then yields the following result (Alder and Wainwright, 1960):

$$(\beta P/\varrho) - 1 = (N\langle v^2\rangle t)^{-1} \sum_{\text{collision}} b_{ij}, \qquad (2.36)$$

where b_{ij} is given by Eq. (2.34). For hard particles, an alternative formula for calculating P can be obtained by a relationship connecting the right-hand side of Eq. (2.36) with the collision rate (Alder and Wainwright, 1960; Hoover and Alder, 1967):

$$[(\beta P/\varrho) - 1]/B_2\varrho = \Gamma/\Gamma_0, \tag{2.37}$$

where Γ_0 is the low-density limiting form of the collision rate and B_2 represents the second virial coefficient.

Still another expression for P [see Eq. (3.7) in the next section] can be obtained for hard-core particles; this requires evaluation of $g(r)$ at $r = \sigma$. Note that the quantities $\langle v^2 \rangle$, b_{ij}, and Γ used in Eqs. (2.36) and (2.37) are directly measurable quantities during the course of the "experiment." Thus, the pressure calculation from Eq. (2.36) or (2.37) is advantageous compared to the third method [Eq. (3.7)], in which $g(\sigma)$ has to be extrapolated to $r = \sigma$ from data on the adjacent $g(r)$. The data on $g(r)$ in turn must be obtained by numerical differentiations of the measured cumulative pairs at different values of r. Because of its rapid convergence rate to the true P, the pressure calculations by the first method [(2.36)] are found to be superior to calculations done by the other two (Hoover and Alder, 1967).

2. *Soft-Core Potentials*

A potential $\phi(r)$ belonging to this class is bounded (except at $r = 0$) and has a smoothly varying functional form. The potential is usually cut off at $r = r_c$. Contributions to P by $\phi(r)$ for $r > r_c$ can be included by the methods described in Section II,A. For this type of potential, no practical way of reducing Newton's equations (2.31) to a set of simple algebraic equations analogous to Eqs. (2.33) and (2.34) has yet been devised. For the soft-core potential, however, it is easier to directly solve the corresponding Newton equations by using a suitable difference scheme, in which the time increment t is a parameter for the difference equations.

This has been done by Gibson *et al.* (1960) by using the following two-step difference formula for the original Newton equation (2.31):†

$$\mathbf{v}_i(t + \Delta t) = \mathbf{v}_i(t) + \Delta t m^{-1}\mathbf{f}_i[\mathbf{r}_1(t), \mathbf{r}_2(t), \ldots, \mathbf{r}_N(t)], \tag{2.38a}$$

$$\mathbf{r}_i(t + \Delta t) = \mathbf{r}_i(t) + \Delta t \mathbf{v}_i(t + \Delta t). \tag{2.38b}$$

† Instead of $\mathbf{v}_i(t)$ and $\mathbf{v}_i(t + \Delta t)$ in these equations, Gibson *et al.* use $\mathbf{v}_i(t - (\Delta t/2))$ and $\mathbf{v}_i(t + (\Delta t/2))$, respectively. The procedures are mathematically equivalent if the velocities evaluated from the above equations at time t are interpreted as the corresponding velocities of Gibson *et al.*'s procedure at $t - (\Delta t/2)$.

Starting from initial positions $\mathbf{r}_i(0)$ and velocities $\mathbf{v}_i(0)$, a computer program evaluates $\mathbf{v}_i(t + \Delta t)$ from Eq. (2.38a). This is substituted into Eq. (2.38b) to obtain the position of particle i at time $t + \Delta t$. Positions and velocities for the other particles are calculated in the same way. This procedure repeats at $t + 2\,\Delta t$ and so on. An alternate difference scheme for solving Eq. (2.31) is Rahman's predictor–corrector method with the following predictor formula for position $\mathbf{r}_i(t + \Delta t)$ of particle i (Rahman, 1964):

$$\bar{\mathbf{r}}_i(t + \Delta t) = \mathbf{r}_i(t - \Delta t) + 2\,\Delta t \mathbf{v}_i(t). \tag{2.39a}$$

With the predicted position $\bar{\mathbf{r}}_i(t + \Delta t)$, the corresponding predicted acceleration $\bar{\boldsymbol{\alpha}}_i(t+\Delta t)$ $\{= m^{-1}\mathbf{f}_i[\mathbf{r}_i(t+\Delta t), \mathbf{r}_2(t+\Delta t), \ldots, \mathbf{r}_N(t+\Delta t)]\}$ is evaluated. Using $\bar{\mathbf{r}}_i(t + \Delta t)$ and $\bar{\boldsymbol{\alpha}}_i(t + \Delta t)$, a new position and a new velocity for particle i are calculated from

$$\mathbf{v}_i(t + \Delta t) = \mathbf{v}_i(t) + \tfrac{1}{2}\,\Delta t[\bar{\boldsymbol{\alpha}}_i(t + \Delta t) + \bar{\boldsymbol{\alpha}}_i(t)], \tag{2.39b}$$

$$\mathbf{r}_i(t + \Delta t) = \mathbf{r}_i(t) + \tfrac{1}{2}\,\Delta t[\mathbf{v}_i(t + \Delta t) + \mathbf{v}_i(t)]. \tag{2.39c}$$

The corrected position $\mathbf{r}_i(t + \Delta t)$, Eq. (2.39c), is used again to predict the new, improved $\bar{\boldsymbol{\alpha}}_i(t + \Delta t)$ and $\mathbf{r}_i(t + \Delta t)$. This predictor–collector cycle terminates when a desired accuracy in position at $t + \Delta t$ is reached. Then the program advances one time step (Δt), and another predictor–corrector cycle repeats. Rahman finds that the one or two cycles usually gives sufficiently accurate $\mathbf{r}_i$ and $\mathbf{v}_i$. Verlet (1967) employs a simple centered difference scheme for the Newton equation,

$$\begin{aligned}\mathbf{r}_i(t + \Delta t) = {} & -\mathbf{r}_i(t - \Delta t) + 2\mathbf{r}_i(t) \\ & + (\Delta t)^2 m^{-1}\mathbf{f}_i[\mathbf{r}_1(t), \mathbf{r}_2(t), \ldots, \mathbf{r}_N(t)].\end{aligned} \tag{2.40}$$

In much the same manner as was done previously in the hard-core calculations, the computer program can be divided into two steps to reduce calculation time. In the first step, the computer calculates interparticle forces between any pair among all $N(N - 1)/2$ possible pairs. Since most of the pairs are separated by distances larger than the cutoff distance r_c of $\phi(r)$, the corresponding forces are identically zero. Since the computer will be wasting most of its time in calculating this useless information, the first step is repeated only after every n executions of the second step in the program. The second step keeps track of the positions of neighbors of each particle within a distance r_M from the particle. The r_M is chosen so large that a particle initially outside the r_M can scarcely migrate within r_c during n executions of the second step in the computer program. This two-step method is reported to cut the computing time by an order of 10 (Verlet, 1967). The same initial positions as used in the Monte Carlo calculations are also used in the molecular-dynamic calculations. For studying steady-state phenomena, initial velocities can be picked arbitrarily, the usual choice being the Gaussian distribution in order to make the system achieve equilibrium faster.

The pressure is calculated by using Eq. (2.14b), where the ensemble average is replaced by a time average. The suitable finite-system correction to P needs to be added to agree with the corresponding infinite-system result. Temperature is derived from the total kinetic energy average as follows:

$$T = (m/dk)\langle \sum_{i=1}^{N} v_i^2 \rangle / N. \tag{2.41}$$

Other thermodynamic quantities, such as the configurational specific heat $[(\partial U/\partial T)_\varrho]$ and $(\partial P/\partial T)_\varrho$, can be calculated as average fluctuations in kinetic energy or potential energy or both (Lebowitz *et al.*, 1967). The calculated values, however, are less precise than the values obtained by direct numerical differentiation.

Errors can occur from three sources: machine error, error due to the finite-difference scheme, and roundoff error due to the finite number of digits used to represent $\mathbf{r}_i$ and $\mathbf{v}_i$. The last source of error can be neglected since a present-generation computer contains a sufficient number of bits to represent a quantity of interest. The second source of error can be empirically corrected by repeating the calculation with different time increments Δt. A good check to detect this error as well as the machine error (which can also occur in the hard-core calculations discussed previously) is to print out the total energy as the calculation progresses. The total energy must be strictly a constant but can have small fluctuations around the average. Therefore, if a continuous gain or loss of energy occurs in the course of the calculation, it must be traceable to these errors. In a problem such as the slowing down of an initially energetic particle, a relatively small Δt must be chosen initially; this value, however, can be made larger as the system approaches equilibrium.

C. Additional Remarks

It is appropriate here to give a few remarks on the relationship between the Monte Carlo and the molecular-dynamic methods. An ensemble satisfying the conditions (constant total energy and momentum) of the molecular-dynamic system is the microcanonical ensemble with an added restriction of constant total momentum $\mathbf{P}$, which can be called the $NVE\mathbf{P}$-ensemble [or, as referred to by Wood (1968a), the "molecular-dynamic" ensemble]. Under identical conditions (i.e., same V, N, m, boundary conditions, etc.), there is a question of whether or not a dynamically calculated equilibrium quantity is identical to the corresponding quantity averaged over the $NVE\mathbf{P}$-ensemble. Thus far, there has been no rigorous proof of this, but it has been assumed to be generally true (the famous ergodic hypothesis). At the low-density limit for the hard-core potential however, the equality can be established. The correction to the order $1/N$ for the second virial coefficient and the collision rate for the $NVE\mathbf{P}$-ensemble in this limit are (Hoover and Alder, 1967)

$$B_2(NVE\mathbf{P}) = B_2(NVE)[1 + (1/N)], \tag{2.42}$$

$$\Gamma_0(NVE\mathbf{P}) = \Gamma_0(NVE)[1 + (4d + 1)/8Nd], \tag{2.43}$$

where $B_2(NVE)$ and $\Gamma_0(NVE)$ are B_2 and Γ_0 calculated using the microcanonical ensemble, which is, due to the absence of energy fluctuations, identical to the NVT-ensemble in the case of the hard-core particles. The quantity $B_2(NVE)$ is related to the second virial coefficient $B_2(\infty)$ of the infinite system by (Oppenheim and Mazur, 1957):

$$B_2(NVE) = B_2(\infty)[1 - (1/N)]. \tag{2.44}$$

Equations (2.42) and (2.43) imply that $B_2(NVE\mathbf{P})$ is identical to an order of $1/N$ to $B_2(\infty)$ and that the Monte Carlo $(\beta P/\varrho) - 1$ for the NVT-ensemble is identical to the corresponding dynamic value if it is scaled by a factor $N/(N - 1)$. If this scaling is presumed to hold at any density, we obtain a relationship linking the molecular-dynamic pressure [Eq. (2.36) or (2.37)] with the NVT-Monte Carlo pressure [Eq. (3.7)] (Hoover and Alder, 1967):

$$[(P/\varrho kT) - 1]_{\text{molecular dynamic}} = [N/(N - 1)][(P/\varrho kT) - 1]_{NVT\text{-Monte Carlo}}. \tag{2.45}$$

The scaling relation (2.45) (therefore, the ergodic hypothesis within this context) has been shown empirically to be valid within the accuracy of the numerical data over the whole density range for the 12 hard-disk (Hoover and Alder, 1967) and the 72 hard-disk (Wood, 1970) systems. Table I shows the 12-disk results of Hoover and Alder at selected densi-

TABLE I

COMPARISON OF THE EQUATION-OF-STATE DATA FOR 12 HARD DISKS WITH PERIODIC BOUNDARY CONDITION[a,b]

ϱ_0/ϱ	$\beta P/\varrho$		
	Molecular-dynamic[c]	NVT-Monte Carlo[d]	$NVE\mathbf{P}$-Monte Carlo
1.1	21.9	20.62(9)	22.40
1.4	7.42	6.86(8)	7.39
1.5	6.33	5.91(5)	6.36
1.6	5.56	5.23(7)	5.61
1.8	4.26	4.07(6)	4.34
2.0	3.54	3.37(6)	3.59
3.0	2.10	2.03(3)	2.12

[a] From Hoover and Alder (1967).

[b] The molecular-dynamic $\beta P/\varrho$ is calculated from Eq. (2.36); the NVT-Monte Carlo $\beta P/\varrho$ (Wood, 1963) is obtained from Eq. (3.7), and Eq. (2.45) is used to obtain the $NVE\mathbf{P}$-Monte Carlo $\beta P/\varrho$ from the NVT-Monte Carlo values.

[c] Numbers in this column are accurate within 1%.

[d] Numerals in parentheses indicate the standard deviation in the last digit.

ties. A similar attempt to establish an equivalence between a dynamic average and an ensemble average of an equilibrium quantity for the soft-core particles would be a worthwhile thing to do. This could be done by separately carrying out the usual molecular-dynamic calculation and computing, for example, its pressure with the Monte Carlo value of pressure obtained for the NVT-ensemble with an additional restraint by fixing the center of mass of the system to make the total momentum zero. Thermodynamic variables obtained for this new ($NVT\mathbf{P}$-) ensemble can be expressed in terms of the corresponding variables for the $NVE\mathbf{P}$-ensemble (Lebowitz *et al.*, 1967).

Precise comparisons of relative efficiency under identical conditions between the molecular-dynamic and the Monte Carlo methods are not yet in the literature. If only the standard deviations are used as a principal yardstick to measure the efficiency, in the case of the hard-core particles in a high-density solid phase, the molecular-dynamic method seems to be more efficient than the existing Monte Carlo methods. The latter, however, could be improved by devising a more efficient way of sampling important configurations. Assuming that both methods take about the same order-of-magnitude calculation time, the relative efficiency of the two methods can be seen from the values of $\beta P/\varrho$ for hard disks at $\varrho_0/\varrho = 1.1$; Alder *et al.* (1968) obtained, about 3000 collisions per particle later, $\beta P/\varrho = 21.951 \pm 0.005$ for the 72-disk system, while the Monte Carlo calculations of Wood (1963) at the same density give $\beta P/\varrho = 20.62 \pm 0.09$ after about 36,800 moves per particle for the 12-disk system. The NVT-ensemble Monte Carlo procedure for estimating pressure is handicapped at high density, since it requires extrapolation of $g(r)$ to its contact value $g(1)$. At high density in the neighborhood of $r = 1$, $g(r)$ is a rapidly decreasing function of r. The error involved in this extrapolation shows a drawback of the NVT-Monte Carlo procedure. In the case of the Lennard-Jones particles, however, Verlet's calculations seem to indicate that the methods have approximately equal efficiency (Wood, 1968a).

III. Applications

A. Communal Entropy and Melting Transition

A substance in a solid phase is characterized by its mechanical strength to withstand shear flow. As it is heated, however, this property disappears suddenly and the material melts. From a thermodynamic viewpoint, the phase change takes place because the fluid phase due to its lower chemical potential becomes more stable relative to the solid phase. At the melting point, chemical potentials and pressures between the two phases must exactly balance each other so that neither solid nor fluid can dominate at the expense of the other phase. Therefore, any successful statistical-mechanical theory of melting should make use of thermodynamic information from both phases.

Starting in 1953 (Metropolis *et al.*, 1953), computers have been increasingly applied to simulating and understanding solid–fluid and liquid–gas transitions in systems containing a finite number of particles with a given interaction force. Initial investigation along this line by a Monte Carlo technique was first carried out by Rosenbluth and Rosenbluth (1954). Their work, however, suggested the absence of a solid–liquid transition for a system of 256 hard spheres. As pointed out later by Wood and Jacobson (1957), this was due to the relatively slow speed of the machine then available and resulting poor statistical averages. Later, by employing a relatively faster machine with a smaller system

(32 spheres), Wood and Jacobson demonstrated that the two phases occur as two distinct branches. This was further substantiated by Alder and Wainwright's (1957) molecular-dynamic work based on 32 and 108 hard spheres. Advances in computer technology in later years not only added further evidence supporting the solid–fluid transition for spheres and disks (Hoover and Alder, 1967; Wood, 1970), but also made it possible to study a more realistic system, such as a system of particles interacting with the Lennard-Jones potential (Wood and Parker, 1957; Hansen and Verlet, 1969) or the square-well potential (Rotenberg, 1965a).

In studying the melting transition by a computer-generated "experiment," the single most important factor which must be taken into consideration is the effect of particle number on the thermodynamic variables. This is so because thermodynamic phase transition is accompanied by a large fluctuation in one of the thermodynamic variables, such as large density fluctuations. A "computer simulation" of two coexisting phases is indeed a difficult task to achieve, since the particle number used in the "experiment" has to be sufficiently large in order to recreate configurations with several different regions with low and high local number densities. Each part of this region is smaller than the size of the system, yet must be large enough to include many particles. Unless this condition is fulfilled, a finite system will likely tend to stay in a single phase rather than break up into two phases. Systems containing as many as 500 hard spheres (Alder *et al.*, 1968) proved to be insufficient to recreate the coexisting phases.

The situation for a two-dimensional system is, however, less formidable. In their molecular-dynamic study on a system of 870 disks, Alder and Wainwright (1962) showed that fluid and solid phases can coexist in the transition region and that the isotherm has a van der Waals loop. Molecular-dynamic studies of Hoover and Alder (1967) confirmed this for a system containing as few as 72 disks. This favorable situation in case of the hard-disk system is due to the smaller number of particles on the surface of a box than inside it, the ratio being approximately proportional to $N^{-1/2}$ in a two-dimensional problem, the ratio for a three-dimensional system behaving approximately as $N^{-1/3}$.

Locations of the melting and the freezing densities are not, however, provided by the computer studies described above since they do not give essential information on the entropy constant. This constant enables one to evaluate, for example, the solid-state entropy from the *same* reference state as used in the fluid-branch entropy. If an ideal-gas state is chosen as the reference state, entropies of the fluid can be measured without difficulty from a low density where the ideal gas is a sufficiently good approximation. This is achieved by integrating the thermodynamic relationship $\beta P = \varrho^2\, \partial A/\partial\varrho$ on the fluid branch; i.e.,

$$A = A_{\text{ideal}} + \int_0^{\varrho} d\varrho \{[(\beta P/\varrho) - 1]/\varrho\}, \tag{3.1}$$

where A_{ideal} corresponds to the free energy for the ideal gas at ϱ, i.e.,

$$A_{\text{ideal}} = 1 - \ln \varrho. \tag{3.2}$$

Evidently a similar approach cannot be used in the case of the solid branch, since this branch terminates at a fairly dense metastable state. Hence, the use of an entropy constant (and also the free-energy constant) for a solid referenced to an ideal-gas state has remained uncertain until recently.

This difficulty has been overcome by using the single-occupancy system as suggested by Hoover and Ree (1967). The single-occupancy system is a system with a constraint on the volume in which a particle can move around. The constraint assigns each particle to its own Wigner–Seitz cell with volume v, and the particle's center is confined within the cell. Figure 1 shows a representative state in the 12-particle single-occupancy

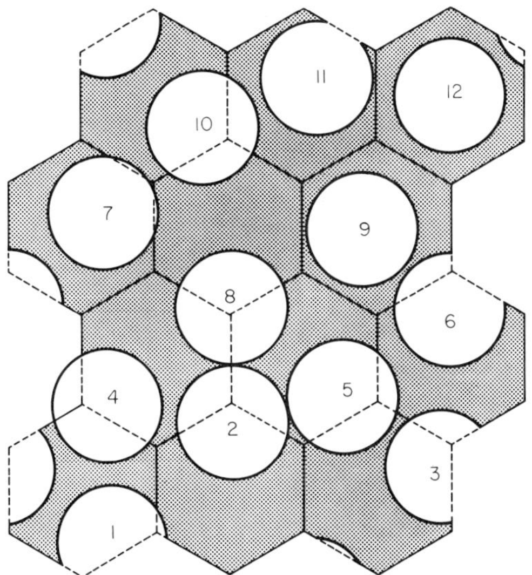

FIG. 1. A single-occupancy system of 12 hard disks in a periodic box, in which the center of each disk is confined in a hexagonal cell. Notice that both the nearest-neighbor disk (5) and the second-neighbor disk (8) of a disk numbered (2) can interact with 2.

system with a hexagonal cell in a two-dimensional periodic box. For the single-occupancy system at high densities, a particle with a repulsive core would predominantly collide with neighbor particles, and only very rarely would it travel to make a collision with its own wall. Since the particle cannot "feel" the presence of the wall, thermodynamic properties at high densities could be satisfactorily approximated by those without the wall constraint. This fact, plus the fact that the artificial wall forces the single-occupancy system to remain in a solid phase at *all* densities, offers a practical way to extend the real solid isotherm to lower fluid densities by using the isotherm of the artificial solid. The Helmholtz free energy follows a similar relation as Eq. (3.1) but applied to the single-occupancy system; i.e.,

$$A_{so} = A_{ideal} + \int_0^{\varrho} d\varrho \{[(\beta P_{so}/\varrho) - 1]/\varrho\}, \tag{3.3}$$

where the subscript "so" signifies the single-occupancy thermodynamic quantities, and A_{ideal} is the same as in Eq. (3.2). Within the "experimental" accuracy, the quantity A_{so} at solid densities can be identified with the free energy A. In this way, we have entirely bypassed a previously encountered difficulty in finding a reversible path linking the solid and the fluid branches. The melting pressure and coexistence densities of solid and fluid can then be fixed by the use of a Maxwell's double tangent construction drawn using A and A_{so}.

Additional information which can be gained from a study of this type is the communal entropy. The entropy difference $\Delta S = S - S_{so}$ between the unconstrained system and the constrained system discussed above is identical to the definition of the communal entropy given by Kirkwood (1950). Interest in this quantity started even earlier with Hirschfelder *et al.* (1937), who suggested that the communal entropy might account for a major portion of entropy of melting, $\sim Nk$. However, further experimental data on the entropy of melting had shown that it depends strongly on pressure (Rice, 1938), unlike the ideal-gas prediction. Communal entropy, therefore, is gradually excited as the system expands further upon melting.

Investigations of the difference between the constrained system and the unconstrained system can be done by using the configurational part Q_N of the partition function. The following formulation is suitable for systems of particles with repulsive cores and without attractive tails. We indeed apply it later to systems of hard spheres and hard disks. For these systems, only the kinetic energy of hard spheres can contribute to the internal energy E, which is equal to that of an ideal gas, $(3/2)NkT$, since the Boltzmann factor $\exp(-\beta\phi_{ij})$ of the interaction potential ϕ_{ij} between the two particles i and j is either 0 or 1, depending on whether the edges of the particles are inside or outside each other's cores. Consequently, the configurational part of the Helmholtz free energy, $A = E - TS$, is due to the contribution by entropy alone. Therefore, the configurational canonical partition function Q_N for the unconstrained system is directly related to S as follows:

$$Q_N = (1/N!) \int d\mathbf{r}^N \exp(-\beta\Phi_N) = \exp(S/k), \tag{3.4}$$

where Φ_N represents the net potential energy of N particles located at $\mathbf{r}_N = (\mathbf{r}_1, \mathbf{r}_2, \ldots, \mathbf{r}_N)$. For the single-occupancy system, the effect of the cell-wall can be represented by the factor Φ_{cw}, which is zero if the center of each particle is located within its own cell, and otherwise is equal to infinity; i.e.,

$$Q_{\text{so}} = (1/N!) \int dr_N \exp(-\beta\Phi_N - \beta\Phi_{\text{cw}}) = \exp(S_{\text{so}}/k). \tag{3.5}$$

The communal entropy $\Delta S \equiv S - S_{\text{so}}$ is, therefore, related to Q and Q_{so} by

$$\Delta S = k \ln(Q_N/Q_{\text{so}}). \tag{3.6}$$

The pressures P $(= -\partial A/\partial V)$ for both the constrained and the unconstrained systems can be obtained through use of the virial theorem, Eq. (2.17):

$$\beta P/\varrho = 1 + b\varrho g(1), \tag{3.7}$$

$$\beta P_{\text{so}}/\varrho = 1 + b\varrho g_{\text{so}}(1), \tag{3.8}$$

where b is the second virial coefficient. For hard spheres, b is equal to $2\pi\sigma^3/3$, and for hard disks, $\pi\sigma^2/2$. The functions $g(1)$ and $g_{\text{so}}(1)$ represent the radial distribution functions for the unconstrained and the constrained systems at sphere or disk diameter. The diameter (sphere, disk, rods, etc.) will be used, hereafter, as the unit of length in discussing the hard-core particles. The fluid-phase equations of state for hard spheres and disks are rather well known. They can be represented by Padé approximants (Ree and Hoover, 1964a, 1967). Among the various Padé approximants which can be constructed from the known first seven virial coefficients, the following symmetric Padé approximants $P(3, 3)$

seem to describe the "experimental" data most accurately:

$$P(3,3) = (\beta P/\varrho) - 1 = [b\varrho + 0.063499(b\varrho)^2 + 0.017327(b\varrho)^3] \times [1 - 0.561501 b\varrho + 0.081316(b\varrho)^2]^{-1}, \tag{3.9}$$

$$P(3,3) = (\beta P/\varrho) - 1 = [b\varrho - 0.202080(b\varrho)^2 + 0.005589(b\varrho)^3] \times [1 - 0.984085 b\varrho + 0.242916(b\varrho)^2]^{-1}, \tag{3.10}$$

for spheres and disks, respectively. The above Padé approximants are constructed so that the first six coefficients of the density expansions will reproduce a like number of the exact virial coefficients. For a fluid, Padé approximants for entropy can be similarily constructed from the density series of S obtained by using the virial expansion of P in Eq. (3.4) (Hoover and Ree, 1968); i.e.,

$$Q_N(\text{fluid}) = (\varrho e)^N \exp\{-Nb\varrho[1 - 0.11075186 b\varrho + 0.00469232(b\varrho)^2] \times [1 - 0.42325186 b\varrho + 0.04130870(b\varrho)^2]^{-1}\}, \tag{3.11}$$

$$Q_N(\text{fluid}) = (\varrho e)^N \exp\{-Nb\varrho[1 - 0.2781515 b\varrho + 0.0059612(b\varrho)^2] \times [1 - 0.6691537 b\varrho + 0.0901912(b\varrho)^2]^{-1}\}, \tag{3.12}$$

for spheres and disks, respectively. The Padé approximants on the fluid side described in Eqs. (3.9)–(3.12) fit the "experimental" data within 1% and $0.01Nk$ for the pressure and the entropy, respectively. The solid-phase data (the face-centered lattice for spheres) on the pressure are also available from the molecular-dynamic calculations of Alder *et al.* (1968). Hence, the solid entropy can be evaluated very accurately by interpolation except for the unknown constant of integration.

The single-occupancy pressure P_{so} is calculated by constructing a Monte Carlo "experiment" which samples the configurational space with a probability $\exp(-\beta\Phi_N - \beta\Phi_{cw})$. The difference between the single-occupancy "experiment" and the Monte Carlo procedure described in Section II,A lies in the occurrence of the cell-wall potential Φ_{cw} around Wigner–Seitz cells. The twelve faces of the Wigner–Seitz cell for the face-centered lattice are formed by planes perpendicular at midpoints of the twelve lines joining the center of the cell and centers of its twelve nearest neighbors. A particle is chosen randomly out of N particles in a box with a periodic boundary. If an arbitrary displacement of this particle within volume $(2\delta)^d$ (d is the number of dimensions) puts a particle outside its own cell, Φ_{cw} prohibits this move. Otherwise, the particle is placed to the new position with probability proportional to

$\exp[-\beta\Phi_N(\text{new}) + \beta\Phi_N(\text{old})]$, as explained in the previous section. In actual calculations for the single-occupancy hard-sphere system with the face-centered-cubic structure, any possible overlap of a particle with its third neighbors is ignored. Although such overlaps can occur in principle for densities greater than 0.19245 of the close-packed density ϱ_0, any contribution by the third-neighbor interactions to the thermodynamic variable is actually negligible at solid phases. Note that the hexagonal cubic lattice structure will also have fourth- and fifth-neighbor interactions. The radial distribution function at the contact distance of two particles, $g_{\text{so}}(1)$, in Eq. (3.8) is obtained by extrapolation using a quadratic polynomial fit to three data points of $g(r)$ immediately adjacent to the contact distance.

Figure 2 shows the single-occupancy pressure for spheres. The Monte Carlo data in the figure correspond to results for an infinite-system pressure extrapolated from the data for the 32-, 108-, 256-, and 500-particle systems. Similar plots for disks obtained by analysis of the 12-, 72-, and 780-disk results are shown in Fig. 3. If the Mayer f-function $\{\equiv [\exp(-\beta\phi)] - 1\}$ is used to express Q_{so}, the first two terms of P_{so} $(\equiv kT\varrho^2\, \partial A/\partial\varrho)$ in its low-density expansion can be calculated analytically:

$$\beta P_{\text{so}}/\varrho = 1 + 4.44288(\varrho/\varrho_0)^{4/3}, \tag{3.13a}$$

$$\beta P_{\text{so}}/\varrho = 1 + 2.30940(\varrho/\varrho_0)^{3/2}, \tag{3.13b}$$

for spheres and disks, respectively. The Monte Carlo results agree with the above analytic expressions at the lowest density $(0.050\varrho_0)$. In fact, the low-density Monte Carlo values of P_{so} can be shown to be independent of particle number. In contrast to this, note that, for the unconstrained system, $(\beta P/\varrho) - 1$ lies below the infinite-system limit by a factor $(N-1)/N$ [see Eq. (2.45)]. Both Figs. 2 and 3 show that P_{so} lies slightly below P for all densities. For the constrained system of hard spheres, there is a cusp in P_{so} at $\varrho = 0.625\varrho_0$. This apparently corresponds to a density where the cell wall begins to prevent spheres from escaping. Absence of a similar phenomenon in P_{so} of the contrained hard-disk system may be related to the absence of the long-range order in the width of the singlet distribution function in two dimensions [see, for example, Frenkel's book (1955)].

By numerical integration of P_{so}, the communal entropy can be calculated. The result listed in Tables II and III for the single-occupancy systems of spheres and disks, respectively, are values of S_{so} extrapolated

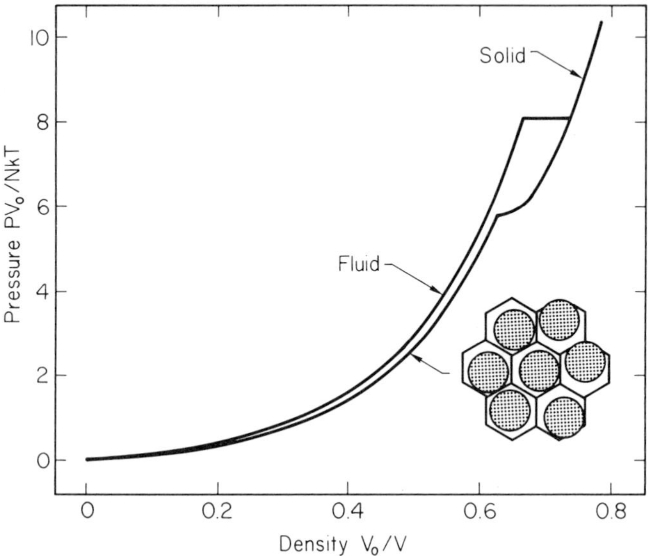

FIG. 2. Equations of state of unconstrained and constrained (single-occupancy) hard-sphere systems (Hoover and Ree, 1968).

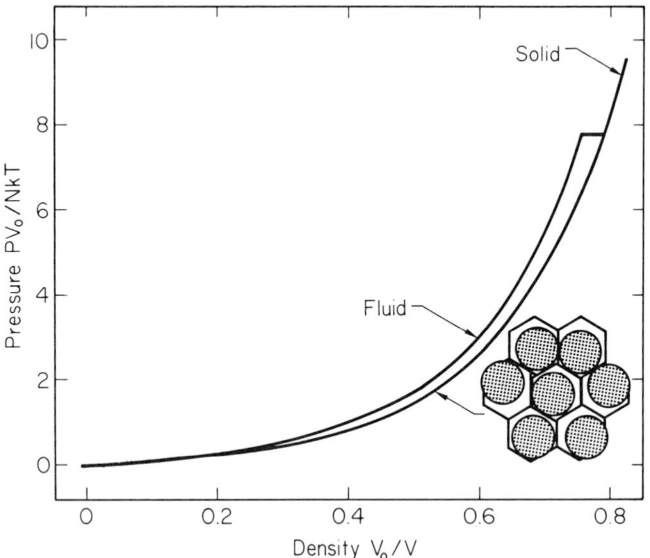

FIG. 3. Equations of state of unconstrained and constrained hard-disk systems (Hoover and Ree, 1968).

TABLE II

Monte Carlo Results on the Single-Occupancy Entropy S^e_{so}, Relative to an Ideal Gas at the Same Density and Temperature, for Hard Spheres[a]

ϱ/ϱ_0	$S^e(\text{Padé})/Nk$	S^e_{so}/Nk [b]	$\Delta S/Nk$ [b,c]
0.00	0.000	−1.000	1.000
0.20	−0.726	−1.428	0.702
0.40	−1.864	−2.352	0.488
0.50	−2.702	−3.101	0.399
0.60	−3.843	−4.155	0.312
0.65	−4.577	−4.806	0.229
0.67	−4.911	−5.058	0.148
0.70	−5.463	−5.440	0.044
0.72	−5.870	−5.703	0.009
0.74	−6.313	−5.978	0.000

[a] The excess entropy S^e(Padé) calculated from Eq. (3.11) for the hard-sphere fluid, and the communal entropy ΔS are shown. These data are taken from Hoover and Ree (1968).

[b] The expected error is 0.015 at the highest density.

[c] The communal entropy is the same as the difference S^e(Padé) − S^e_{so}, except for the two-phase region, where it is calculated from data on the coexistence tie-line.

TABLE III

Monte Carlo Results on the Single-Occupancy Entropy S^e_{so}, Relative to an Ideal Gas at the Same Density and Temperature, for Hard Disks[a]

ϱ/ϱ_0	$S^e(\text{Padé})/Nk$	S^e_{so}/Nk [b]	$\Delta S/Nk$ [b,c]
0.00	0.000	−1.000	1.000
0.20	−0.424	−1.146	0.722
0.40	−1.035	−1.515	0.480
0.50	−1.461	−1.832	0.371
0.60	−2.025	−2.292	0.267
0.70	−2.818	−2.977	0.159
0.75	−3.354	−3.430	0.076
0.78	−3.742	−3.735	0.011
0.80	−4.036	−3.956	0.000

[a] The excess entropy S^e(Padé) [Eq. (3.12)] for the hard-disk fluid, and the communal entropy ΔS. These data are taken from Hoover and Ree (1968).

[b] The expected error is 0.01 at the highest density.

[c] The communal entropy is the same as the difference S^e(Padé) − S^e_{so} except for the two-phase region, where it is calculated from data on the coexistence tie-line.

to an infinite system based on a recipe that Q_{so} for a finite system follows a relation v_f^{N-1}/ϱ. This empirical relation gives v_f, free volume, which is nearly independent of N. This expression was suggested by the results of Salsburg and Wood (1962).

Using the data given in Tables II and III, we can now conclude that hard spheres melt at $(0.736 \pm 0.003)\varrho_0$ and freeze at $(0.667 \pm 0.003)\varrho_0$. The pressure at the coexistence region is determined to be $(8.27 \pm 0.13) \times \varrho_0 kT$. Similar analysis for hard disks shows that the coexistence range stretches between $0.761\varrho_0$ and $0.798\varrho_0$ with the coexistence pressure $8.08\varrho_0 kT$. The latter result is consistent with the tie-line given by Alder and Wainwright (1962) for the 870-disk system. Difference between the above result and Alder–Wainwright's result (a coexistence line between $0.762\varrho_0$ and $0.790\varrho_0$ with $P = 7.72\varrho_0 kT$) can be ascribed to the number dependence in the 780-disk system. A substantial agreement found between the melting data given in this section with Alder–Wainwright's result, which was obtained without the cell constraint, is reassuring with respect to an observation that any possible contribution to entropy lost by the single-occupancy constraint cannot exceed more than $0.01Nk$. As expected, large local density fluctuations, characteristic of two dimensions, are expected to have a negligible influence on thermodynamic quantities. In the above argument, the effect of vacancies on melting points was ignored. An approximate theoretical analysis employing a single-occupancy system can be used to predict the number of vacancies by means of a Monte Carlo method (Squire and Hoover, 1969). At the melting point, the fraction of vacancy is found to be approximately 10^{-5}, which in turn contributes negligibly to entropy, not exceeding $0.001Nk$.

The numerical values for the communal entropy listed in Tables II and III are shown in Fig. 4. This figure includes a similar plot for the constrained hard-rod system (Hoover and Alder, 1966). As was expected, the communal entropy does not suddenly vanish at the freezing density, but rather decreases approximately linearly as density is decreased to about 0.6 of the close-packed density. This approximate linearity of ΔS versus ϱ is a common feature present in all three constrained systems. Therefore, if this feature is taken into consideration of a cell model, the resulting calculations may give improved thermodynamic values.

Melting data given above for hard spheres and disks agree within "experimental" uncertainties with those predicted by Ross and Alder's empirical rule (Ross and Alder, 1966). This rule predicts that the coexistence tie-line could occur at the fluid side corresponding to the highest density at which the solid branch in a Monte Carlo or molecular-

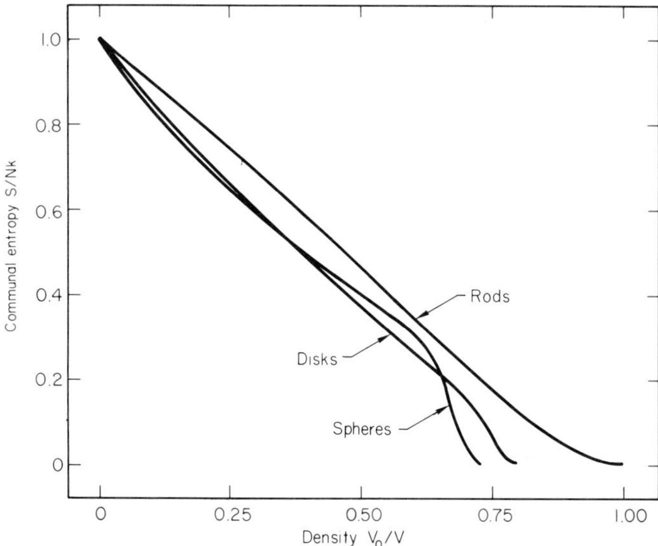

FIG. 4. Communal entropies for hard spheres, disks, and rods (Hoover and Ree, 1968).

dynamic system makes a jump to the fluid-phase isotherm. Ross and Alder have shown that the coexistence pressure derived from this rule for 108 argonlike particles closely agrees with Simon's equation for melting (Simon and Glatzel, 1929) at various melting temperatures. Their Monte Carlo results also confirmed that Lindemann's law (Lindemann, 1910) for melting holds over a 100-fold increase in melting temperature and a 4-fold increase in density from the normal condition. This "law" states that melting takes place at a volume corresponding to a 10% linear expansion from the close-packed volume.

For a system of the Lennard-Jones particles, the solid–fluid transition has been recently determined by Hansen and Verlet (1969), who used a method similar to that discussed above for the hard-sphere case. Near the triple point, a straightforward application of the above method may result in an additional difficulty due to the attractive interaction, which causes a large pressure fluctuation. This difficulty was overcome by introducing a "charging parameter" for the attractive potential energy, which was turned on *after* the single-occupancy system went from a fluid phase to a region where a solid would be a stable phase without the cell constraint. Furthermore, Hansen and Verlet determined a liquid–gas tie-line using a method which inhibits a large density fluctuation from taking place. This will be reviewed in Section D in detail. Thermodynamic properties of particles with r^{-12} repulsion have been calculated recently by Hoover *et al.* (1970) and Hansen (1970).

Table IV lists the solid and fluid transition densities (Hansen and Verlet, 1969) for the Lennard-Jones particles, together with the data on argon (Hansen and Verlet, 1969; Michels *et al.*, 1958; Gosman *et al.*, 1969) at temperatures ranging from 6.7% above the triple point (83.8°K) to 4.5% above the critical point (150.9°K). In Table IV, temperature and length are expressed in units of the Lennard-Jones parameters, ε/k and σ_{LJ}, respectively. The reduced freezing density ϱ_r(fluid) and melting density ϱ_r(solid) calculated from various simple rules are also listed in Table IV. Numbers given in the column under $S(k) = 2.85$ in Table IV are the freezing densities obtained by Hansen and Verlet (1969) under the assumption that the structure factor $S(k)$ at the freezing density is independent of temperature and is equal to 2.85. This value is also the value of $S(k)$ for the hard-sphere fluid at its freezing point. This correspondence rule gives the freezing densities, which are in almost perfect agreement with the Monte Carlo data for all temperature regions except in the neighborhood of the triple point, where the difference is still very small, amounting to only about 1–2%. Numbers quoted under "variable core" are the transition densities obtained using the hard-sphere transition densities (discussed earlier in the present section) together with the temperature-dependent diameter σ obtained by using the Lennard-Jones potential energy in the following formula given by Barker and Henderson's (1967a) perturbation theory of liquids:

$$\sigma = \int_0^{\sigma_{LJ}} dx \{1 - \exp[-\beta\phi_{ij}(x)]\}. \tag{3.14}$$

Small deviations in the transition densities at low temperatures may be caused by attractive interaction, which may become important at these temperatures. However, it is noteworthy that the transition densities predicted from this rule agree remarkably well at high temperatures ($T_r \gtrsim 1.35$, or $T = 160$°K). Ross and Alder's (1966) rule on freezing density also seems to work well at the highest temperature ($T_r = 2.74$, $T = 328$°K) in the table. For this comparison, the reduced density at the freezing point (1.08) is chosen as an average value of the two data points ($\varrho_r = 1.05$ and $\varrho_r = 1.11$) used in Wood and Parker's (1957) calculations on the 32 Lennard-Jones particles. Further tests of this rule remain to be done at still lower temperatures, where the attractive part of the potential energy may shift the transition data. The column

TABLE IV

COMPARISON OF FREEZING ϱ_r(fluid) AND MELTING ϱ_r(solid) DENSITIES AT THE SOLID–FLUID TRANSITION RANGE FOR THE LENNARD-JONES PARTICLES, TOGETHER WITH EXPERIMENTAL DATA ON ARGON[a]

T_r	ϱ_r(fluid)					ϱ_r(solid)			
	Monte Carlo[b]	$S(k) = 2.85$[c]	Variable core[d]	Ross–Alder.	Argon[b,f]	Monte Carlo[b]	Lindemann[g]	Variable core[d]	Argon[b,g]
0.75	0.871	0.864	1.008	—	0.856	0.987	0.990	1.112	0.967
1.15	0.936	0.935	1.033	—	0.947	1.024	1.015	1.140	1.028
1.35	0.964	0.964	1.044	—	0.982	1.053	1.043	1.150	1.056
2.74	1.113	1.116	1.108	1.08	—	1.179	1.192	1.220	—

[a] Temperature and density are reduced by dividing them by ε/k and $(\sigma_{LJ})^{-3}$, respectively. The Lennard-Jones parameters are $\varepsilon/k = 119.8°K$ and $\sigma_{LJ} = 3.408$ Å.

[b] Hansen and Verlet (1969).

[c] Freezing is assumed to take place whenever the first peak of $S(k)$ becomes 2.85, where $S(k)$ is the reduced structure factor. See Hansen and Verlet (1969). Numbers quoted in this column may be inaccurate as much as ± 0.06 due to "readoff" error made during translation of a graphic representation in Hansen and Verlet's (1969) work.

[d] Hard-sphere melting data of Hoover and Ree (1968) are used with the temperature-dependent diameter Eq. (3.14) of Barker and Henderson (1967b). The errors in numbers quoted in this column are estimated to be less than ± 0.005.

[e] See the text for Ross and Alder's rule. Wood and Parker's (1957) data are used in conjunction with this rule. A possible error of ± 0.03 may occur.

[f] Michels *et al.* (1958). See data quoted in Hansen and Verlet (1969).

[g] The melting is assumed by Hansen (1969a) to take place when the root-mean-square displacement of an atom from its lattice position reaches one-seventh of the nearest-neighbor distance. The data quoted in this column are read from a graph in Hansen's paper. Aside from a possible error in translation, there are large (~10%) statistical fluctuations in the original figure. The last two digits in the numbers quoted in this column may be in error.

under "Lindemann" lists densities at which the root-mean-square displacement of a particle from its lattice position reaches one-seventh the nearest neighbor lattice spacing at constant temperature. The ratio 1/7 is assumed by Hansen (1969a) to be a constant in the Lindemann law of melting. Although values in this column agree well with the Monte Carlo values, one must be cautious in drawing conclusions, because of the large uncertainties (approximately 10%) in Hansen's data on the mean-square displacement.

B. Singlet Distribution Function

The singlet distribution function $s(\mathbf{r})$ describes the probability of finding a particle at distance $\mathbf{r}$, with $\mathbf{r}$ being measured from the center of a Wigner–Seitz cell belonging to the particle. More specifically, the mathematical expression for $s_{so}(\mathbf{r})$ of the single-occupancy system is given as follows:

$$s_{so}(\mathbf{r}) = Q_{so}^{-1} \int d\mathbf{r}^{N-1} \exp(-\beta\Phi_N - \beta\Phi_{cw}). \tag{3.15}$$

The mean-square half-width $\langle r^2 \rangle_v$ of $s_{so}(\mathbf{r})$ follows from Eq. (3.15) as

$$\langle r^2 \rangle_v = \int_v d\mathbf{r}[r^2 s_{so}(\mathbf{r})], \tag{3.16}$$

where the integration is performed over the Wigner–Seitz cell of the particle.

In the case of hard spheres in a solid phase, particles will usually collide with other particles, and collisions with their own cell walls are relatively rare. Therefore, the difference between $s_{so}(\mathbf{r})$ and $s(\mathbf{r})$, the singlet distribution function without the wall constraint, is negligible within the cell wall. Outside the cell wall, $s_{so}(\mathbf{r})$ vanishes identically but $s(\mathbf{r})$ does not. However, its value is very small. The running of a computer "experiment" on $s(\mathbf{r})$ or $s_{so}(\mathbf{r})$ has not yet received as much attention as have thermodynamic quantities or radial distribution functions. At the time of writing,[†] the only published "experimental" data on this subject are limited to the ratio of the mean-square displacement for the face-centered-cubic, hard-sphere solid to that of the hexagonal close-

[†] We expect that additional information will be published soon. Further detailed analysis on the singlet distribution function and its various moments for the hard-sphere solid will be published soon by Young and Alder (1971).

packed crystal at the close-packed limit (Alder *et al.*, 1969). This ratio (1.02 ± 0.01) can be regarded as predicting that the face-centered-cubic crystal is more stable than the hexagonal close-packed crystal, if the mean-square displacement is used as a measure of the free volume. Aside from providing information relevant to possible stability among various crystals, $s(\mathbf{r})$ can serve as an additional criterion which a successful theory of solids must reproduce reasonably well. Although a three-dimensional crystal can have a finite width for $s(\mathbf{r})$, this is not the case for the unconstrained system in one and two dimensions. This point is elegantly described in Frenkel's book (1955). More recently, Mermin (1968) has shown the impossibility of a crystalline long-range order in a two-dimensional system for power-law potentials of the Lennard-Jones type. In essence, this occurs in the limit of infinite N and is caused by long-wavelength density fluctuations. As we have demonstrated in Section II,A, conclusive evidence exists for the first-order thermodynamic transition for hard disks. The hard-disk system serves as a good example that a thermodynamic phase transition (of, at least, the first-order type) is not necessarily accompanied by a long-range crystalline order. This, however, does not at all rule out the possibility of a long-range order in higher-order distribution functions. A further theoretical investigation to find out the possible relationship between the nature of a thermodynamic transition and the presence of an order in particle distribution functions will be very useful.

For the constrained system, the divergence of $\langle r^2 \rangle$ is prevented by the cell walls, which restrict particles to stay always within their own cells. We confine our attention to only the constrained system and calculations of its $s_{\text{so}}(\mathbf{r})$ and $\langle r^2 \rangle_v$. In order to do this, the Monte Carlo program described in Section III,A needs to be slightly modified so that $s_{\text{so}}(\mathbf{r})$ is evaluated in the fixed center-of-mass coordinates. This is essential in the present case, since the Monte Carlo calculations performed in the floating center-of-mass frame, as has been commonly used, give different-looking $s_{\text{so}}(\mathbf{r})$. This peculiarity can occur only in the Monte Carlo calculations and is not present in the corresponding molecular-dynamic calculation, which, by conservation of the total momentum, automatically constrains its center of mass to remain fixed. While the differences in shapes of $s_{\text{so}}(\mathbf{r})$ resulting from the two different ways of computing are only minor at low densities, they gradually become more pronounced as density is increased. Only small differences occur at low densities since particles will collide mostly with the cell walls. Small differences in the two systems are attributable to number dependence. At high solid densities, however, particles will not be able to "see" the cell walls. Therefore, the instantaneous center of mass generated by moving a randomly selected particle a certain amount (as described in the Monte Carlo procedure in Section II,A) will drift around with a low-frequency motion in comparison with the number of jumps taken in a realization of the corresponding Markov chain. If the long "time" average of $s_{so}(\mathbf{r})$ is measured from a center of the Wigner–Seitz cell, the resulting $s_{so}(\mathbf{r})$ will look nearly flat.

One possible method of "playing the Monte Carlo game" in the fixed center-of-mass system uses a perfect lattice configuration as its initial configuration. Each subsequent move is performed by selecting two particles arbitrarily and moving them simultaneously by an equal amount, but in opposite directions. The two particles are chosen randomly out of N enumerated particles by choosing two random integers i $(1 \leq i \leq N)$ and i' $(1 \leq i' \leq N - 1)$. The first integer (i) corresponds to particle i; the second particle is chosen to be particle $i + i'$ if $i + i' \leq N$; otherwise, particle $i + i' - N$ is chosen as the second particle. The rest of the program is nearly identical to the previous one used in Section III,A. Using this Monte Carlo procedure, single-occupancy systems of 72, 896, and 1400 hard disks in periodic containers are examined (Ree, 1970). For the 72-disk system in a hexagonal crystal configuration for the Wigner–Seitz cell, the lattice sites consist of eight rows each containing regularly spaced 9-disk sites. The 896- and 1400-disk systems use 32 and 40 rows, respectively. To save computation time, only the singlet distribution function averaged over its angular orientation has been examined. Therefore, the resulting angle-independent $s_{so}(r)$ cannot tell possibly interesting orientation-dependent behavior of $s_{so}(\mathbf{r})$. After the end of each move in the Monte Carlo program, the program evaluates r^2 for the two particles. These distances are "scored" in appropriate counters, each of which stores the scores belonging to separations lying between $\bar{r}^2$ and $\bar{r}^2 + \Delta r^2$. The maximum distance of $\bar{r}$ is stretched to the nearest cell wall. Numerical differentiation, making use of the total accumulated scores in each of the counters, yields the desired $s_{so}(r)$. As was usually done, $\bar{r}^2$ instead of $\bar{r}$ was used to identify each counter. While this procedure saves some computing time by not computing square roots, it amounts to relatively a small portion of the net computing time and has a disadvantage that the resulting data on $s_{so}(r)$ are given in uneven intervals of r, necessitating further manual interpolations if the data are to be displayed in evenly spaced intervals of r.

Figures 5a, 5b, and 5c show $s_{so}(r)$ at the low-fluid density $(0.3\varrho_0)$,

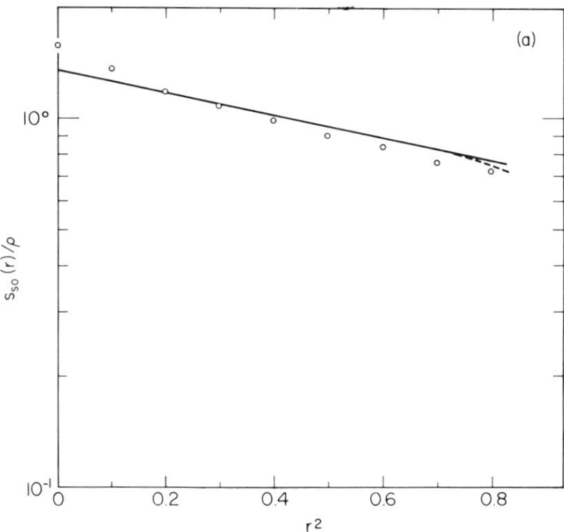

FIG. 5a. Plot of the singlet distribution function in units of disk diameter, $s_{so}(r)/\varrho$, for the single-occupancy hard disks versus the square of the distance, r^2, measured from the center of the Wigner–Seitz cell. Here, $\varrho = 0.3\varrho_0$. (——) $s_{so}(r)$ from approximation 1 (IO1); (- - -) $s_{so}(r)$ from approximation 2 (IO2). The Monte Carlo data on $s_{so}(r)$ shown by the open dots are 1400-disk data for $0.3\varrho_0$.

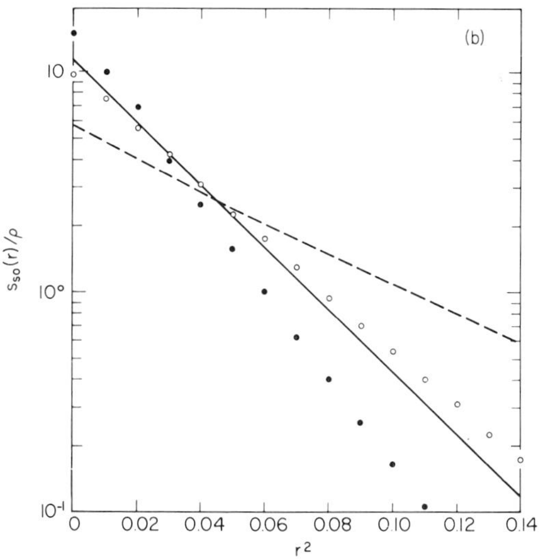

FIG. 5b. Plot of $s_{so}(r)/\varrho$ versus r^2 for $\varrho = 0.8\varrho_0$. (——) Approximation 1 (IO1); (- - -) approximation 2 (IO2). Open dots are 896-disk Monte Carlo data and black dots are 72-disk Monte Carlo data for $\varrho = 0.8\varrho_0$.

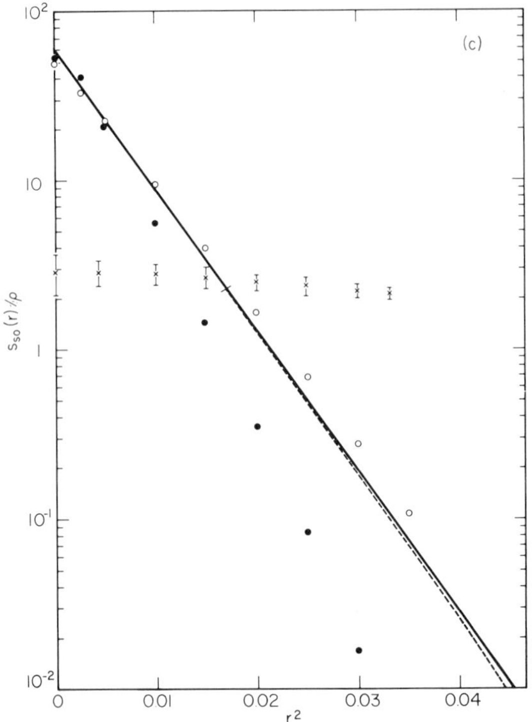

FIG. 5c. Plot of $s_{so}(r)/\varrho$ versus r^2 for $\varrho = 0.9\varrho_0$. (——) Approximation 1 (IO1); (- - -) approximation 2 (IO2). Open dots are the 896-disk Monte Carlo data and black dots are 72-disk Monte Carlo data for $\varrho = 0.9\varrho_0$. Also, the Monte Carlo data on $s_{so}(r)$ at $= 0.9\varrho_0$ performed on a single-occupancy system of 72 hard disks using the floating center-of-mass system are indicated by $\times$. Notice the large error bars in this case and an almost flat appearance of $s_{so}(r)$.

the near-melting density ($0.8\varrho_0$), and the high-solid density ($0.9\varrho_0$). The hard-disk diameter is chosen to be unit of length in these figures. At $\varrho = 0.3\varrho_0$, N dependence of $s_{so}(r)$ is only noticeable at distances closer ($\lesssim 3.5$) to the center of the cell. The N-dependent behavior, although small in magnitude, is in the direction that would make $s_{so}(r)$ flatter. For example, $s_{so}(r{=}0.0722)/\varrho$ decreases from 1.69, 1.65, to 1.64 (with the standard deviations equal to 0.01), as the number of hard disks increases from 72, 896, to 1400 disks. Apparently the largest system (1400 disks) studied here is not yet large enough so that a limiting form for $s_{so}(r)/\varrho$ as $N \to \infty$ can be extracted with confidence from the present result. Figure 5c also contains $s_{so}(r)/\varrho$ for the 72-disk system when the center of mass of the system is not fixed. As was mentioned earlier,

$s_{so}(r)/\varrho$ in this case has a flat-looking appearance. At $\varrho = 0.8\varrho_0$, which lies close to the density where melting would occur for disks without the cell restriction, number-dependent behavior of $s_{so}(r)$ is particularly significant (see Fig. 5b). Moreover, at this density, the Markov chain used in the Monte Carlo program takes a length of time approximately an order of magnitude longer to achieve an accuracy of $s_{so}(r)$ comparable to that obtained at $0.3\varrho_0$ and $0.9\varrho_0$. These facts agree with an observation that a long-wavelength cooperative motion of the whole system exerts a significant influence on the behavior of $s_{so}(r)$, particularly in the neighborhood of the phase-transition point.

Mean-square widths of $s_{so}(r)$ are tabulated in Table V for systems with various sizes. Agreement at $\varrho = 0.3\varrho_0$ among the 72-, 896-, and 1400-disk systems are good. A small increase in $\langle r^2 \rangle_v$ as the system size becomes bigger is attributable to the presence of the N-dependent fluctuation effect even at this low density. In Figs. 5a–5c and Table V, $s_{so}(r)$ and $\langle r^2 \rangle_v$ calculated from a theory recently proposed by Fixman (1969) are also listed. In the low-density region, agreement between the approximate theory 1 or 2 and the "experiment" is rather good. In the case of

TABLE V

THE MONTE CARLO AND APPROXIMATE THEORETICAL VALUES OF THE MEAN-SQUARE HALF-WIDTH $\langle r^2 \rangle_v$ FOR A SINGLE-OCCUPANCY SYSTEM OF HARD DISKS, GIVEN IN UNITS OF DISK DIAMETER[a]

ϱ/ϱ_0	$\langle r^2 \rangle_v$			
	N	Monte Carlo[b]	Approximation 1	Approximation 2
0.3	72	0.392 ± 0.001	0.4122	0.4086
	896	0.395 ± 0.001		
	1400	0.396 ± 0.001		
0.8	72	0.0218 ± 0.0004	0.0308	0.0626
	896	0.0343 ± 0.0003		
0.9	72	0.0037 ± 0.0001	0.0052	0.0052
	896	0.0058 ± 0.0001		

[a] Approximation 1 and Approximation 2 follow from the use of Eq. (3.26) in conjunction with the truncated form $(\varepsilon + \gamma r^2)$ for the effective potential $v_e(r)$ or the untruncated expression (3.28). These data are taken from Ree (1970). Here, N is the number of hard disks employed in the Monte Carlo "experiment."

[b] Error is estimated as twice the standard deviation in the Monte Carlo averages.

the hard-sphere system, Fixman's approximation 1 predicts the pressure within "experimental" accuracy of the molecular-dynamic data at high density (Alder *et al.*, 1968). On the other hand, approximation 2, which gives inferior data on the pressure, nevertheless is superior for describing $s_{s0}(r)$ and $\langle r^2 \rangle_v$. For hard spheres at $0.9\varrho_0$ in the face-centered configuration, for example, approximation 2 gives $0.00173\sigma^2$ for $\langle r^2 \rangle_v$. The corresponding molecular-dynamic value is $0.00170\sigma^2$ with 1% accuracy. This value is obtained by Young and Alder (1971) by extrapolating the available finite-system data for $\langle r^2 \rangle_v$ to that for the infinite system. Since $\langle r^2 \rangle$ at high densities for hard spheres does not depend on the presence or absence of the wall constraint, a good agreement observed between the two serves to justify further examination of Fixman's theory. This theory is reviewed in the following discussion. The mean-square half-width depends strongly on the volume accessible to a particle (large *free volume*, for example, implying a large $\langle r^2 \rangle_v$). Therefore, entropy, which is proportional to the logarithm of the free volume, would depend on the magnitude of $\langle r^2 \rangle_v$. Therefore, calculations of $\langle r^2 \rangle_v$ for different types of crystals may yield certain correlations among the stabilities of these crystals.

C. High-Density Behavior of Hard Spheres and Disks

In a low-density situation, a thermodynamic variable such as pressure has a power-series expansion in density or fugacity. The coefficient of ϱ^n or z^n appearing in the expansion has a graph-theoretical expression which involves only n-particle clusters. Systematic inclusion of increasingly higher-order terms in the expansion enables us to evaluate thermodynamic quantities at gradually denser states. This procedure is limited in practice by a mathematical difficulty of evaluating individual coefficients (see Section III,G). For example, even in case of a simple potential such as the hard-sphere potential, only the first seven coefficients for pressure (Ree and Hoover, 1967) have been evaluated. Nevertheless, when used in a proper Padé approximant (Baker, 1965), the seven coefficients are proved to be sufficient to describe the hard-sphere and disk fluids accurately.

A similar well-formulated expansion does not exist at a high-density solid phase. In view of this lack of a well-defined, theoretically rigorous formulation, at present we must study high-density behavior essentially in an empirical manner. This can be done at least for hard-core particles.

In this case, a natural parameter characterizing high density is α, which measures available free volume per particle relative to v_0:

$$\alpha = (v/v_0) - 1. \tag{3.17}$$

Alternatively, we can employ a parameter

$$\eta = (v/v_0)^{1/d} - 1, \qquad d = 1, 2, \text{ or } 3. \tag{3.18}$$

The parameter η is the nearest-neighbor distance minus the hard-core diameter, which is used as a unit of length. Thus, it characterizes the magnitude of an average reduced length which a particle can move before colliding with other particles. Although either α or η can be used in studying high-density behavior, η rather than α appears in many of the currently available approximate theories on the free volume. At any rate, the advantage of using one over the other is not clearly known at present. One factor affecting the choice would be the rate of convergence when the pressure is expanded in these parameters:

$$\beta P/\varrho = (d/\alpha) + c_0 + c_1\alpha + \ldots, \tag{3.19}$$

$$\beta P/\varrho = (1/\eta) + \bar{c}_0 + \bar{c}_1\eta + \ldots, \tag{3.20}$$

where, from Eqs. (3.17) and (3.18), $\bar{c}_0 = c_0 - [(d-1)/2]$ and $\bar{c}_1 = c_1 d + [(d^2 - 1)/12]$.

Using Eq. (3.19), analysis of the high-precision pressure measurement (with 0.01% error) for hard spheres and disks was carried out by Alder *et al.* (1968). It is perhaps worthwhile to remark again that the high-precision data resulted from the molecular-dynamic calculations of the pressure by means of the virial theorem (2.36). This procedure looks particularly attractive near the close-packed density. Numerical values of c_0 and c_1 are extracted using them from the intercept and slope at $\alpha = 0$ of the plot of $(\beta P/\varrho) - (d/\alpha)$ versus α. These values, together with values of $\bar{c}_0$ and $\bar{c}_1$, are listed in Tables VI and VII for hard spheres and disks, respectively. Note that coefficients of the first two terms (η^{-1} and η^0) in the η-expansions are relatively insensitive to the dimensionality of the systems. These data serve a useful purpose in checking out the corresponding values derived from various free-volume theories. This sort of comparison permits us to examine the adequacy of various approximations made in the theories. Numerical values for these coefficients corresponding to various approximate theories are also listed in Tables VI and VII. In the following discussions, we shall take up one approximate theory at a time and make comments relevant to it.

TABLE VI

Coefficients of High-Density Expansions of Pressure [Eqs. (3.19) and (3.20)] and Entropy [Eqs. (3.21) and (3.22)] for Hard Spheres in Which the Expansion Parameter Is either the Reduced Free Length $\eta = (\varrho_0/\varrho)^{1/3} - 1$ or the Reduced Free Volume $\alpha = (\varrho_0/\varrho) - 1$

	η-Expansion			α-Expansion		
	$\bar{s}_0$	$\bar{c}_0$	$\bar{c}_1$	s_0	c_0	c_1
Experiment[a]	1.86 ± 0.04	1.56 ± 0.02	2.3 ± 0.2	-0.22 ± 0.04	2.56 ± 0.02	0.56 ± 0.08
IO1[b]	1.080	1.556	~2.3	−0.999	2.556	~0.56
IO2[c]	1.140	1.40 ± 0.04	—	−0.939	2.40 ± 0.04	—
Cell Cluster[d]	1.8629[h]	1.449	—	−0.2165	2.449	—
	1.8637[i]	1.458	—	−0.2157	2.458	—
SCFV[e]	0.000	1.000	0.000	−2.079	2.000	−0.222
LJD[f]	2.079	1.125	0.219	0.000	2.125	−0.149
Correlated[g]	2.020	1.333	1.168	−0.059	2.333	0.167

[a] Alder *et al.* (1968). The data on the α-expansion quoted above are taken from this reference.
[b] Independent-oscillator, approximation 1: Fixman (1969) and Ree (1970).
[c] Independent-oscillator, approximation 2: Ree (1970).
[d] Cell-cluster theory: see Rudd *et al.* (1969) for hard spheres and Salsburg *et al.* (1967) for hard disks.
[e] Self-consistent free-volume theory: Wood (1952).
[f] Lennard-Jones–Devonshire cell theory: see Buehler *et al.* (1951) for hard spheres and Alder *et al.* (1963) for hard disks.
[g] Correlated cell theory: see Squire and Salsburg (1961) and Alder *et al.* (1963).
[h] Face-centered-cubic lattice: three-particle cluster for $\bar{s}_0$ (s_0) and two-particle cluster for $\bar{c}_0$ (c_0).
[i] Hexagonal close-packed lattice: numbers of clusters used are the same as the ones used in the face-centered-cubic lattice.

TABLE VII

COEFFICIENTS OF HIGH-DENSITY EXPANSIONS OF PRESSURE [Eqs. (3.19) and (3.20)] AND ENTROPY [Eqs. (3.21) and (3.22)] FOR HARD DISKS EXPANDED IN EITHER THE REDUCED FREE LENGTH $\eta = (\varrho_0/\varrho)^{1/2} - 1$ OR THE REDUCED FREE VOLUME $\alpha = (v/v_0) - 1$ [a]

	η-Expansion			α-Expansion		
	$\bar{s}_0$	$\bar{c}_0$	$\bar{c}_1$	s_0	c_0	c_1
Experiment	1.44 ± 0.01	1.40 ± 0.01	1.6 ± 0.1	0.05 ± 0.01	1.90 ± 0.01	0.67 ± 0.07
IO1	0.908	1.511	2.5 ± 0.2	−0.478	2.011	1.1 ± 0.1
IO2	0.910 ± 0.03	1.51 ± 0.05	—	−0.476 ± 0.003	2.01 ± 0.05	—
Cell Cluster	1.397	1.40	—	0.011	1.90	—
SCFV	0.000	1.000	0.000	−1.386	1.500	−0.125
LJD	1.386	1.056	0.078	0.000	1.556	−0.086
Correlated	1.327	1.389	1.780	−0.059	1.889	0.765

[a] Refer to footnotes *a–g* respectively of Table VI for the literature on the experiment and approximate theories listed in the first column.

All of the theories produce the correct leading term d/α. In fact, the asymptotic expression for the equation of state for a *finite* system of hard spheres or disks near close-packing has been shown by Salsburg and Wood (1962) to be exactly d/α. However, except for a one-dimensional system of hard rods (Tonks, 1936), a proof such as Salsburg and Wood's has not yet been extended to a thermodynamic infinite system. In the case of a lattice-gas system with a hard-core repulsion, the relevant mathematics used for the high-density analysis becomes much simpler. In fact, analytic expressions are available for the high-density expansion carried out in several different systems (Alder *et al.*, 1963; Gaunt and Fisher, 1965; Ree and Chesnut, 1966; Bellemans and Nigam, 1967). All of these expressions for P, however, behave as a logarithm of free volume near close-packing, in contrast to the reciprocal of free volume as empirically found in a continuum system of spheres or disks. A hybrid system whose particles have a lattice-gaslike nearest-neighbor interaction, yet moving in a continuous configurational space, has been proposed by Hoover (1966) as an alternative system. For this rather artificial system, Hoover was able to prove that its pressure behaves as d/α near the close-packed limit even as N goes to infinity.

The constant s_0 given in Tables VI and VII is the deviation of the "experimental" or theoretical entropy from the entropy $[\approx(\ln v_0) - d \times \ln(2\alpha/d)]$ of the Lennard-Jones–Devonshire (LJD) theory (Buehler *et al.*, 1951) in the close-packed limit:

$$s_0 = \lim_{\alpha\to 0} [(S/Nk) - (\ln v_0) - d \ln(2\alpha/d)]. \tag{3.21}$$

In calculating the free volume available to a particle, the LJD theory makes the assumption that neighboring particles are fixed at lattice sites. The analogous expression $\bar{s}_0$ for the η-expansion is

$$\bar{s}_0 = \lim_{\eta\to 0} [(S/Nk) - (\ln v_0) - d \ln \eta], \tag{3.22}$$

which is also equal to the deviation of the entropy at the close-packed limit from the entropy given by the self-consistent free volume (SCFV). The free volume v_f and the equation of state (Wood, 1952) for d-dimensional hard spheres are given in the SCFV approximation by

$$v_f = v_0 \eta^d, \tag{3.23}$$

$$\beta P/\varrho = (1 + \eta)/\eta. \tag{3.24}$$

Except for the multiplicative constant v_0 in Eq. (3.23), a similar result was also obtained on an intuitive basis by Hirschfelder *et al.* (1937). The SCFV theory calculates v_f without the requirement of the fixed-neighbor assumption which was made in the LJD theory. However, it relies on the assumption that the configurational probability density is represented by a product of single-particle probability densities (Kirkwood, 1950). In deriving Eqs. (3.23) and (3.24), however, an additional restriction, namely to only the nearest-neighbor interactions, is imposed.

The entropy in the expressions (3.21) and (3.22) for the "entropy constants" s_0 and $\bar{s}_0$ has to be obtained near the close-packed density. The "experimental" S is evaluated by integrating the pressure from the ideal-gas limit to the close-packed limit. The pressure data required for this purpose are available for both solid (Alder *et al.*, 1968) and fluid [Eqs. (3.9) and (3.10)] branches. These, together with the melting data obtained in Section III,A, comprise the necessary information required for evaluation of S. As can be seen in Tables VI and VII, the numerical values of these constants vary widely for different approximate theories. Therefore, s_0 or $\bar{s}_0$ can be used as another good criterion to judge the relative success of a theory.

The SCFV s_0 or $\bar{s}_0$ is less than the corresponding exact value for both hard spheres and disks (see Tables VI and VII). This implies that the self-consistency requirement places too strong a restriction on available free volume near close-packing. These tables also show that the LJD s_0 or $\bar{s}_0$ underestimates the free volume in the case of hard spheres and overestimates it for hard disks near close-packing. However, the amount of error would be small. Pressure deviations $(\beta P/\varrho) - (1/\eta)$ for various theories, and the molecular-dynamic data of Alder *et al.*, are illustrated in Figs. 6 and 7. The relatively poor performances for the LJD and the SCFV theories become apparent in these figures.

In the case of the hard-disk system, the correlated cell (CC) model shows very good agreement with the molecular-dynamic data (see Fig. 7 and Table VII). In contrast to the LJD theory, which assumes fixed neighbors, the CC model introduces some correlation of a center particle with its neighbors by allowing some (not all) of its nearest neighbors to move away by an equal distance as the center particle moves toward them. Thus, the correlated motion gives an increased free volume. The CC model calculations were carried out by Squire and Salsburg (1961) for hard spheres and by Alder *et al.* (1963) for hard disks. For a two-dimensional system, for example, the CC model assumes that an entire row of particles arranged in a hexagonal close-packed configuration slides past the two neighboring rows. Particles in each row are kept at a constant

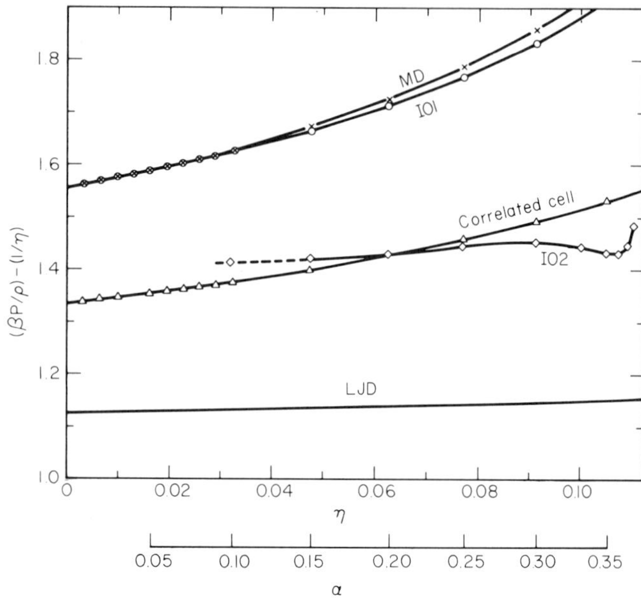

FIG. 6. Pressure differences for hard spheres ($\beta P/\varrho$ minus its limiting high-density form) plotted against the high-density expansion parameter η. MD: Molecular-dynamic data (Alder *et al.*, 1968); IO1: independent oscillator approximation 1 (Fixman, 1969); IO2: independent oscillator approximation 2 (Ree, 1970); correlated cell: correlated cell model (Alder *et al.*, 1968); LJD: Lennard-Jones–Devonshire cell model (Alder *et al.*, 1968).

distance from each other. It is noteworthy that the CC model for hard disks has a loop with a cusp in its P versus ϱ plot and this loop appears in *approximately* the same position as the observed van der Waals loop for the 780-disk system (Alder and Wainwright, 1962). The CC model for hard spheres does not give equally good agreement with the molecular-dynamic data (see Table VI and Fig. 6). The CC model isotherm for hard spheres exhibits a slight inflection point around $v/v_0 \approx 1.54$ (Squire and Salsburg, 1961). At present, any possible relation between the singularities obtained in the CC model calculations of hard spheres and disks and a fluid–solid transition is limited only to the speculative stage.

Tables VI and VII show that the cell-cluster theory gives remarkably good numerical values for the hard-sphere s_0 and c_1 (Rudd *et al.*, 1969). For hard disks, the cell-cluster c_1 is good, while its s_0 is not (Salsburg *et al.*, 1967). The cell-cluster theory (Salsburg *et al.*, 1967; Stillinger *et al.*, 1965) is a formally exact theory which uses the LJD configurational

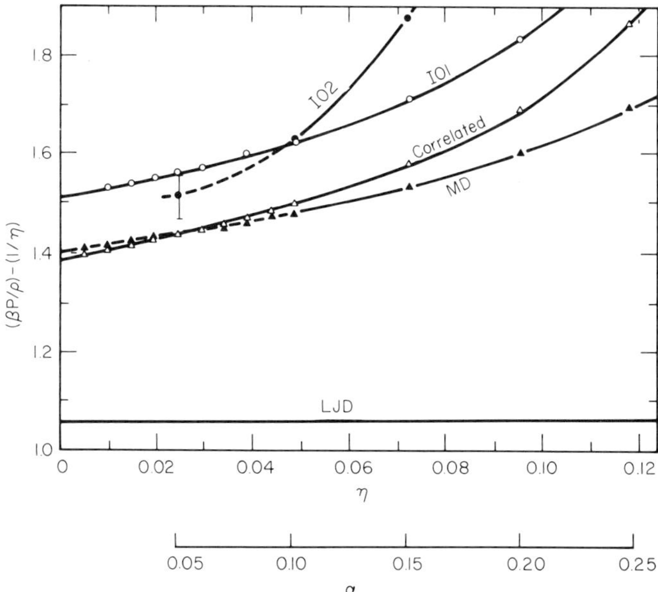

FIG. 7. Pressure differences for hard disks ($\beta P/\varrho$ minus its high-density form). See the caption to Fig. 6 for the notation used in this plot.

partition function as the zeroth-order contribution and which introduces higher-order corrections due to two- and higher-particle clusters. While the cell-cluster theory is attractive from a theoretical viewpoint because of its systematic (though not unique) formalism, its success depends crucially on the rapidity of convergence of the successive corrections, since the higher-order corrections not only involve considerations of a larger number of topologically distinct clusters of a given number of particles, but also each integral associated with these clusters becomes harder to evaluate. In this respect, good agreements observed for the cell-cluster s_0 and c_0 for hard spheres are rather encouraging and indicate that inclusion of only two- and three-particle clusters is sufficient to give accurate estimates of s_0 and c_0. Although the next-order correction will be much harder to evaluate, its evaluation will be probably worthwhile to do, since it may provide some insight into the rate of convergence of the cell-cluster theory applied to hard spheres, and may also affect s_0 by contribution of the higher-order clusters which may play an important role in deciding the relative stability of hard spheres in the hexagonal close-packed and the face-centered-cubic crystals near the close-packed density. The *small* discrepancy (0.0008) found in values of s_0 (Table VI) for the two crystals appears to favor the hexagonal close-packing over the face-centered-cubic arrangement. Considering the small magnitude of s_0, this calculation probably does not contradict an empirical argument (Alder *et al.*, 1969) that supports the opposite view. This argument is partly based on the molecular-dynamic pressure of the 216 hard spheres in the face-centered configuration, which is about 0.1% lower near the melting point than the corresponding pressure for the hexagonal close-packed spheres, showing possibly different locations of the melting tie-line for these crystals. As further evidence to support this observation, Alder *et al.* give a value approximately 2% larger for the mean-square width for the face-centered hard sphere near close-packing. The difference in c_0 for the two crystals so far has not been detected in the molecular-dynamic measurements for hard spheres in two crystal phases. In contrast to the hard-sphere case, there is a relatively large discrepancy for hard disks between the molecular-dynamic s_0 ($= 0.05 \pm 0.01$) and the cell-cluster s_0 ($= 0.11$), which includes contributions up to the five-particle clusters. It is worthwhile to see if the large discrepancy still persists in the cell-cluster calculations using other ways of sorting out cell clusters. If so, this would imply that the rate of convergence of the cell-cluster theory is slow for hard disks. Therefore, entropy calculated by use of the low-order cluster contributions alone would not give a reliable estimate and might lead to a wrong conclusion for hard disks.

The independent oscillator model (IO) (quoted in Tables VI and VII and Figs. 6 and 7) is one of the two interesting models proposed recently by Fixman (1969) in order to account for thermodynamics of the single-occupancy system of hard spheres (see Section III,A). Since Fixman's work is relatively recent, we will describe it here in some detail. Fixman replaces the Boltzman factor of a pair potential ϕ_{ij} by its Gaussian average over a position vector $\mathbf{r}_j$ of particle j measured from the center of the Wigner–Seitz cell j; i.e.,

$$\begin{aligned} \exp(-\beta\phi_{ij}/2) &\approx (t/\pi)^{d/2} \int d\mathbf{r}_j \exp[-(\beta\phi_{ij}/2) - tr_j^2] \\ &\equiv 1 + h(\mathbf{R}_j - \mathbf{r}_i), \end{aligned} \tag{3.25}$$

where $\mathbf{R}_j$ is a displacement vector of the center of the Wigner–Seitz cell j measured from the center of the cell i. The term tr_j^2 occurring in the exponent in Eq. (3.25) corresponds to the potential energy of an independent oscillator vibrating around the center of the cell j. The spring constant t is fixed by requiring that the singlet distribution function $s_{s0}(\mathbf{r})$ resulting from replacing $\exp(-\beta\phi_{ij}/2)$ in Eq. (3.15) by Eq. (3.25) has an identical mean-square displacement $\langle r^2\rangle_v$ as that of an independent oscillator with spring constant t; i.e.,

$$d/2t = \int_v d\mathbf{r}[r^2 s_{\mathrm{so}}(\mathbf{r})] \equiv \langle r^2\rangle_v, \qquad \mathbf{r}_i \equiv \mathbf{r}, \tag{3.26}$$

where the integration is carried out over the Wigner–Seitz cell i. The singlet distribution function for the independent oscillator model is obtained in a straightforward manner and is given as follows:

$$s_{\mathrm{so}}(\mathbf{r}) = \{\exp[-\beta v_{\mathrm{e}}(\mathbf{r})]\} \Big/ \int_v d\mathbf{r} \exp[-\beta v_{\mathrm{e}}(\mathbf{r})], \tag{3.27}$$

where $v_{\mathrm{e}}(\mathbf{r})$ is an "effective potential" energy which particle i feels when it is located at a distance $\mathbf{r}$ from its cell center and is given by

$$v_{\mathrm{e}}(\mathbf{r}) = \sum_j \ln[1 + h(\mathbf{R}_j - \mathbf{r})]. \tag{3.28}$$

The parameter t can be determined by devising a suitable iterative scheme for Eq. (3.26), whose right-hand side contains t implicitly through the effective potential (3.28) appearing in $s_{\mathrm{so}}(\mathbf{r})$. Using this t, the pressure can be easily determined from the relation $P/T = \partial S/\partial V$, and the entropy

$$S = Nk \ln\left[\int_v d\mathbf{r} \exp[-\beta v_{\mathrm{e}}(\mathbf{r})]\right].$$

In practice, integration in Eq. (3.26) is rather laborious to carry out because of its dependence on the orientation vector $\mathbf{r}$. Therefore, Fixman replaces $v_{\mathrm{e}}(\mathbf{r})$ by its first two nonvanishing terms $\varepsilon + \gamma r^2$ in its Taylor expansion. This step is essentially equivalent to spherically averaging out the orientation dependence in $v_{\mathrm{e}}(\mathbf{r})$. A considerable simplification which results after the truncation allows one to give analytic expressions for s_0 and c_0 (Ree, 1970) for spheres in the face-centered-cubic or hexagonal close-packed lattices, disks in the hexagonal close-packed lattice, and rods in regularly partitioned cells:

$$s_0 = n_d\{\ln\tfrac{1}{2}[1 + \mathrm{erf}(b)]\} + d\{\ln[\pi^{1/2}/(2b)]\} + \ln \varrho_0, \tag{3.29}$$

$$\begin{aligned} c_0 = {} & [-10\mathrm{b} - 24b^3 \\ & + (3 + 28b^2)(b^2 + 0.5)^{1/2}]/\{2b^2[2b - (b^2 + \tfrac{1}{2})^{1/2}]\}, \quad \text{spheres,} \end{aligned} \tag{3.30a}$$

$$\begin{aligned} = {} & [-15b - 18b^3 \\ & + 5(1 + 6b^2)(b^2 + \tfrac{2}{3})^{1/2}]/\{4b^2(3b - (b^2 + \tfrac{2}{3})^{1/2}]\}, \quad \text{disks,} \end{aligned} \tag{3.30b}$$

$$= 1, \qquad \text{rods,} \tag{3.30c}$$

where n_d is the coordination number (= 2, 6, 12, respectively, for $d = 1, 2, 3$), and b is a root of the following equation:

$$(2/\sqrt{\pi})[\exp(-b^2)]/[1 + \mathrm{erf}(b)] = -b + [b^2 + (2d/n_d)]^{1/2}. \quad (3.31)$$

The resulting roots of Eq. (3.31) for spheres, disks, and rods are

$$\begin{aligned} b &= 1.00166, && \text{for spheres,} \\ &= 0.79571, && \text{for disks,} \\ &= 0.38934, && \text{for rods.} \end{aligned}$$

Tables VI and VII give values of s_0, c_0, and c_1 calculated from the independent oscillator model with the truncation in v_e (which is hereafter called the IO1 model, to distinguish it from the independent oscillator model without the truncation, called the IO2). As Fig. 6 demonstrates, the IO1 model for hard spheres seems to reproduce the "experimental" coefficients c_0 and c_1 within the numerical accuracy of the molecular-dynamic "experiments." If the hard-sphere value of b (= 1.00166) is replaced by unity, the corresponding expression for c_0 takes a simpler form, $-4.3 + 5.6\sqrt{1.5} = 2.5587$, which again agrees with the experimental value. Complete calculations (IO2) without the truncation procedure (Ree, 1970), however, show significantly different high-density behavior for the pressure and entropy [see the IO2 model in Tables VI and VII, and Figs. 6 and 7]. The difference between the IO1 and the IO2 models becomes large near the two-phase region, as shown in Fig. 8. In the case of the IO2 model, there exist two separate branches in A and P. The branch that extends from the ideal-gas end to approximately 75% of the close-packed density can be called a "fluidlike" branch, which would be unstable without the cell restriction in the single-occupancy system. In contrast to the fluidlike branch, the isotherm of the second branch ("solidlike" branch), which appears at approximately $0.73\varrho_0$ and ends at the close-packed density, results from the first-neighbor interaction alone. That is, particles do not "sense" the cell wall. Within a narrow density interval ($0.73\varrho_0 - 0.75\varrho_0$), P and A are multivalued, very much resembling a similar metastable region observed in the molecular-dynamic or the Monte Carlo calculation for a finite hard-sphere or disk system (Alder and Wainwright, 1957; Wood and Jacobson, 1957). The instability may be due to a mathematical artifact attributable to the expansion scheme based on Hermite polynomials, and therefore no conclusion of physical significance can be attached. Excluding this

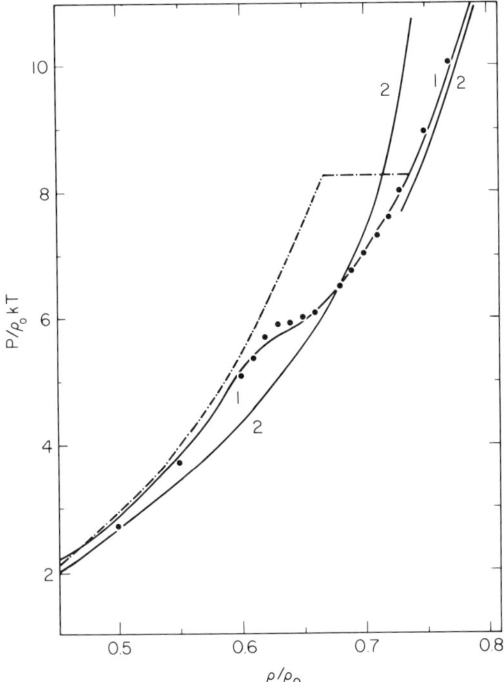

FIG. 8. The single-occupancy pressure for hard spheres plotted against density. The large dots are the Monte Carlo values. The numbers 1 and 2, respectively, indicate that the pressure is evaluated under the independent oscillator approximation 1 and the independent oscillator approximation 2. The line (—·—) shows the pressure of the hard-sphere fluid as well as that of the coexisting phase without the cell restriction (Ree, 1970).

possibility, however, occurrence of the instability is noteworthy in relation to two "experimental" facts. First, the melting density of hard spheres is $0.736\varrho_0$, which lies close to the observed end point $(0.73\varrho_0)$ of the solid-like branch. This suggests a possible relationship between the two. Second, the end point of the solidlike branch may be mimicking the "experimentally" observed "cusp" at $0.625\varrho_0$ in the hard-sphere equation of state (Hoover and Ree, 1968).

In contrast to spheres, the isotherm for disks does not show two distinct branches; instead, the hard-disk system shows a continuous van der Waals-like transition from the fluidlike to the solidlike phase around $0.82\varrho_0$. This again is close to the melting density of hard disks, $0.798\varrho_0$. For rods, no such transition takes place in the IO2 model. The main reason for the different thermodynamic behavior between the IO1 and

IO2 models lies in a term in the expression for $v_e(\mathbf{r})$ which changes its sign depending on whether the position vector $\mathbf{r}$ is inside or outside of the exclusion spheres of its neighbors. The IO2 model takes this into account, while the IO1 model does not. Since the IO1 model gives remarkably good solid equation-of-state data which no other cell theory has so far been able to match, there may exist some still unaccounted reason which favors the truncation procedure. Fixman's second (coupled oscillator) model corrects the effective potential (3.28) of the IO models by introducing correlation of neighboring particles. The coupled oscillator model, when used in conjunction with data on $\langle r^2 \rangle_v$ from the IO1, generally gives an improved value for entropy; in particular, s_0 is reduced from the IO1 value (-1.00) to -0.44. However, the pressure calculation seems to favor the IO1 model over the coupled oscillator model around intermediate and high densities ($\varrho \gtrsim 0.6\varrho_0$).

From the above considerations, it is clear that at present there is no approximate theory suitable for simultaneously describing high-density thermodynamic behavior of *both* hard spheres and disks. A theory which is found to be suitable for hard spheres, for example, turns out to give a less satisfactory answer if it is applied to the hard-disk system. This lack of self-consistency in each theory described here serves as a further challenge to those interested in working out a new theory or refining those currently available.

D. Liquid–Gas Transition and Fluid Equation of State

In contrast to a fluid–solid transition, which was found in Section III,A to originate primarily from a geometric problem of packing particles in a given space, an attractive-particle interaction is required in bringing about a liquid–gas transition. The potentials used most frequently for the computer "experiments" and theoretical work relevant to the present problem are the square-well potential $\phi_{\text{SW}}(r)$ and the Lennard-Jones potential $\phi_{\text{LJ}}(r)$. These potentials can be represented as follows:

$$\begin{aligned} \phi_{\text{SW}}(r) &= \infty, && \text{if} \quad r < 1, \\ &= -\varepsilon, && \text{if} \quad 1 \leq r \leq \lambda, \\ &= 0, && \text{if} \quad \lambda < r, \end{aligned} \tag{3.32}$$

$$\phi_{\text{LJ}}(r) = 4\varepsilon(r^{-12} - r^{-6}), \tag{3.33}$$

where the distance r (in the case of the square-well particle) is reduced by using the hard-core diameter σ of the square-well particle as a unit of distance. For the Lennard-Jones particle, r is expressed in units of σ_{LJ} at which $\phi_{LJ}(r)$ vanishes identically. The parameter ε in Eqs. (3.32) and (3.33) measures the depth of the square well or the Lennard-Jones potentials. The dimensionless temperature T_r $(= kT/\varepsilon)$ will be used in the discussion given below. We shall first review work on the square-well particles and then discuss the Lennard-Jones particles. Discussions concerning the liquid–gas transition of the latter will be centered around Hansen and Verlet's recent work (1969), which fixed the coexistence curve of the Lennard-Jones particles.

The square-well particles at subcritical temperatures were studied by Rotenberg (1965a) for the three-dimensional system of mostly 256 particles at three different temperatures ($T_r = 0.5$, 1, and 3.333). In addition to this paper, there is Alder's unpublished work, a part of which appears in a plot of a figure in Barker and Henderson's paper (1967a). These authors take the attractive-range parameter λ in Eq. (3.32) equal to 1.5. This choice of λ makes the range of the square well enclose center of repulsive core of the second particle when its core is in contact with the core of the first. In this way, the maximum number of particles whose centers are within attractive range of a central particle is limited to, in case of three dimensions, the twelve first-neighbors plus the six second-neighbors.

The highest isotherm ($T_r = 3.333$) obtained by Rotenberg has a qualitatively similar look as the corresponding hard-sphere isotherm ($T_r = \infty$), exhibiting the disconnected fluid and solid branches occurring in a density region which is approximately equal to the corresponding metastable region for hard spheres. The other two of Rotenberg's isotherms ($T_r = 0.5$ and 1), however, have a distinctly different look. First, each of these isotherms has *two* van der Waals-like loops. The first loop shows a broad, shallow minimum with negative pressure and occurs at lower densities than the second. In the case of the $T_r = 0.5$ isotherm, for example, the negative pressures associated with this minimum stretch roughly from $0.59\varrho_0$ to $0.2\varrho_0$. The phase transition tie-line for the $T_r = 0.5$ isotherm will, therefore, be fairly broad and similar in character to a liquid–gas coexistence line of a real gas (such as argon) and will be identified as such. As temperature is increased, this loop becomes narrower and eventually disappears at the critical temperature. The location of the critical point for square-well particles is now accurately established (Alder and Hecht, 1969); i.e., $T_r = 1.26 \pm 0.01$, $\varrho_c/\varrho_0 = 0.241 \pm 0.01$, $P_c/kT_c\varrho_0 = 0.28 \pm 0.01$. The preliminary analysis of the square-well particles at subcritical temperatures is briefly reported in Alder and Hoover's review article (1968), in which they note that the square-well computer system "most probably" obeys the van der Waals equation very accurately. The resulting coexistence curve has a parabolic shape, unlike the cubic-shaped coexistence curve of a real fluid (Fisher, 1966). The most likely source causing this discrepancy is the finiteness of the molecular-dynamic system, which suppresses otherwise possible long-range density fluctuations. The closer the system gets to the critical point, the slower is the density–density correlation (for example, it decays exponentially at large distances). The long-range character of various correlation functions is responsible for such anomalous critical behavior as a cubic shape of the coexistence curve. To properly take the density fluctuations into account in the computer "experiments" would probably require a system which would be too massive to be computationally practicable, even considering the present technology of high-speed computers. Nevertheless, further computer study along this line using larger particle number ($<10^5$), preferably in two dimensions (so as to reduce relative contribution from the boundary effect), is within the realm of possibility and would be helpful for further understanding of the observed discrepancy. This lack of correspondence of a real system with its computer analog is a little disappointing, since, in the case of the two-dimensional Ising model, only four to six spins making up a row in a two-dimensional network are enough to predict the logarithmic divergence of the constant-volume specific heat (Kramers and Wannier, 1941).

The second loop, which occurs around $\varrho_0/\varrho \approx 1.18$, is puzzling in several ways. First, it occurs at densities well within the melting density for hard spheres ($\varrho/\varrho_0 \simeq 0.736$). Second, presence of the loop for the $T_r = 0.5$ and $T_r = 1$ isotherms, and its absence for the $T_r = 3.333$ isotherm, indicate that there is also a critical point (unlike the solid–fluid isotherm, which possibly does not have one) at some temperature between $T_r = 1$ and $T_r = 3.333$. Furthermore, the shape of this loop in the isotherm is quantitatively affected by the number of particles used in the "experiments." This sensitivity of particle number is demonstrated by Lado and Wood (1968). As the number of square-well particles is increased to 256, 500, and 864, its $\beta P/\varrho_0$ at $T_r = 0.5$ and $\varrho_0/\varrho = 1.185$ increased from an initially negative value (-2.82 ± 0.04) to -0.46 ± 0.04, and finally to a positive value (1.10 ± 0.02). By linear extrapolation, $\beta P/\varrho_0$ is found to be 3.24 for an infinite system. Dependence of the pressure on particle number can be examined further from the following formula for the pressure, which results directly from the virial theorem (2.17) applied to square-well particles:

$$\beta P/\varrho = 1 + B_2\varrho\{g(1) - g(\lambda^-)[1 - \exp(-1/T_r)]\lambda^3\}, \qquad (3.34)$$

This formula shows that the pressure depends on the value $g(1)$ at the contact distance of two cores as well as the value $g(\lambda^-)$ just inside $(r = \lambda^-)$ the square well. At a given density and temperature, $g(1)$ increases while $g(\lambda^-)$ decreases, as particle number (thus, V) is increased. From Eq. (3.34), this then eventually brings the pressure to a positive number. At present, the nature and thermodynamic significance (relevant to the infinite system) of this loop remains unclear. However, it should be noted that the range ($\lambda = 1.5$) of a square-well particle chosen for Alder's and Rotenberg's "experiments" is larger than $\sqrt{2}$ and hence is just enough to enclose its second-nearest neighbors at high solid densities. The observed loop at the high-density isotherm may be intimately related to the choice of the parameter which allows the second-neighbor enclosure (Wood, 1968a).

We now turn our attention to the Lennard-Jones particles. Thermodynamic behavior at temperatures above the critical point for a system of particles interacting with the Lennard-Jones potential has been examined using the Monte Carlo method by Wood and Parker (1957) and Fickett and Wood (1960). Further studies of the Lennard-Jones system have been provided by the work of Singer (1966) and McDonald and Singer (1967a,b, 1969). In addition to the above Monte Carlo work, the molecular-dynamic calculations of thermodynamic properties of the Lennard-Jones particles are also carried out by Verlet (1967). In his review on the Monte Carlo method, Wood (1968a) compiled data on pressure and internal energy together with rather detailed analyses of the Monte Carlo calculations of the Lennard-Jones system based mostly on work by him and his colleagues above the critical temperature ($T = 150.9$°K). Accordingly, the interested reader should refer to Wood's review article for the supercritical equilibrium properties. We remark here that data from both the Monte Carlo (Wood and Parker, 1957) and the molecular-dynamic (Verlet, 1967) calculations at one isotherm ($T_r = 2.74$) are in substantial agreement with one another. Moreover, the Lennard-Jones system describes remarkably well the equilibrium properties of argon at supercritical temperatures, except in the neighborhood of the critical point, where suppression in the computer "experiments" of long-range density fluctuations found in a real system such as argon quite likely brings in the observed discrepancy between the two.

In the discussions here, we shall concentrate our attention on equilibrium properties of the Lennard-Jones particles below the critical temperature, focusing mostly on the liquid–gas transition line. Using $\varepsilon/k = 119.76°\text{K}$, $v^* \equiv N\sigma_{\text{LJ}}^3 = 23.79 \text{ cm}^3 \text{ mole}^{-1}$ for the parameters in the Lennard-Jones potential (3.33), Wood's (1968a) 32-particle Monte Carlo isotherm at $T = 126.7°\text{K}$ ($T_r = 1.0579$) shows a van der Waals loop. The loop has a deep negative pressure at the minimum, located at roughly 60 cm³ mole⁻¹, followed by a broad maximum pressure (≈30 atm) centered around approximately 190 cm³ mole⁻¹ (see Fig. 9).

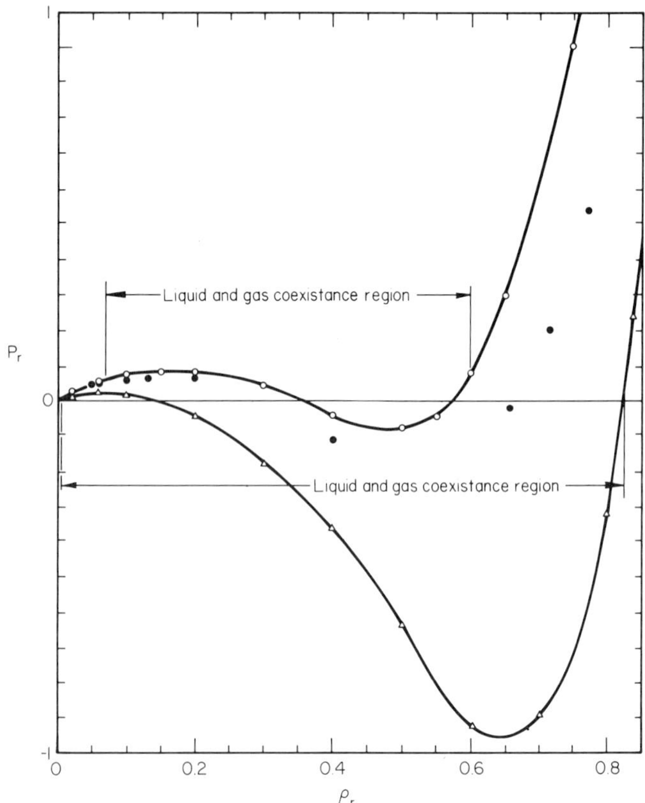

FIG. 9. Equations of state of the homogenized Lennard-Jones fluid expressed in reduced units, $\sigma_{\text{LJ}} = \varepsilon/k = 1$, at the reduced temperatures $T_r = 0.75$ (lower curve) and $T_r = 1.15$ (upper curve) obtained by Hansen and Verlet (1969). The liquid–gas coexistence region for each isotherm is also indicated. The dots indicate the isotherm ($T_r = 1.0579$) obtained by Wood (1968a) for the 32 Lennard-Jones particles without the homogenizing restriction, Eq. (3.35).

Any reliable estimate of the corresponding vapor pressure by a Maxwell equal-area construction requires accurate data on the pressure over the density range where the loop occurs. Woods (1968a) points out that the pressure near the liquid end of the tie-line results after extensive cancellation of the positive value (1) and the negative contribution coming from the second term in the right-hand side of Eq. (2.14b). Therefore, the resulting pressure is small and uncertain within a standard deviation, which is approximately equal in magnitude to the pressure itself. Considering this, therefore, the corresponding vapor-pressure estimates cannot be reliably made. The main cause of the large pressure fluctuations within the coexistence range is due to a tendency of the system to separate into regions of different densities. A similar situation can be seen in a "snapshot" picture of Alder and Wainwright's (1962) 870-hard-disk system in the case of a solid–fluid transition.

One can avoid the liquid–gas coexistence region by simply heating at constant volume (say, at a gas density) to a supercritical temperature, then compressing isothermally or isobarically to a liquid density range, followed by an isochoric cooling to the original temperature. This "experiment," when performed on a computer, involves several steps and consequently has not yet been attempted. A simpler method to remove the difficulty due to the observed large fluctuations has been provided by Hansen and Verlet (1969), who recently proposed a method using a multiple-occupancy system, somewhat analogous to the single-occupancy system treated in Section III,A, but whose cells can accommodate not one but several particles. Verified for feasibility by using the Lennard-Jones particles, Hansen and Verlet's method determines vapor pressure and liquid and gas densities at the coexistence region, as well as the entropy constant in the liquid phase referred to the ideal gas. Their method can be useful at temperatures not very close to either the critical point or the triple point. Removal of the large fluctuation is done as follows: (a) Consider a volume V which has been partitioned into M identical cells having volume v_M ($\equiv V/M$); (b) both upper and lower limits of the permissible number n_M of particles in each cell are fixed; that is,

$$\bar{n}_M - \Delta < n_M < \bar{n}_M + \Delta, \tag{3.35}$$

where $\bar{n}_M$ is the average number (N/M) of particles in each cell. Both of the parameters M and Δ can change as temperature changes, but they do not depend on density. The condition (3.35) inhibits fluctuation in the number of particles beyond a certain preset number Δ from the average number in each cell. At a low gas density, the cell volume v_M becomes large; therefore, the Verlet and Hansen "homogenized" fluid excludes by the restriction (3.35) many configurations which would otherwise have been permissible. However, with a suitable choice of M and Δ, the forbidden configurations can be made to occur relatively infrequently in the Monte Carlo calculations; therefore, their contributions can be ignored in calculating the pressure. Since the value of Δ is also chosen to be large enough, almost no configuration at liquid densities can violate the condition [Eq. (3.35)]. A small number of configurations violating the restrictions makes no statistically significant difference in evaluating the pressure. The only difference in the homogenized and ordinary fluids occurs in the liquid–gas transition range, in which the isotherm for the homogenized fluid is smooth because of the restriction [Eq. (3.35)] which prevents the system from separating into two phases. An optimum value for Δ can be estimated by choosing its value sufficiently larger than the standard deviation of n_M; i.e.,

$$\Delta > \langle (n_M - \bar{n}_M)^2 \rangle^{1/2} = [2\bar{n}_M \, \partial\varrho/\partial(\beta P)]^{1/2}, \tag{3.36}$$

where the angular brackets indicate the configurational average of the quantity they enclose. In practice, the reduced isothermal compressibility $\partial\varrho/\partial(\beta P)$ in Eq. (3.36) is replaced by its value at the lowest liquid density for any given isotherm. Since $\partial\varrho/\partial(\beta P)$ in the liquid region becomes largest at the lowest liquid density, a homogenized isotherm corresponding to a suitable choice of the parameters Δ and $\bar{n}_M$ satisfying Eq. (3.36) will not show any noticeable change from the corresponding liquid without any restriction. As can be seen from Eq. (3.36), the optimum choice of the parameters becomes increasingly difficult to obtain near the critical point ($T = 150.9$°K, or $T_{\rm rc} = 1.26$ for argon), where $\partial\varrho/\partial(\beta P)$ becomes very large. Therefore, from Eq. (3.36), Δ has to become so large that no advantage can be gained from Hansen and Verlet's procedure. The difficulty of choosing suitable parameters also occurs near the triple point ($T = 83.8$°K, or $T_{\rm r} = 0.7$ for argon), where M has to be chosen large to reduce the pressure fluctuations, but too large a value of M is accompanied by a value for Δ too small to describe the isotherm properly at gas densities. In their study of 864 Lennard-Jones particles with the interaction parameters $\sigma_{\rm LJ} = 3.405$ Å and $\varepsilon/k = 119.8$°K, Hansen and Verlet (1969) use the following values for Δ and M: $T_{\rm r} = 1.15$, $M = 27$, $\Delta = 12$, $\bar{n}_M = 32$; and $T_{\rm r} = 0.75$, $M = 64$, $\Delta = 2.5$, $\bar{n}_M = 13.5$. The condition (3.36), when applied to the above two cases, becomes: $T_r = 1.15$, $12 > [2 \times 32 \times 0.41]^{1/2} = 5.1$; and $T_r = 0.75$, $2.5 > [2 \times 13.5 \times 0.081]^{1/2} = 1.5$.

The decrease in compressibility at $T_r = 0.75$, which lies chose to the triple-point temperature of argon, does not make the inequality (3.36) any stronger since it is offset by an increase in cell number (therefore, small Δ) to prevent density fluctuations from taking place in the two-phase region. The resulting "homogenizing" condition ($2.5 > 1.5$) is so weak that, according to Hansen and Verlet, phase separation can still take place even with the cell constraint. This occurs at a relatively rapid rate compared with the total steps required in a Monte Carlo average; some cells are quickly filled up to maximum particle number ($n_M + \Delta$), and at the same time, particles in some other cells are depleted to the minimum permitted value ($n_M - \Delta$). However, since this occurs within the two-phase region and with statistically tolerable fluctuations in pressure, it does not present a serious difficulty, at least up to $T_r \gtrsim 0.75$.

A short description of the Monte Carlo program suitable for the present purpose is in order. This is different from the earlier one for the single-occupancy system only by the restriction (3.35). Thus, for a given set of ϱ_r, T_r, M, and Δ, a particle is chosen randomly out of N particles and is given an arbitrary displacement within a cubic box of side length 2δ whose center matches with the center of the particle. If the new position of the particle takes it away from the given cell, the program checks whether the resulting deletion of the particle from the original cell violates the lower bound in Eq. (3.35) on n_M. If so, no such move is allowed. Next, the program checks to see whether or not the new cell into which the particle is to migrate can afford to accomodate the particle according to the upper bound on n_M, Eq. (3.35). If it cannot, the program does not allow the move. If neither condition is violated, then the remaining Monte Carlo steps (and evaluation of the pressure) are similar to the earlier program discussed in Section III,A.

The Helmholtz free energy of the "homogenized" fluid follows from Eq. (3.1) using the Monte Carlo data on pressures. The data on the liquid–gas tie-line can then be obtained from the free-energy data by a Maxwell double-tangent construction. In Fig. 9, Hansen and Verlet's data on the two isotherms ($T_r = 0.75$ and $T_r = 1.15$) for the Lennard-Jones "homogenized" fluid are plotted, where P_r is the reduced pressure, i.e., $P\sigma_{LJ}^3/\varepsilon$. The figure exhibits characteristic van der Waals loops. For comparison purposes, Wood's (1968a) isotherm at $T_r = 1.0517$ for 32 Lennard-Jones particles is shown by dots in the same plot. From the free-energy data, obtained by numerical integration of the equations-of-state data, Hansen and Verlet give the following values for the liquid and gas densities at the coexistence tie-line, and the vapor pressure: For $T_r = 0.75$, $\varrho_r(\text{gas}) = 0.0035$ (0.0047), $\varrho_r(\text{liquid}) = 0.825$ (0.818), $P_r = 0.0025$ (0.0031); while for $T_r = 1.15$, $\varrho_r(\text{gas}) = 0.073$ (0.093), $\varrho_r(\text{liquid}) = 0.606$ (0.579), $P_r = 0.0597$ (0.0664); where values in parentheses are the experimental data on argon of Michels *et al.* (1958) (at $T_r = 1.15$) and the argon data quoted in Hansen and Verlet's paper. Possible errors in the Monte Carlo "experimental" values of the coexistence densities are not given by these workers. Assuming then that the Monte Carlo values may be inaccurate only in the last digits of the quoted numbers, we note that agreement between the Monte Carlo and argon data on $\varrho_r(\text{liquid})$ is very good. The somewhat larger percentage deviations from the experimental data found in the case of the vapor pressure and $\varrho_r(\text{gas})$ can be regarded as satisfactory in view of their rather small values plus our lack of knowledge of an accurate pair potential.

The coexistence curve of the Lennard-Jones particles obtained by Hansen and Verlet is shown in Fig. 10; estimated critical data based on

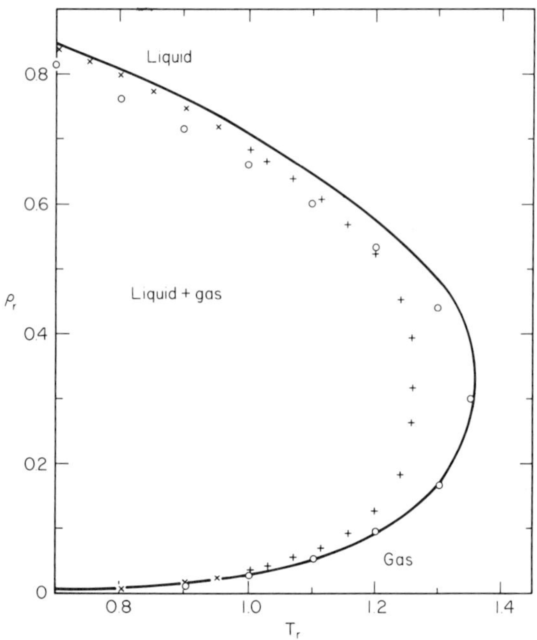

FIG. 10. The coexistence curve for the Lennard-Jones particles obtained by Hansen and Verlet (1969). Coexistence densities obtained from Barker and Henderson's perturbation theory (1967a) are indicated by ○, while the experimental points of Michels *et al.* (1958) and Gosman *et al.* (1969) are indicated by + and ×, respectively.

the higher-order Percus–Yevick (PY2) theory (Verlet and Levesque, 1967) for the Lennard-Jones system was used in plotting this curve. Experimental data on argon due to Michels *et al.* (1958) and the argon data compiled by Gosman *et al.* (1969) are indicated by $+$ and $\times$, respectively. The flatter appearence of the experimental coexistence curve near the reduced critical temperature (which is about 0.1 lower than the coexistence curve of Hansen and Verlet) is most probably caused by large density fluctuations (Levesque and Verlet, 1969). As has been mentioned previously, this effect has not been taken into consideration by Verlet and Hansen's finite-system Monte Carlo calculations. In the same figure, the coexistence curve from Barker and Henderson's (1967a) perturbation theory is shown by open circles. The gas side of the coexistence curve obtained from the perturbation theory agrees especially well with the Monte Carlo data all the way up to the critical density. In addition, Barker and Henderson's critical-point data $(P_r, T_r, \varrho_r) = (0.14, 1.35, 0.30)$ agree well with the estimated values for the corresponding numbers (0.13-0.17, 1.32-1.36, 0.32-0.36) from Verlet and Levesque's estimates. Small deviations from the Monte Carlo results, amounting to approximately 0.03–0.04 in liquid density of the coexistence curve at low temperatures, could be corrected if the next-order correction is included in the perturbation theory. One should note here that the perturbation approach also fails to account for the occurrence of the long-range correlation near the critical point. Finally, we note that, in contrast to several formulas available to describe the solid–fluid transition discussed in Section III,A, there is no convenient formula describing the liquid–gas transition.

E. Radial Distribution Function—Theories and "Experiments"

We are concerned here with determination of the radial distribution function $g(r)$ by a computer "experiment." We use the "experimental" $g(r)$ to test the adequacy of several approximate expressions for $g(r)$. Particular emphasis is placed on numerical calculations using an improved Born–Green–Yvon (BGY) equation (Yvon, 1935; Born and Green, 1946). This equation, which hereafter will be called the BGY2 equation and which has been analyzed previously for hard spheres (Lee *et al.*, 1968; Lee, 1970; Ree *et al.*, 1971), should serve as a suitable example for the second class of problems (as classified in Section I) which can be handled by a today's high-speed computer.

Before proceeding further, we briefly remark on several recent integral equations which presumably improve the earlier ones. Further details on the radial distribution function are given elsewhere in this book (Baxter, Chapter 4, this volume).

An improved Percus–Yevick (PY) equation is proposed by Verlet (1966). This equation (hereafter, termed the PY2 equation for convenience) was extensively studied numerically by Levesque (1966). Similar improvement for the convolution-hyper-netted-chain (CHNC) equation, termed the CHNC2 equation, can equally be constructed (Rushbrooke, 1968). In addition to these, there is an alternative second-order equation of the PY theory which was proposed recently by Wertheim (1967). On the one point, Wertheim's equation, which hereafter will be called the PYII equation, appears to be superior to the PY2 equation due to (a) the symmetric appearance of particle coordinates in the equation, and (b) a superior value for the fifth virial coefficient for hard-core particles; this equation, however, seems to indicate slightly inferior behavior at low temperatures (Rowlinson, 1968) for particles with the Lennard-Jones-like potential.

All of the above second-order equations (BGY2, PY2, PYII, CHNC2) result from truncating in one way or another the higher-order correlation functions appearing in the second stage of a formally exact infinite hierarchy of integral equations. The success or failure of the distribution-function method in describing thermodynamic properties depends crucially on the assumption that a satisfactory approximation truncating the next-order distribution function can be devised at an early stage of the infinite hierarchy. This is so because the effort involved in solving more refined theories such as the BGY3, PY3, PYIII, and CHNC3 in a nontrivial case appears too complicated to treat even by present-day computers. Notwithstanding this, a successful theory of the radial distribution function, if found in the future, would give valuable information on its asymptotic behavior at large distances. This information, which is relevant to the study of the critical point, is not obtainable by means of the presently available Monte Carlo or molecular-dynamic methods employing a finite number of particles.

The BGY hierarchy can be expressed as an infinite set of the following intergodifferential equations [see, for example, Hill (1956)]:

$$-\boldsymbol{\nabla}_1 g_{12} = g_{12}\boldsymbol{\nabla}_1 u_{12} + \varrho \int d\mathbf{r}_3[(\boldsymbol{\nabla}_1 u_{13})g_{123}], \tag{3.37a}$$

$$-\boldsymbol{\nabla}_1 g_{123} = g_{123}\boldsymbol{\nabla}_1(u_{12} + u_{13}) + \varrho \int d\mathbf{r}_4[(\boldsymbol{\nabla}_1 u_{14})g_{1234}], \tag{3.37b}$$

$$-\boldsymbol{\nabla}_1 g_{1234} = \text{equation involving } g_{12345}, \tag{3.37c}$$

where ∇_1 is the gradient operator acting on particle 1, and u_{12} is the reduced interparticle potential energy $[\phi_{12}/(kT)]$ between particles 1 and 2. For a fluid, the pair (g_{12}) and triplet (g_{123}) distribution functions can be represented as functions of only interparticle distances $r_{ij} \equiv | \mathbf{r}_i - \mathbf{r}_j |$. Thus, the abbreviated notation g_{12}, g_{123}, and g_{1234} is used in the above equations to denote $g(r_{12})$, $g(r_{12}, r_{13}, r_{23})$, and $g(r_{12}, r_{13}, r_{14}, r_{23}, r_{24}, r_{34})$, respectively.

The most widely known truncation in the BGY hierarchy is Kirkwood's (1935) superposition approximation, which replaces g_{123} in Eq. (3.37a) by a triplet product of the pair distribution functions:

$$g_{123} = g_{12} g_{13} g_{23}. \tag{3.38}$$

A less known, though apparently promising, approximation for g_{123} is

$$g_{123} = g_{ij} g_{jk}, \qquad (r_{ij}, r_{jk}) \leq r_{ik}, \qquad (i, j, k) = (1, 2, \text{ or } 3). \tag{3.39}$$

This simpler approximation becomes exact for the one-dimensional system of hard rods (Salsburg *et al.*, 1953), and use of this approximation in Eq. (3.37a) results in a two-dimensional hard-disk $g(r)$ which is comparable in accuracy to one obtained from the PY equation (Chae *et al.*, 1969). When functions such as these are employed in Eq. (3.37a), the resulting $g(r)$ for a dense fluid behaves poorly, although the low-density behavior of $g(r)$ is good. A logical question is: Will inclusion of the second equation (3.37b) in the hierarchy, with a suitable truncation for g_{1234}, lead to a promising result? An approximation for g_{1234} suitable to answer this question was proposed by Fisher and Kopeliovich (1960). It has the following form:

$$\begin{aligned} g_{1234} &= g_{123} g_{124} g_{234} g_{314} / g_{12} g_{13} g_{14} g_{23} g_{24} g_{34}, \\ &= \chi_{1,23} \chi_{1,24} \chi_{2,34} \chi_{3,14} g_{12} g_{13}, \end{aligned} \tag{3.40}$$

where

$$\chi_{i,jk} = g_{ijk} / g_{ij} g_{ik}. \tag{3.41}$$

Any approximation on a distribution function $g_{123\ldots n}$ [such as Eqs. (3.38), (3.39), and (3.40)] should meet a symmetry condition and an asymptotic condition. The symmetry condition requires invariance of $g_{123\ldots n}$ under the exchange of particle positions; for example,

$$g_{123\ldots n} = g_{213\ldots n} = g_{n32\ldots 1}. \tag{3.42}$$

The asymptotic condition requires the distribution function to be expressed as a product of lower-order distribution functions, if a batch of particles are separated sufficiently far away from batches containing the other particles. For example,

$$g_{123\ldots n} = g_{123\ldots i} g_{i+1\ldots j} g_{j+1\ldots n}, \tag{3.43}$$

$$g_{12} = 1 \qquad \text{if} \quad r_{12} \to \infty. \tag{3.44}$$

Stell (1964) proposed a generalized superposition approximation for $g_{123\ldots n}$ which can satisfy the symmetry and the asymptotic conditions and which reduces to the results of Kirkwood's and of Fisher and Kopeliovich's for $n = 3$ and 4, respectively.

In the following, we truncate g_{1234} by the approximation (3.40) and use it in Eq. (3.37b). In numerically integrating the resulting coupled integrodifferential equations (3.37a) and (3.37b), it is computationally less burdensome if the integrals in Eqs. (3.37a) and (3.37b) are transformed slightly so that they will vanish for large particle separations. If an integral gives a small number, though not necessarily so for its integrand, cancellations of positive and negative factors in the integral should occur during the integration process. This must be taken into consideration in the present problem. To avoid this complication, we introduce the following identity:

$$\int d\mathbf{r}_3[(\nabla_1 u_{13}) g_{13}] = 0. \tag{3.45}$$

Adding Eq. (3.45) to Eqs. (3.37a) and (3.37b), and expressing Eq. (3.37b) by means of $\chi_{i,jk}$, we obtain

$$\nabla_1 \ln[g_{12}\exp(u_{12})] = -\varrho \int d\mathbf{r}_3\{(\nabla_1 u_{13}) g_{13}[\chi_{1,23} - 1]\}, \tag{3.46}$$

$$\begin{aligned} &\nabla_1 \ln[\chi_{2,13} g_{12} \exp(u_{12} + u_{13})] \\ &\qquad = -\varrho \int d\mathbf{r}_4\{(\nabla_1 u_{14}) g_{14}[(\chi_{1,34}\chi_{3,24}\chi_{1,24}/g_{24}) - 1]\}, \end{aligned} \tag{3.47}$$

where we used Eq. (3.40) together with the following relationships, which follow directly from Eq. (3.41) and the symmetry condition (3.42) for $n = 3$:

$$\chi_{i,jk} = \chi_{i,kj}, \tag{3.48}$$

$$\chi_{i,jk}/g_{jk} = \chi_{k,ij}/g_{ij}. \tag{3.49}$$

Notice that, if particle 1 is relatively far away from particles 2 and 3, g_{24} and the $\chi_{i,jk}$ in the integrand of Eq. (3.47) should approach unity, since particle 4 has to stay relatively close to particle 1 in order that $\nabla_1 u_{14}$ be nonnegligible. This situation, however, makes the integrand vanish, which is the desired result.

The set of Eqs. (3.46) and (3.47) has to be further reduced to a form convenient for practical computation. The first simplification comes from the symmetry properties (3.48) and (3.49) of $\chi_{i,jk}$ under particle exchanges. By using these properties, for example, it is necessary to know values of $\chi_{1,23}$ only within the region $r_{12} \geq r_{13} \geq r_{23}$. In a real computation, this region is broken up into a network of finite mesh-points with discrete interval Δr, and each $\chi_{i,jk}$ corresponding to (r_{ij}, r_{ik}, r_{jk}) is stored in a location in the computer memory assigned for the configuration (r_{ij}, r_{ik}, r_{jk}). Secondly, the integration in Eq. (3.47) is found to be best performed in the usual spherical coordinates, since its Jacobian of transformation stays finite for any position of particle 4, while for most of the other choices it does not. If this coordinate system is adopted, however, $\chi_{i,jk}$ in the integrand of Eq. (3.47) does not usually fall right on the configurations stored in the computer memory, and, therefore, has to be interpolated by means of data on neighboring points of $\chi_{i,jk}$. The third consideration which requires our attention is a suitable integration path for the gradient ∇_1 in Eq. (3.47). The best choice is a straight-line path which carries particle 1 at its present position to infinity at a fixed angle between r_{12} and r_{23} (see Fig. 11). This choice of path makes the constant of integration vanish for Eq. (3.47), due to the asymptotic conditions imposed on g_{12} and $\chi_{1,23}$.

For the hard-sphere potential, our numerical calculations have an additional advantage in that the derivative of the corresponding Mayer function f_{ij} has a delta-function property; i.e., choosing the diameter $\sigma = 1$,

$$f_{ij} = \exp(-u_{ij}) - 1, \tag{3.50}$$

$$\partial f_{ij}/\partial r_{ij} = \delta(r_{ij} - 1). \tag{3.51}$$

This is indeed a very helpful property. That is, use of Eq. (3.51) reduces the number of integrations required in Eqs. (3.46) and (3.47) by one. Furthermore, each triplet configuration for $\chi_{i,jk}$ in Eqs. (3.46) and (3.47) has a geometry with one pair of particles, say particles 2 and 3, at contact. Therefore, instead of a three-dimensional network for storing $\chi_{1,23}$, only a two-dimensional network of r_{12} and r_{13} is required. A further improvement of the computer program results after analytically integrating part of the region in which the $\chi_{i,jk}$ in the integrands of Eqs. (3.46) and

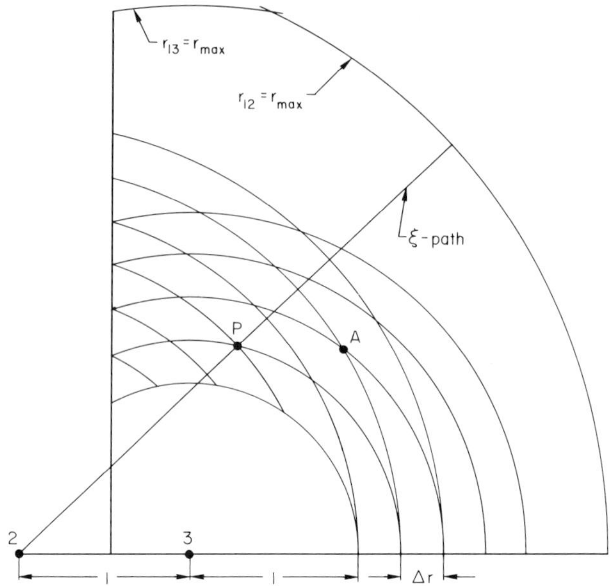

FIG. 11. Geometric region in which particle 1 is located ($r_{23} = 1 \leq r_{13} \leq r_{12} \leq r_{max}$). The right-hand side of Eq. (3.53) is to be evaluated at every intersection (such as A or P) of the two concentric circles whose origins lie on particles 2 and 3. The straight line (ξ-path) shows the integration path for the left-hand side of Eq. (3.53).

(3.47) vanish by the restriction of a hard core ($r_{jk} < 1$). Applying the procedures outlined thus far to Eqs. (3.46) and (3.47), we obtain the following expressions:

$$\ln[g_{12}\exp(u_{12})] \equiv G_{12} = (\lambda/4\pi) \underset{1\quad 2}{\overset{3}{\wedge}}$$

$$-(\lambda/2)\int_{r_{12}}^{r_{max}} dr_{12}r_{12}^{-1}\int_{\max(1,r_{12}-1)}^{\min(r_{12}+1,r_{max})} dr_{23}[r_{23}\cos\theta_{213}(\chi_{1,23}-1)],$$

$$r_{13} = 1, \tag{3.52}$$

$$(\partial/\partial\xi)\left\{[\ln(\chi_{2,13}g_{12}\exp(u_{12}+u_{13}))] - (\lambda/4\pi)\left[\underset{1\quad 2}{\overset{4}{\wedge}} + \underset{1\quad 3}{\overset{4}{\wedge}} + \underset{1\ \ 2\ \ 3}{\overset{4}{\wedge}}\right]\right\}$$

$$\equiv (\partial/\partial\xi)T_{2,13} = (\lambda/2\pi r_{12})\int_{\max(r_{12}-1,1)}^{\min(r_{12}+1,r_{max})} dr_{24}[r_{24}\cos\theta_{214}]$$

$$\times\int_0^\pi d\phi[(\chi_{1,34}\chi_{3,24}\chi_{1,24}/g_{24}) - 1], \qquad r_{14} = 1, \quad r_{34} \geq 1, \tag{3.53}$$

where $r_{\max}$ is chosen to be large, so that the region outside of $r_{\max}$ contributes negligibly to the integral, and $\partial\xi$ is the differential element along a constant angle θ_{123} (see Fig. 11). The quantity λ and the angle θ_{ijk} are defined as follows:

$$\lambda = 4\pi\varrho g(1), \tag{3.54}$$

$$\theta_{ijk} = \cos^{-1}[(r_{ij}^2 + r_{jk}^2 - r_{ik}^2)/2r_{ij}r_{jk}]. \tag{3.55}$$

The graphic notation such as 3 1 2 and 4 1 2 3 in Eqs. (3.52) and (3.53) represent the usual Mayer–Montroll graphs (see, for example, Uhlenbeck and Ford, 1962), where the black dots for particle 4 represent integration over all configurations of particle 4. A straight line between two particles denotes the Mayer f-function (3.50). For hard spheres, the quantity 4 1 2 3 has been evaluated by several investigators (Weissberg and Prager, 1962; Rowlinson, 1963; Powell, 1964; Ree *et al.*, 1966), while the integral 3 1 2 can be evaluated in a straightforward fashion.

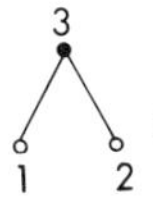

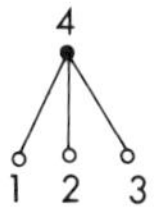

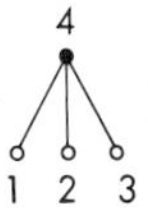

Details of each iterative cycle to solve Eqs. (3.52) and (3.53) are as follows:

(a) Assume the guess functions $g_{12}^{(1)}$ and $\chi_{2,13}^{(1)}$ over a discrete network of r_{12} and r_{13} ($r_{23} = 1$) such that $1 \leq r_{13} \leq r_{12} \leq r_{\max}$. The value of $\chi_{2,13}^{(1)}$ for r_{12} and r_{13} lying outside of this range can be generated from similar configurations within this range by using the symmetry relationships (3.48) and (3.49).

(b) Substitute the nth guess functions into the right-hand side of Eq. (3.52), and evaluate G_{12} in the left-hand side of Eq. (3.52). The $(n+1)$th guess function for g_{12} then follows from the following linear combination of $g_{12}^{(n)}$ and $\exp(G_{12})$ with a suitable mixing parameter θ_g:

$$g_{12}^{(n+1)} = \theta_g[\exp(G_{12})] + (1 - \theta_g)g_{12}^{(n)}, \qquad 0 \leq \theta_g \leq 1. \tag{3.56}$$

(c) Next, evaluate the right-hand side of Eq. (3.53) for all discrete mesh points spanning the surface defined by $1 \leq r_{13} \leq r_{12} \leq r_{\max}$.

(d) Using the data points obtained in the previous steps, integrate Eq. (3.53) along the line ξ from $\xi = r_{12}$ to $r_{\max}$ (see Fig. 11). This requires another interpolation, since the right-hand side of Eq. (3.53) has been evaluated only at discrete mesh points in (r_{12}, r_{13}), while the ξ-integration requires values of the right-hand side of Eq. (3.53) at a discrete interval $\Delta\xi$.

(e) Finally, calculate the desired solution for the $(n+1)$th guess on $\chi_{2,13}$ by a linear combination with the preceding solution $\chi^{(n)}_{2,13}$, with a mixing parameter θ_χ; i.e.,

$$\ln \chi^{(n+1)}_{2,13} = -G_{12} + (\lambda/4\pi)\left[\overset{4}{\underset{1\quad 2}{\wedge}} + \overset{4}{\underset{1\quad 3}{\wedge}} + \overset{4}{\underset{1\ \ 2\ \ 3}{\wedge}} \right] + \theta_\chi T^{(n+1)}_{2,13} + (1-\theta_\chi)\, T^{(n)}_{2,13}, \qquad 0 < \theta_\chi < 1. \tag{3.57}$$

In the above iterative procedure, step (d) requires data on the right-hand side of Eq. (3.53) along the straight-line path of particle 1 with a fixed angle θ_{123}. This, however, proved to be a computationally hazardous thing to do. The interpolation procedure in step (d) uses only the radial component $\partial T_{2,13}/\partial\xi$ at neighboring mesh points, ignoring contributions by the azimuthal component $\partial T_{2,13}/\partial\theta_{123}$. Neglecting this part would make little or no difference as long as Δr is chosen sufficiently small. But we must note that step (d) may cause a small discrepancy at small values of θ_{123} ($\neq 0$), where the interpolation formula is likely to perform poorest. The iteration cycle stops when $g^{(n)}_{12}$ and $\chi^{(n)}_{2,13}$ achieve a preset accuracy ε. An economical way of doing the iterative scheme outlined here is to repeat this procedure twice—once using a large mesh Δr to speed up the convergence, and next, using the resulting solution g_{12} and $\chi_{1,23}$ as initial guess functions for the second iterative procedure with smaller Δr. The two-step method cuts calculation time approximately in half.

The above iterative scheme uses an external parameter λ together with internal parameters ($r_{\max}$, Δr, θ_g, θ_χ, ε). Although in principle the internal parameters should not affect the final solutions, this is not true in the present problem, mainly on account of a finite memory capacity and the limited speed of the computer. Therefore, optimum choice of some parameters is imperative. For example, Table VIII shows the dependence of the density,

$$\varrho = \lambda/4\pi g(1), \tag{3.58}$$

on the mesh sizes at $\lambda = 10$ and 15, along with the other internal parameters used in the computation where $\theta \equiv \theta_g \equiv \theta_\chi$. Extrapolations of

TABLE VIII

DENSITIES CALCULATED FROM THE RELATION[a] $\varrho = \lambda/6B_2g(1)$ AT $\lambda = 10$ AND 15 AND AT VARIOUS MESH SIZES Δr USED IN INTEGRATING THE BGY2 EQUATION, TOGETHER WITH THE EXTRAPOLATED VALUES AT $\Delta r = 0$

Δr [b]	ϱ/ϱ_0	
	$\lambda = 10$[c]	$\lambda = 15$[d]
0.1	0.297	0.367
0.05	0.297	0.368
0.04	0.298	0.369
0.00 (extrapolated)	0.298	0.372

[a] Here, B_2 is the second virial coefficient. The data in this table are from Ree *et al.* (1971).

[b] Distance is expressed in units of sphere diameter.

[c] Mixing parameter $\theta = 0.6$, maximum cutoff parameter $r_{\max} = 6$.

[d] $\theta = 0.7$, $r_{\max} = 7$.

ϱ to $\Delta r = 0$ are made under the assumption of a monotonic dependence of ϱ on Δr. Even though solutions of the BGY2 equations at still smaller Δr are obviously desirable, they would take an enormous amount of computation. The computation time required to complete one iterative cycle by a CDC 6600 computer in the above iterative scheme is roughly estimated to follow the relation:

$$t = 0.13 r_{\max} (10\,\Delta r)^{10(\Delta r/3)-4.14} \quad \text{min.} \tag{3.59}$$

For $\Delta r = 0.04$ and $r_{\max} = 7$ used in case of $\lambda = 15$, the above formula predicts 37 min per iterative cycle. Fortunately, for a fixed Δr, the behavior of g_{12} with respect to the iteration number is sufficiently smooth and an extrapolation procedure using approximately the first half of the total number of iterations required turned out to be sufficient in correctly predicting the final solutions.

In Figs. 12a,b and 13a,b, the BGY2 $g(r)$ is compared with the BGY1 $g(r)$ obtained from Eq. (3.46) using the Kirkwood superposition approximation (3.38) along with the hard-sphere solution (Wertheim, 1963; Thiele, 1963) of the Percus–Yevick (PY) equation at $\varrho = 0.298\varrho_0$ and $0.372\varrho_0$. These theories are also compared with the Monte Carlo "experimental" data. The densities $0.298\varrho_0$ and $0.372\varrho_0$ chosen for the

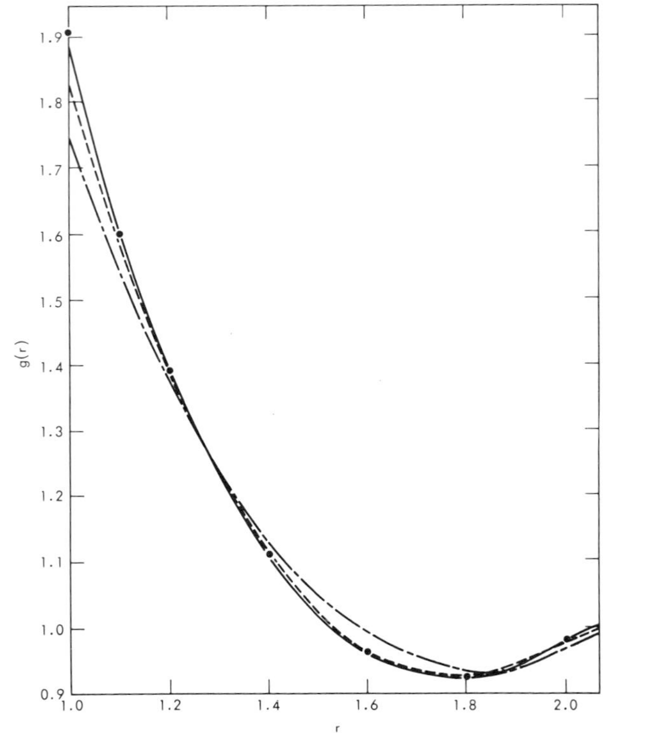

FIG. 12a. Radial distribution functions of hard spheres at $\varrho/\varrho_0 = 0.298$ from the Monte Carlo "experiment" (dots) and the BGY1 (— - —), the BGY2 (——), and the PY (- - -) theories

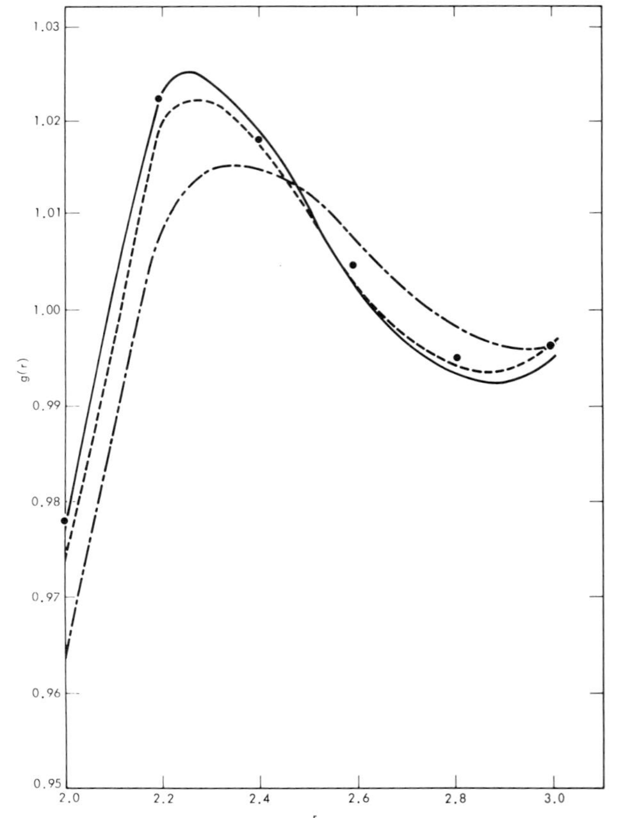

FIG. 12b. Radial distribution functions as in Fig. 12a but for $r = 2.0$–3.0.

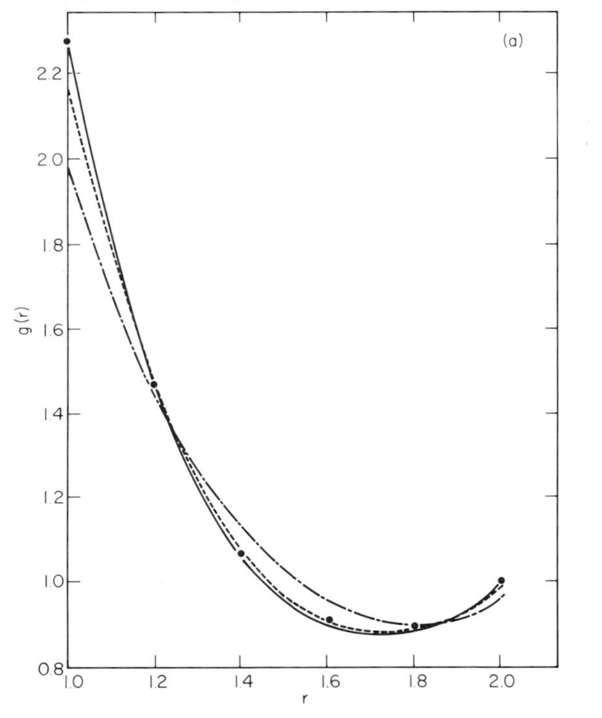

FIG. 13a. Radial distribution functions of hard-spheres at $\varrho/\varrho_0 = 0.372$ from the Monte Carlo "experiment" (dots) and the BGY1 (— - —), the BGY2 (——), and the PY (- - -) theories for $r = 1.0$–2.0 (Lee, 1970; Ree *et al.*, 1971).

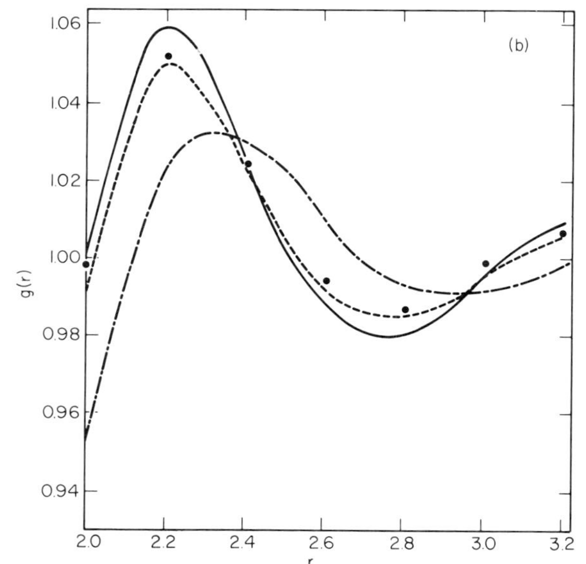

FIG. 13b. Radial distribution functions as in Fig. 13a but for $r = 2.0$–3.2.

"experiments" are made to match with the densities predicted from the BGY2 solutions for $\lambda = 10$ and 15, respectively. At $\varrho = 0.298\varrho_0$, use of the BGY2 $g(r)$ improves significantly over the corresponding quantity calculated form the BGY1 equation. Moreover, the BGY2 $g(r)$ agrees with the Monte Carlo value of $g(r)$ within their numerical accuracies. At $\varrho = 0.372\varrho_0$, the BGY2 $g(r)$ agrees with the Monte Carlo $g(r)$ for $r \lesssim 1.4$; however, at large distances ($r \gtrsim 1.6$), the PY $g(r)$ apparently behaves in a manner superior to the BGY2 $g(r)$. However, discrepancies between the BGY2 $g(r)$ and the Monte Carlo $g(r)$ are minute, amounting to 1.4% or less. An almost perfect agreement of the BGY2 $g(r)$ with the corresponding Monte Carlo value at $r = 1$ is especially encouraging for hard-core particles, for calculation of the pressure by the virial theorem (3.7) requires only the knowledge of $g(1)$. In this respect, it will be useful to know if the good agreement is maintained even at higher densities. To avoid a large amount of computation time involved in this case, investigation on a calculationally simpler two-dimensional hard-disk system may prove to be more advantageous.

The Monte Carlo data on $g(r)$ plotted in Figs. 12a,b and 13a,b are obtained by using a system of 500 hard spheres which are initially placed in faced-centered lattice points in a periodic cubic box having the same side lengths. The particles are arranged in ten equidistant layers, each of which contains ten rows consisting of five particles in each row. In the Monte Carlo calculations, the side length of the box is chosen to be the unit of length. Accordingly, the hard-sphere diameter is made to vary in order to represent a system at a desired density. The first 100,000 configurations generated in the Monte Carlo procedure are thrown away in the calculation of equilibrium properties so that the system can achieve spatial equilibrium configurations. The calculation of $g(r)$ follows using configurations generated thereafter. At each Monte Carlo step, pair separations between a randomly chosen particle and all other particles are measured. A set of counters which accumulates the number of pairs with specified r^2 is increased by the number of times that such pairs fall within the range $(r^2, r^2 + \Delta r^2)$ specified by one of its counters [say $G(r^2)$]. At $\varrho = 0.298\varrho_0$ and $0.373\varrho_0$, for example, these spacings are chosen to be approximately equal to $\Delta r^2 = 0.138$ and 0.116, respectively. The radial distribution function $g(r)$, after $\bar{m}$ Monte Carlo steps, is calculated by using the following formula:

$$g(\bar{r}) = N(N-1)^{-1}\varrho^{-1}G(r^2)/(2\pi\bar{r}\,\Delta r^2\bar{m}), \tag{3.60}$$

where introduction of the correction factor $N(N-1)^{-1}$ is expected to make the present Monte Carlo calculations agree with similar calculations performed by the molecular-dynamic method [see Eq. (2.45)]. The distance $\bar{r}$ in Eq. (3.60) is calculated from the following relationship:

$$2\pi\bar{r}\,\Delta r^2 = (4\pi/3)[(r^2 + \Delta r^2)^{3/2} - r^3]. \tag{3.61}$$

The Monte Carlo values of $g(r)$ plotted in Figs. 12a,b and 13a,b are obtained after 1.8×10^6 Monte Carlo steps for the case $\varrho = 0.298\varrho_0$, and 2.5×10^6 steps for the case $\varrho = 0.372\varrho_0$. These data have an error of about 0.4% or less.

Before concluding the discussion of $g(r)$, we mention the existence of the molecular-dynamic $g(r)$ for 108 hard spheres at $\varrho_0/\varrho = 3, 2, 1.7$, and 1.6 (Alder and Hecht, 1969). These data are judged to have approximately 99% accuracy, probably reflecting the relatively slow earlier machine (IBM 704) used for the calculations. For hard disks, there are data on $g(r)$ carried out on the *NVT*-ensemble by Chae *et al.* (1969), and the data by Wood (1970), who used the *PVT*-ensemble. In the latter paper, the interested reader can also find discussions on the number dependence of $g(r)$ and a comparison of the calculation of $g(r)$ by means of the *NVT*-ensemble and of the *NPT*-ensemble.

F. Coulomb System

Discussions on the Monte Carlo calculations of the equilibrium properties for a Coulomb system will be given here. The Monte Carlo data will be compared with data obtained from calculations using several approximate theories of the radial distribution function.

Probably the earliest reported work on the machine "experiments" on the Coulomb system is the thesis work by Shaw (1963), who dealt with two-component, hard-sphere ions. This work has been further reviewed in some detail by Poirier (1966), who used Shaw's data to assess the validity of various approximations made in the original derivation of the Debye–Hückel theory (Debye and Hückel, 1923). The above work was followed by work of Barker (1965) and of Brush *et al.* (1966). The latter work is concerned with the one-component plasma with a uniform, neutralizing background charge. However, it is probably one of the most extensive works in the available literature and covers various thermodynamically interesting ranges using relatively long Monte Carlo chain lengths. Consequently, Brush's *et al.* work will be discussed in some detail.

Conventional Monte Carlo or molecular-dynamic methods described in previous parts of this section have only limited success when directly applied to a system of classical, charged particles whose interaction potential is a Coulomb potential:

$$\phi(r) = (Ze)^2/r, \tag{3.62}$$

where Z denotes the number of electronic charges e in each ion. The main difficulty with the Coulomb system occurs with problems of how to handle a relatively long-ranged tail in the Monte Carlo calculations of the finite Coulomb system. Unless the density is dilute enough or the temperature is high enough so that contributions by the Coulomb forces can be approximated as a small perturbation from an ideal state, a suitable long-range correction to the total potential energy by all "image" particles outside the box must be incorporated into a Monte Carlo program to simulate an infinite thermodynamic system.

One additional effect which requires our attention is that of the uniform, negative background charge which is introduced to conserve the charge neutrality. The background charge exerts a stabilizing influence on the ions, reinforcing the confining effect of the wall of the periodic box. The effect of background and image charges can be taken into account in the potential-energy calculations by using Ewald's method (Ewald, 1921; Barker, 1965) of computing lattice sums. This has been carried out by Brush *et al.* (1966). The resulting expression for the potential energy $U(\mathbf{r}^N)$ of N particles is

$$U(\mathbf{r}^N) = (NE_\mathrm{m}/2) + \sum_{i>j}^{N} \psi(|\,\mathbf{r}_i - \mathbf{r}_j\,|), \tag{3.63}$$

where E_m is equal to $-2.837297\varrho^{1/3}$ and is the Madelung energy of a simple cubic crystal with a lattice spacing equal to the edge length L of the periodic box. The quantity $\psi(\mathbf{x})$ is the Ewald potential and is defined by

$$\psi(\mathbf{x})/kT = \Gamma[\psi_1(x) + \psi_2(\mathbf{x})], \tag{3.64}$$

$$\psi_1(x) = x^{-1}\operatorname{erfc}(\pi^{1/2}x/L) - L^{-1}, \tag{3.65}$$

$$\begin{aligned}\psi_2(\mathbf{x}) = {} & L^{-1} \sum_{\mathbf{n}}{}' \{|\mathbf{n} - (x/L)\,|^{-1} \operatorname{erfc}[\pi^{1/2}\,|\,\mathbf{n} - (\mathbf{x}/L)\,|]\} \\ & + L^{-1} \sum_{\mathbf{n}}{}' \{(\pi n^2)^{-1}[\exp(-\pi n^2)] \exp[i(2\pi\mathbf{n}\cdot\mathbf{x})/L]\},\end{aligned} \tag{3.66}$$

where $\mathbf{n}$ is a three-dimensional index defining location of the center of an image of the periodic box. The prime on the summations in Eq. (3.66) denotes exclusion in the summations of $\mathbf{n} = (0, 0, 0)$ corresponding to the periodic box. In the above expressions, x and Γ are dimensionless quantities and are defined as follows:

$$x = r(4\pi\varrho/3)^{1/3}, \tag{3.67}$$

$$\Gamma = [(Ze)^2/kT](4\pi\varrho/3)^{1/3}. \tag{3.68}$$

Using the reduced distance x, the potential energy (3.62) can be written in a dimensionless form by dividing it by kT, i.e.,

$$\phi(x)/kT = \Gamma/x. \tag{3.69}$$

The parameter Γ is a controllable input variable which roughly measures the ratio of the Coulomb potential energy to the average kinetic energy. Thus, depending on its magnitude, the system will behave either ideally ($\Gamma \lesssim 0.01$) or show a freezing phenomenon ($\Gamma \gtrsim 100$). In Fig. 14, relative contributions of $\psi_1(x)$ and $\psi_2(\mathbf{x})$ to the Ewald potential, Eq. (3.64), are shown. As can be seen in this figure, over a wide range of x, $\psi_2(\mathbf{x})$ behaves in a manner similar to the Coulomb potential x^{-1}. For large L, we see from Eq. (3.65) that $\psi_1(x)$ can be approximated by x^{-1} except for those values of L near the edge of the box. In contrast to this, $\psi_2(\mathbf{x})$ is angle-dependent. However, its angle dependence is minor and is nearly constant throughout the box. Furthermore, the magnitude of $\psi_2(\mathbf{x})$ is relatively small compared with $\psi_1(x)$ and, therefore, makes a minor contribution to the Ewald potential (see Fig. 14). Brush *et al.* use an approximate expression for $\psi_2(\mathbf{x})$ in the actual Monte Carlo calculations, in which the approximation results from truncation of a three-dimensional Taylor expansion of $\psi_2(\mathbf{x})$.

By using the Ewald potential energy as the interaction potential for the Coulomb particles, Brush *et al.* generated 50 NVT-ensemble Monte Carlo chains as discussed in Section II,A. Each Monte Carlo chain corresponds to different combinations of parameters N (32–500 particles) and Γ (0.05–100). In addition to these data, Brush *et al.* evaluated the pair distribution functions of 500 particles interacting with the true Coulomb interaction [Eq. (3.69)] (that is, instead of the Ewald potential) at $\Gamma = 10$, 14, and 16. These calculations are carried out using the minimum-image distance convention discussed in Section II,A. Comparison of the Monte Carlo $g(x)$ obtained at $\Gamma = 10$ for particles in-

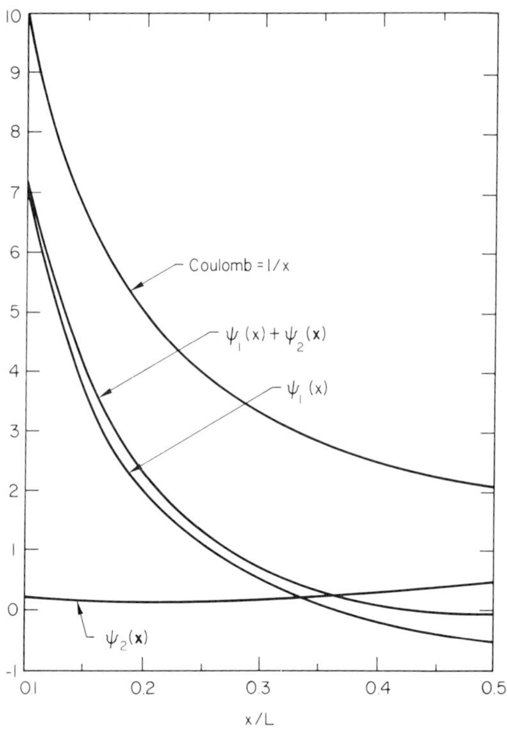

FIG. 14. The Ewald potential $\psi_1(x) + \psi_2(\mathbf{x})$ [Eq. (3.64)] and its dominant term $\psi_1(x)$ [Eq. (3.65)] at $\Gamma = 1$. The Coulomb potential is plotted for comparison. Here, L is the edge length of the box and x is the dimensionless distance defined by Eq. (3.67). (Brush *et al.*, 1966.)

teracting with the Ewald potential shows no significant difference from those obtained for the true Coulomb particles when the minimum-image distance convention is used. At $\Gamma = 14$, a small difference appears at large distance ($x \gtrsim 1.8$). For still larger $\Gamma \gtrsim 16$, the difference between the two becomes significant. This indicates that the 500 Coulomb particles, through use of the minimum-image distance convention, are probably not large enough to simulate an infinite thermodynamic system when Γ becomes larger than $\Gamma = 14$. In one of their Monte Carlo chains corresponding to $\Gamma = 126$ and $N = 32$, the system in the fluid phase actually underwent a change in phase to a solid in a face-centered-cubic structure. After staying there for a relatively long period ($\sim 10^3$ Monte Carlo moves per particle), the system returned to the original fluid state. Existence of the discrete solid and fluid branches (rather than a continuous van der Waals-like loop) is due to a small number of particles as described in Section II,A in the case of hard spheres and disks. Transition to the body-centered-cubic crystal structure would have been possible had they used the number of particles needed to represent this crystal. The parameter $\Gamma = 126$ at which two phases coexisted would indicate that an infinite thermodynamic Coulomb system in a solid phase would probably undergo melting at Γ_{m}, which lies slightly above $\Gamma = 126$. Van Horn (1969) estimated the value of Γ_{m} to be 170 ± 10. This number is obtained using the assumption of the Lindemann melting rule and agrees reasonably well with the Monte Carlo observation. Brush *et al.* mention that if the minimum-image distance method is used, onset of transition to the solid phase takes place at a different density and the crystal structure of the system is not the face-centered-cubic structure but is the simple cubic crystal! This difference is very likely caused by a finite-number effect. If this is the case, then the Brush *et al.* results imply a sensitive dependence of equilibrium properties of the Coulomb system on the choice of two different (Ewald and Coulomb) potentials. Whether or not this is true remains to be determined and will be a subject interesting to pursue further.

It is possible to make a comparison of the Monte Carlo results of Brush *et al.* with the other available theories. Depending on their approaches, the current approximate theories can be classified into two essentially different types. The first approach is the diagram method of Meeron (1958) and Hirt (1965), who modified the Mayer–Montroll cluster expansion of $g(x)$ so that $g(x)$ for the Coulomb gas can be expanded by means of certain powers of the product of Γ and $\ln \Gamma$. This method has an attractive feature in that $g(x)$ can be calculated analytically. However, this advantage likely occurs at the expense of its inherent disadvantage, which limits the range of its applicability to a small value of Γ. In most situations, consideration of large values of Γ ($\gtrsim 2$) will necessitate evaluations of progressively larger classes of graphs, each of which becomes progressively more difficult to evaluate. Nevertheless, a step in this direction has been recently taken by various workers (del Rio and DeWitt, 1969; Mitchell and Ninham, 1968; Cohen and Murphy, 1969). In particular, del Rio and DeWitt obtained an explicit expression for $g(x)$ correct to the order of $\Gamma^3 \ln \Gamma$, thus improving the Debye–Hückel theory (1923). Moreover, del Rio and DeWitt's work shows a broad maximum in $g(x)$ for $\Gamma \gtrsim 1.8$. One may recall that there is no maximum in the Debye–Hückel $g(x)$. The maximum indicates that the short-range order in $g(x)$ starts to set in at $\Gamma \gtrsim 1.8$, which lies close to the "experimental" value, $\Gamma \approx 2.5$. Figures 15a and 15b show the absence ($\Gamma = 1$) and the presence ($\Gamma = 10$) of the short-range order in the Monte Carlo $g(x)$.

The second approach, which is less rigorous but more computationally feasible, uses the radial distribution functions obtained from the various approximate theories. This has been undertaken by Carley (1963a,b), who used the Percus–Yevick (PY) and the convolution-hypernetted-chain (CHNC) integral equations, and by Hirt (1967), who employed the Born–Green–Yvon (BGY) equation with the Kirkwood superposition approximation (3.38) for the triplet distribution function. For Coulomb particles, these equations can be represented as follows:

PY and CHNC:
$$H(x) = (3/2) \int_0^\infty ds\, s[1 - g(s)] \int_{|s-x|}^{s+x} dt \times [H(t) - tg(t) + t], \tag{3.70}$$

PY:
$$g(x) = \{1 + [H(x)/x]\} \exp[-\Gamma/x], \tag{3.71a}$$

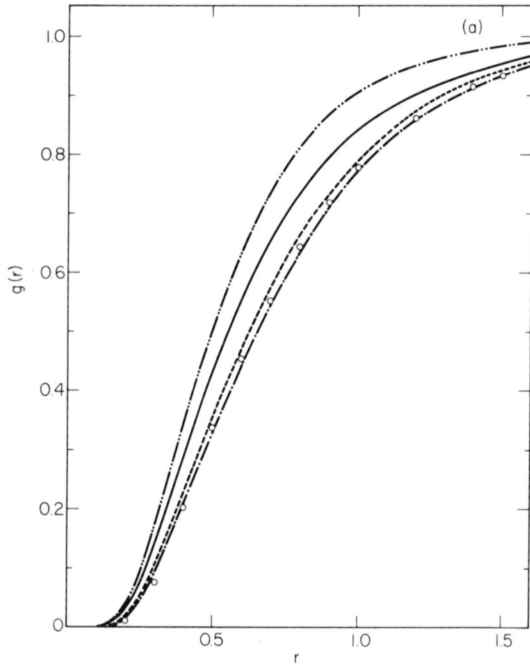

FIG. 15a. The radial distribution functions of the Coulomb gas at $\Gamma = 1$. Open dots: The Monte Carlo results; (——) the asymptotic Debye–Hückel $g(r)$ [Eq. (3.74)]; (- - -) the PY $g(r)$; (—·—·) the BGY $g(r)$; (— ·· — ··) the CHNC $g(r)$. These data are taken from Brush *et al.* (1966), Carley (1963b), and Hirt (1967).

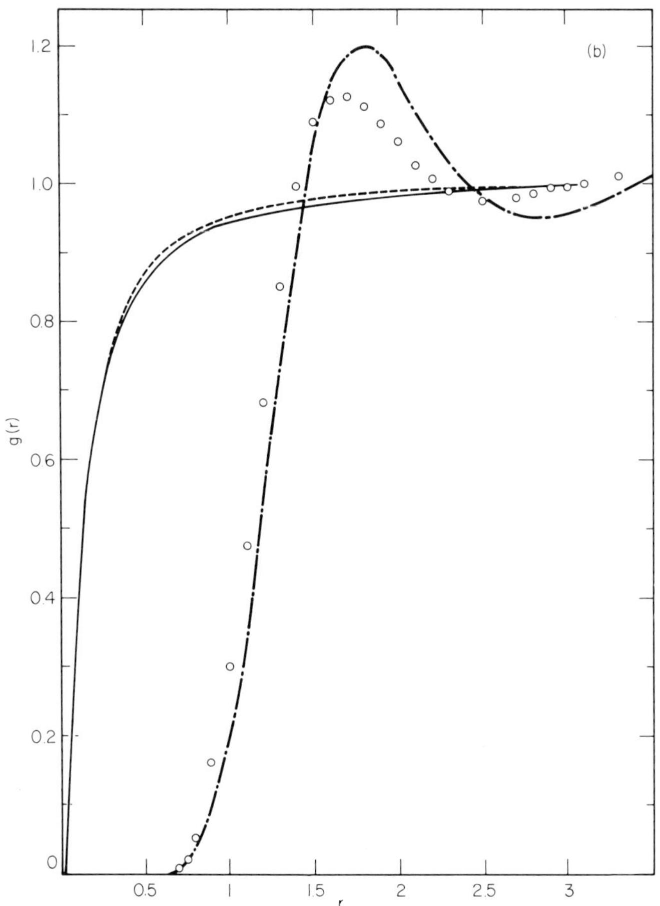

FIG. 15b. The radial distribution functions of the Coulomb gas at $\Gamma = 10$. See Fig. 15a for key.

CHNC:
$$g(x) = \exp\{[H(x)/x] - (\Gamma/x)\}, \tag{3.71b}$$

BGY:
$$\begin{aligned} -\Gamma^{-1} \ln g(x) = x^{-1} &+ 3x^{-1} \int_0^x dt\, \{t^2[g(t) - 1]\} \\ &+ 3 \int_x^\infty dt\, \{t[g(t) - 1]\} + (3/4x) \int_0^\infty ds \\ &\times \{s^{-2}[g(s) - 1]\} \int_{x-s}^{x+s} dt\, \{t[s^2 - (x - t)^2] \\ &\times [g(|\, t\,|) - 1]\}, \end{aligned} \tag{3.72}$$

where the PY $g(x)$ can be obtained from Eqs. (3.70) and (3.71a), and the CHNC $g(x)$ from Eqs. (3.70) and (3.71b). Hirt obtained the BGY equation (3.72) using the original BGY equation (3.37) and an *additional* requirement on $g(x)$ imposed by Gauss's law of electrostatics, which can be expressed mathematically as follows:

$$\int d\mathbf{x}_3\{(\boldsymbol{\nabla}_1\phi_{13})[g_{23}-1]\} = (\boldsymbol{\nabla}_1\phi_{12})4\pi\int_0^{x_{12}} dx\{x^2[g(x)-1]\}. \tag{3.73}$$

In his derivation of Eq. (3.72) from the original BGY equation (3.37), Hirt replaces a term in the equation that is exactly same as the left-hand side of Eq. (3.73) by the right-hand side of Eq. (3.73). The resulting equation (3.72) is not an *exact* equation for $g(x)$. Hence, its solution most likely will not satisfy Eq. (3.73). Therefore, the above-described replacement procedure is suspect for its internal inconsistency. In this respect, we may regard Eq. (3.72) as an augmented BGY equation.

Numerical solutions of Eqs. (3.70)–(3.72) can be obtained by an iterative procedure such as the one described in Section II,E. The resulting numerical data at $\Gamma = 1$ and 10 are plotted in Figs. 15a and 15b, respectively. In the same figures, the following asymptotic solution of the Debye–Hückel (DH) equation is also plotted:

$$g(x) = \exp\{-(\Gamma/x)\exp[-x(3\Gamma)^{1/2}]\}. \tag{3.74}$$

At $\Gamma = 1$, the BGY $g(x)$ agrees quantitatively very well with the Monte Carlo $g(x)$ for all values of x and is better than $g(x)$ calculated from the PY, the CHNC, and the DH theories. While the range of Γ within which the DH $g(x)$ can be applicable is limited to $\Gamma \lesssim 0.1$ (Brush *et al.*, 1966), the BGY $g(x)$ is found to be applicable to a wider range of Γ, reaching close to $\Gamma \lesssim 2.5$. As shown in Fig. 15b, even at $\Gamma = 10$, the BGY $g(x)$ follows the qualitative behavior of the Monte Carlo $g(x)$. The presence of the oscillatory behavior found in the BGY $g(x)$ at Γ beyond a certain critical value has not yet been demonstrated by solutions of all other approximate integral equations (PY, CHNC, etc.) that have been tested so far for the Coulomb system. In the above comparison, we have omitted discussions on values of $g(x)$ calculated from a little more complicated Broyles–Sahlin equation (see Carley, 1963b) or solution of the exact Debye–Hückel equation, which can be solved by means of a direct iteration as was done by Hirt (1967). Readers interested in these equations should review the work of these authors.

Table IX summarizes calculations made by Brush *et al.* (1966), Carley (1963a,b), and Hirt (1967) on the average potential energy U at different

TABLE IX

COMPARISON OF MONTE CARLO DATA ON THE AVERAGE POTENTIAL ENERGY U OF A ONE-COMPONENT COULOMB SYSTEM AT VARIOUS VALUES OF Γ $\{\equiv[(Ze)^2/kT](4\pi\varrho/3)^{1/3}\}$ WITH THE CORRESPONDING DATA OBTAINED FROM THE VARIOUS APPROXIMATE THEORIES

Γ	$-U/NkT$				
	Monte Carlo[a]	BGY[b]	PY[c]	CHNC[c]	DH[c]
0.05	0.0128	—	0.008	0.008	0.009
0.1	0.0270	—	0.020	0.019	0.025
1.0	0.579	0.577	0.539	0.328	0.468
2.0	1.338	—	1.448	0.647	0.925
2.5	1.729	1.742	1.903	0.784	1.128
10.0	8.069	8.192	—	1.875	3.140

[a] Brush *et al.* (1966).
[b] The Born–Green–Yvon theory: Hirt (1967).
[c] See Carley (1963a,b) and Hirt (1967).

values of Γ ranging from 0.05 to 10. The average potential energy of the Coulomb system can be obtained from the formula

$$U/NkT = (3\Gamma/2) \int_0^\infty dx\, x[g(x) - 1]. \tag{3.75}$$

A good agreement of the U resulting from use of the BGY $g(x)$ in Eq. (3.75) with the Monte Carlo U is particularly noteworthy within this range of Γ. From the virial theorem for the Coulomb system, the pressure is directly related to U:

$$PV/NkT = 1 + (U/3NkT). \tag{3.76}$$

Hence, comparison of pressure calculations provides no new information as to the relative merits of the approximate theories.

G. VIRIAL EXPANSIONS

At low densities, experimental data on equilibrium properties such as P and $g(r)$ can be fitted empirically as polynomials in ϱ, i.e.,

$$\beta P/\varrho = 1 + B_2\varrho + B_3\varrho^2 + B_4\varrho^3 + B_5\varrho^4 + \cdots, \tag{3.77}$$

$$g(r) \exp[\beta\phi(r)] = 1 + g^1(r)\varrho + g^2(r)\varrho^2 + g^3(r)\varrho^3 + \cdots. \tag{3.78}$$

From a theoretical point of view, the above expansions in ϱ, which are more popularly known as the *virial expansions*, can be rigorously shown to hold for particles interacting with a classical two-body force. Each of the integrals in the *virial coefficients* B_n and $g^n(r)$ can be given a convenient graph-theoretic meaning (see, for example, Uhlenbeck and Ford, 1962). The following discussion is concerned with presentation of the formal prescriptions for writing down B_n and $g^n(r)$ and their reformulations, which are particularly useful for particles with a repulsive interaction. Two powerful methods of evaluating the virial coefficient which have been developed over the last decade will be reviewed. The first method uses the Monte Carlo integration of the reformulated expressions for the virial coefficients and is especially useful for a repulsive potential. The second method uses Legendre polynomial expansions (Barker and Monaghan, 1962; Henderson and Oden, 1966) to reduce the manifold integrals in the expressions of the virial coefficients into computationally tractable expressions. The second approach has been used primarily for attractive potentials but is applicable to repulsive potentials.

A reason for calculating virial coefficients higher than those experimentally measurable ($\approx B_2$ or B_3 in some instances) is to check the adequacy of approximate theories by using an idealized potential. An improved version of an approximate theory often is accompanied by improved values of the first few B_n. Another justification lies in the possibility that knowledge of many virial coefficients may provide information relating to the radius of convergence of the virial series. Since the current theory on the convergence problem is by no means complete yet, any empirical information regarding this will most certainly be welcome. The convergence of the virial series and various exact lower and upper bounds on B_n, as well as the fugacity coefficients b_n, will be the subject of the last part of the present discussion.

1. *Formulation and Reformulation*

For a system of classical particles with a pairwise additive interaction force, the virial coefficient B_n can be expressed as follows (Mayer and Mayer, 1940):

$$B_n = [(1 - n)/n!] \lim_{V\to\infty} V^{-1} \int d\mathbf{r}^n V_n, \tag{3.79}$$

where V_n is the sum of all n-point *star* graphs formed by placing various combinations of the Mayer f-function [Eq. (3.50)] to link all possible pairs of particles i and j among the assembly of n particles. By using a straight line to represent the Mayer f-function, V_4 and V_5 representing the collection of entire four- and five-point star graphs are given by

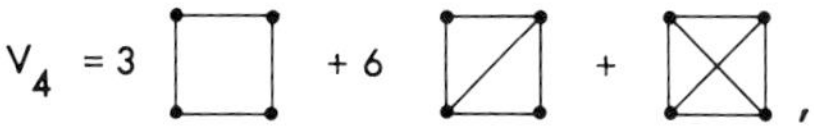

$$V_4 = 3\,[\text{graph}] + 6\,[\text{graph}] + [\text{graph}], \tag{3.80}$$

$$V_5 = 12\,[\text{graph}] + 60\,[\text{graph}] + 10\,[\text{graph}] + 10\,[\text{graph}] + 60\,[\text{graph}] + 30\,[\text{graph}] + 15\,[\text{graph}] + 30\,[\text{graph}] + 10\,[\text{graph}] + [\text{graph}] \tag{3.81}$$

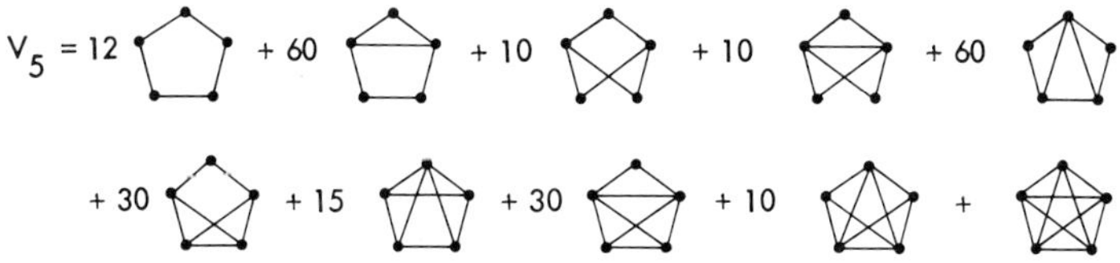

where the number preceding each graph denotes the different ways of labeling that graph.

Likewise, according to Mayer and Montroll (1941), $g^n(r)$ can be expressed as

$$g^{n-2}(r) = (1/n!) \int d\mathbf{r}^{n-2} W_n, \tag{3.82}$$

where the coordinates of particles 1 and 2 are excluded in the integration in the above equation, and the quantity W_n represents the sum of all possible doubly-rooted star graphs formed by connecting particles 1, 2, . . ., n by f-functions, provided that no f_{12} line is drawn. For example, W_4 and W_5 are given by

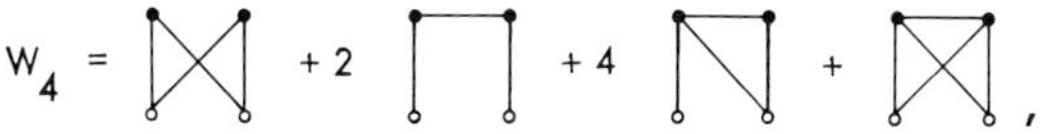

$$W_4 = [\text{graph}] + 2\,[\text{graph}] + 4\,[\text{graph}] + [\text{graph}], \tag{3.83}$$

$$W_5 = 6\,[\text{graph}] + 6\,[\text{graph}] + 12\,[\text{graph}] + 12\,[\text{graph}] + 6\,[\text{graph}] + 6\,[\text{graph}] + [\text{graph}] + 12\,[\text{graph}] + 3\,[\text{graph}] + 12\,[\text{graph}] + 12\,[\text{graph}] + 12\,[\text{graph}] + 6\,[\text{graph}] + 6\,[\text{graph}] + 6\,[\text{graph}] + 3\,[\text{graph}] + 12\,[\text{graph}] + 6\,[\text{graph}] + 3\,[\text{graph}] + 6\,[\text{graph}] + 6\,[\text{graph}] + 3\,[\text{graph}] + 6\,[\text{graph}] + [\text{graph}] \tag{3.84}$$

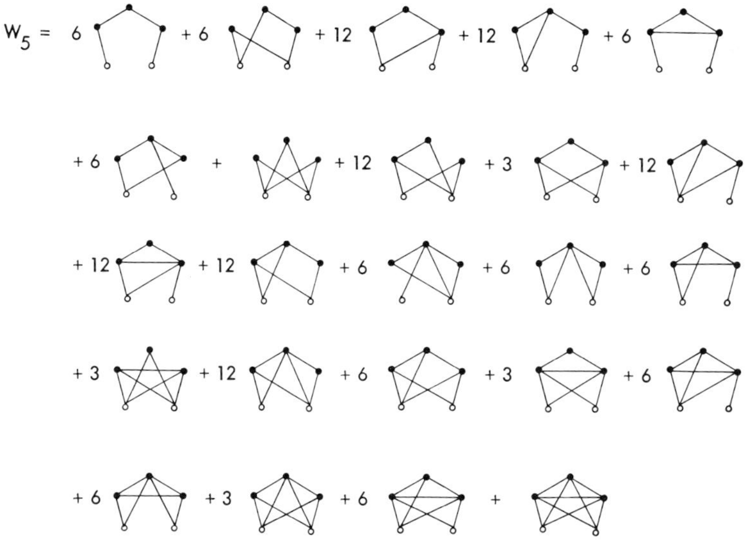

In Eqs. (3.83) and (3.84), the open circles refer to particles 1 and 2 and are known as rooted points, and the black circles refer to particles 3, 4, and 5, which are integrated over all space under the restriction of f-functions. These points are known as field points.

In practice, direct use of expressions such as Eqs. (3.80)–(3.84) to calculate B_n or $g^n(r)$ proved to be unprofitable. For instance, calculations for hard potentials show that the net contribution of positive integrals to $B_n \geq 4$ is roughly equal to that of negative integrals. Separating the positive and the negative terms in B_5 [Eq. (3.79)] gives $B_5/(B_2)^4 = 7.902 - 7.792 = 0.110$ for hard spheres (Ree and Hoover, 1964a) and $6.364 - 6.352 = 0.012$ for hard parallel cubes (Hoover and DeRocco, 1962). The final value of B_n is small in comparison to both the positive and negative contributions—about the same order of magnitude as the contribution of the smallest star integral, namely the complete star integral. The five-point complete star graph [the last graph in Eq. (3.81)] has a contribution to $B_5/(B_2)^4$ of 0.024 for spheres and 0.020 for cubes. Origin of the large cancellation derives from the fact that, in Mayer graphs appearing in B_n and $g^n(r)$, many pairs of points are not connected by Mayer f-functions. In integrations of Eqs. (3.79) and (3.82) for repulsive potentials, these pairs of points have less geometric restrictions, hence correspondingly larger values for individual graphs. This is especially a handicap in a practical calculation, where the numerical accuracy of a desired solution is mostly limited by the time and size of an electronic computer. This disadvantage can be overcome, at least for calculations of lower-order ($\lesssim n = 8$) virial coefficients, by using the $\tilde{f}$-function

$$\tilde{f}_{ij} = f_{ij} + 1, \tag{3.85}$$

and by replacing any unfilled (by an f-function) pair of points in a Mayer graph by means of the identity $1 = \tilde{f}_{ij} - f_{ij}$. When the factors $\tilde{f}_{ij} - f_{ij}$ are multiplied out, a Mayer star can be written as a sum of *modified stars* or *modified doubly-rooted graphs* composed of two types of lines f_{ij} (denoted by a straight line between i and j) and $\tilde{f}_{ij}$ (denoted by a wiggly line between i and j). When this procedure is applied to all of the stars or doubly-rooted graphs occurring in either B_n or $g^n(r)$, many of the modified graphs cancel identically. For example, the final expressions for V_4, V_5, W_4, and W_5 are shown as follows (Ree and Hoover, 1964b; Ree *et al.*, 1966):

$$V_4 = \left[-2 \times 1\right] \quad + \left[1 \times 3\right] \quad , \tag{3.86}$$

$$V_5 = [-6 \times 1]\,\phi + [3 \times 15] + [-2 \times 30] + [1 \times 10] + [1 \times 12], \tag{3.87}$$

and

$$W_4 = [-2 \times 1]\,\phi + [1 \times 1] + [1 \times 2] \tag{3.88}$$

$$W_5 = [-6 \times 1]\,\phi + [3 \times 3] + [3 \times 6] + [1 \times 6] + [-2 \times 3] + [-2 \times 6] + [-2 \times 12] + [-1 \times 3] + [1 \times 1] + [1 \times 6] + [1 \times 6] + [1 \times 6], \tag{3.89}$$

where ϕ represents the complete star graph. In modified graphs, we use the convention of only drawing the *wiggly-line* ($\tilde{f}$) *graph.* The wiggly-line graph consists only of lines which represent $\tilde{f}$-functions in the modified graphs; the lines not drawn are understood to be Mayer f-functions. This convention has the advantage that a single type of wiggly-line graph can be used to represent an infinite class of corresponding modified-star integrals, as is the case, for example, for the graph in Eqs. (3.86) and (3.87). The first multiplicative factor preceding each wiggly-line graph in V_n (modified doubly-rooted graph in W_n) is called the *star* (*doubly-rooted star*) *content* and can be obtained from the following rule:

Count the number of labeled Mayer stars (doubly-rooted stars) that can be formed by successively removing 0, 2, 4, . . . f-functions from the f-functions of any modified star (doubly-rooted modified star); then, subtract the number of labeled Mayer stars (doubly-rooted modified stars) that can be formed by removing 1, 3, 5,. . . f-functions of the same modified star (doubly-rooted star). The resulting number (which can be positive, negative, or zero) is the star (doubly-rooted star) content of the modified Mayer star (doubly-rooted Mayer star).

The second factor in front of each modified graph in Eqs. (3.86)–(3.89) is the number of different ways of labeling that graph.

This reformulation of the original Mayer formulation of B_n and $g^n(r)$ contains three advantages which can be especially useful for repulsives potentials.

(a) Because there are many modified stars or doubly-rooted modified stars with identically zero star content, the number of integrals of topologically different graphs contributing to the virial coefficients is reduced. For example, the number of topologically different graphs contributing to B_4, B_5, B_6, and B_7 in Mayer's formulation is, respectively, equal to 3, 10, 56, and 468. In contrast to this, the corresponding number of integrals which must be evaluated is considerably reduced, being 2, 5, 23, and 171.

(b) In Mayer's formulation, none of the star or doubly-rooted star integrals contributing to B_n or $g^n(r)$ (for hard potentials, at least) could be ignored; each made a nonnegligible contribution. In the reformulated expression, many integrals make either negligible or zero contribution for hard potentials. Therefore, the previously discussed extensive cancellation of positive and negative terms found in Mayer's formulation is significantly reduced. Using the previous hard-sphere and hard-cube potentials as examples, when $B_5/(B_2)^4$ is evaluated using modified-star integrals, separating the positive and negative terms gives $0.158 - 0.048 = 0.110$ for hard spheres and $0.120 - 0.108 = 0.012$ for hard cubes. The number of zero-valued integrals contributing to B_n and $g^n(r)$ increases with n and decreases with dimensionality. For one-dimensional hard rods, all but one (complete star) of the contributing modified-star integrals are zero. Moreover, in contrast to the Mayer representation, which makes the complete star integral contribute least to B_n and $g^n(r)$, the reformulation assigns to the complete star integral the largest contribution among all the modified star integrals contributing to B_n through $n = 7$ for one-, two-, and three-dimensional hard potentials. This reduced cancellation observed for hard-core potentials occurs because the presence of two kinds of lines (f and $\tilde{f}$) requires particles to be overlapping and nonoverlapping. In original Mayer stars, some pairs of particles are restricted to overlap by f-functions, but overlap or nonoverlap of the other pairs (for which no f-function appears) is left unspecified.

(c) The third advantage for the reformulation occurs in the form of several new ways of summing the virial series by successive approximations, which can give an increasingly large number of correct virial coefficients.

Taking the hard parallel squares as an example, contributions to B_n by the five modified stars can be worked out (Ree and Hoover, 1964b).

Using the side of the square as the unit of length, these are

$$\{\phi\}: \qquad B_n = n \quad \text{if} \quad n \geq 2; \; = 0 \quad \text{otherwise}; \tag{3.90}$$

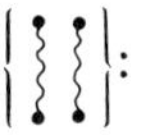

$$: \qquad B_n = -(n-2)(n-3)/2(n-1) \;\; \text{if}\, n \geq 4; = 0 \,\text{otherwise}; \tag{3.91}$$

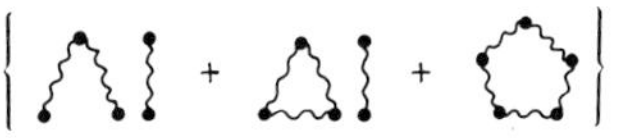

$$: \qquad B_n = 2(n-4)/[3(n-1)(n-2)(n-3)]$$
$$\times[-2(n-2)(n-3)^2+2n-5] \quad \text{if} \;\; n \geq 5; \;\; = 0 \;\; \text{otherwise}. \tag{3.92}$$

The equation of state for hard squares which includes only Eq. (3.90), or both Eqs. (3.90) and (3.91), or all three terms, Eqs. (3.90), (3.91), and (3.92), reproduces, respectively, the first three, four, and five virial coefficients exactly. These equations of state are given as follows:

$$\{\phi\}: \qquad \beta P/\varrho = 1/(1-\varrho)^2, \tag{3.93}$$

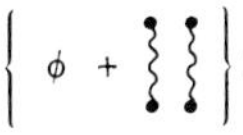

$$|\phi + \;|: \qquad \beta P/\varrho = [1 + \varrho - (3\varrho^2/2)]/(1-\varrho)^2 + \varrho \ln(1-\varrho), \tag{3.94}$$

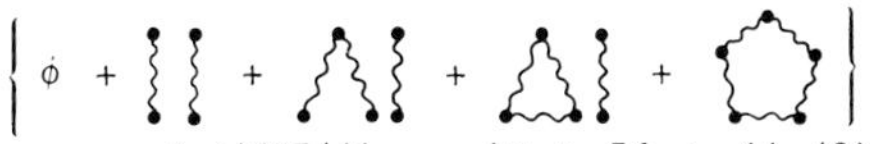

$$|\phi + \; + \; + \; + \;|: \qquad \beta P/\varrho = [1+6\varrho - (23\varrho^2/3)$$
$$-(\varrho^4/6)]/(1-\varrho)^2 + [6 + (4\varrho/3) + (\varrho^2/3)] \ln(1-\varrho). \tag{3.95}$$

The above three equations diverge at the close-packed density ($\varrho = 1$) of hard squares, and the third approximation exhibits a maximum in pressure and predicts negative pressures at high densities. It is as yet unknown whether the feature follows from a true behavior of the virial series, or if use of the higher-order approximation will contain a large positive contribution which can cause a van der Waals loop, or an isotherm which increases monotonically with density.

The present reformulation can also be generalized to the fugacity series for pressure, the number-density series for the surface tension, and the series for the s-particle potential of mean force. This reformulation is of value for more realistic potentials such as the Lennard-Jones potential (except at low temperatures). At low temperatures, the contribution by the attractive portion of a given potential to B_n or $g^n(r)$ overwhelms the contribution by the repulsive portion. As will be shown later, the virial series is consequently expected to be of limited use at low temperatures.

2. *Application—Repulsive Potentials*

In the following discussion, the hard-sphere potential will be used to explain how to incorporate the reformulation given in the preceding subsection into the Monte Carlo calculations of B_n and $g^n(r)$. For hard disks, the 1-, 8-, and 93-modified-star integrals occurring in the expressions of B_5, B_6, and B_7 are zero because configurations corresponding to these modified stars are geometrically inaccessible in two dimensions. Likewise, the 0-, 1-, and 7-modified stars are geometrically inaccessible for hard spheres. Correspondingly, the numbers of modified star integrals necessary for evaluating B_5, B_6, and B_7 are at most 4, 15, and 78, respectively, for hard disks, and 5, 22, and 164 for hard spheres. Similarly, the doubly-rooted star integrals occurring in the expressions such as $g^3(r)$ [Eq. (3.89)] have severe geometric restrictions which vary with separation distance.

In the following, we shall illustrate in some detail how a Monte Carlo method can be used to evaluate $g^3(r)$ by means of its reformulated expression (3.89). Generalization of the method to other B_n or $g^n(r)$ can be done in a straightforward manner.

To make a "trial configuration," place particle 1 at the origin and particle 2 along the x axis. Particles 3 and 4 are next chosen at random within a sphere of radius 1 (using the sphere diameter as the unit of length) with the restriction that $f_{13} = f_{34} = -1$. Next, particle 5 is picked and placed in a random position within a unit sphere centered around particle 2. This completes the buildup step of the trial configuration at specified distance r between particles 1 and 2 in the Monte Carlo calculations. The computer next checks the remaining distances between all pairs of points to see if any modified doubly-rooted graph in $g^3(r)$ can match this random configuration. The total number of configurations, which matches a certain modified doubly-rooted graph generated from the total number of random trial configurations, represents the ratio of integrals corresponding to the modified doubly-rooted graph to that for the trial configuration. For example, the following scheme evaluates the integral corresponding to the sixth modified doubly-rooted graph in the expression for W_5 given in Eq. (3.89):

$\int d\mathbf{r}_3\, d\mathbf{r}_4\, d\mathbf{r}_5$ 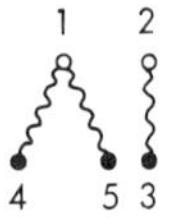 $= (2B_2)^3(-1)^6$

$\times$ [number of occurrences of 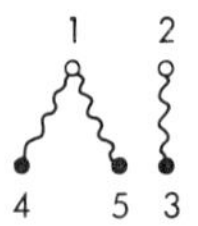

and 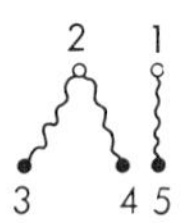]

$\times$ (unlabeling factor $\times$ number of trials)$^{-1}$. (3.96)

In Eq. (3.96), the factor $(-1)^6$ represents the number of f-functions in the graph integral. The unlabeling factor in Eq. (3.96) is defined as the number of ways this graph can be placed in a configuration with the restrictions $f_{13} = f_{34} = f_{25} = -1$. For the graph in Eq. (3.96), the corresponding unlabeling factor is 2. When the testing is over at a given r, r is increased by a fixed amount dr and the occurrence of configurations satisfying modified doubly-rooted stars in $g^3(r)$ is tested again using the same trial configuration. This procedure is repeated until r reaches $r_{\max}$ $(= 2)$, beyond which all integrals for modified doubly-rooted graphs in $g^3(r)$ either vanish or are analytically calculable. Then the whole procedure is repeated using a new random trial configuration until a desired accuracy for $g^3(r)$ is achieved. Of the modified doubly-rooted graphs contributing to $g^3(r)$, only the ninth member (a triangle made up of wiggly lines $\tilde{f}_{34}$, $\tilde{f}_{45}$, and $\tilde{f}_{35}$) in Eq. (3.89) cannot be realized from the trial configuration mentioned above. Thus, the integral corresponding to this graph is evaluated by a separate Monte Carlo program using a different type of trial configuration specified by $f_{13} = f_{14} = f_{15} = f_{23} = f_{24} = f_{25} = -1$. At any rate, this modified doubly-rooted graph integral contributes very little for $r \geq 1$ and vanishes for $r \geq (8/3)^{1/2}$. For hard disks, this integral vanishes identically for $r \geq 1$. Numerical values of $g^3(r)$ by the above scheme would require about 1 hr of CDC 7600 computing time to achieve better than 99% accuracy within the range of $r \leq 2$. At $r = 1$, the 12 modified doubly-rooted graphs in Eq. (3.89) make the following contributions to $g^3(r)$ [arranged in order of appearance in Eq. (3.89)]:

$$\begin{aligned} g^3(1) &= 1.3073 + 0.3083 - 0.4404 - 0.0518 + 0.1029 - 0.1568 \\ &\quad -0.1956 - 0.0031 + 0.0001 - 0.0074 - 0.1360 + 0.2876 \\ &= 1.015. \end{aligned}$$

This shows the relative importance of various graph integrals. Table X compares the Monte Carlo data on $g^3(r)$ (Lee *et al.*, 1966) together with the corresponding results obtained from various approximate theories of fluids.

The virial coefficients for hard spheres and hard disks can be similarly evaluated by means of a suitable Monte Carlo method. For further information, refer to the original articles (Ree and Hoover, 1964a, 1967). The Monte Carlo value of the fifth virial coefficient for hard spheres is given in Table XI, which also lists data on B_5 calculated from the various approximate theories. Note, in particular, in Tables X and XI, the

TABLE X

VALUES OF THE HARD-SPHERE $g^3(r)$ CALCULATED FROM THE EXACT, THE BGY2, THE PYII, THE BGY, THE PY, AND THE CHNC THEORIES OF $g(r)$[a]

r	Exact[b]	BGY2[c]	PYII[d]	BGY[c]	PY[c]	CHNC[c]
0.0	32.990	32.918(6)	32.868	39.355	10.910	57.610
0.2	21.680(9)[e]	21.933(6)[f]	21.627(9)[e]	23.762	8.472	38.645
0.4	13.365(8)	13.522(6)	13.344(8)	12.928	6.138	23.906
0.6	7.414(7)	7.462(6)	7.406(7)	6.061	4.011	13.113
0.8	3.425(6)	3.424(6)	3.422(6)	2.217	2.193	5.785
1.0	1.015(4)	1.005(6)	1.014(4)	0.436	0.790	1.330
1.2	−0.210(4)	−0.214(4)	−0.210(4)	−0.144	−0.106	−0.879
1.4	−0.612(3)	−0.619(2)	−0.613(3)	−0.139	−0.485	−1.523
1.6	−0.531(4)	−0.531(2)	−0.531(4)	0.090	−0.443	−1.264
1.8	−0.175(3)	−0.173(1)	−0.175(3)	0.351	−0.123	−0.616
2.0	0.302	0.302	0.302	0.588	0.321	0.102
2.2	0.558	0.558	0.558	0.663	0.561	0.498
2.4	0.505	0.505	0.505	0.533	0.505	0.493
2.6	0.332	0.332	0.332	0.336	0.332	0.330
2.8	0.164	0.164	0.164	0.164	0.164	0.164
3.0	0.061	0.061	0.061	0.061	0.061	0.061

[a] The sphere diameter is taken as the unit of distance.

[b] Ree *et al.* (1966).

[c] Lee *et al.* (1968).

[d] Values in this column are evaluated by subtracting from the exact $g^3(r)$ a contribution from a missing graph integral in the PYII expression (Wertheim, 1967), which was evaluated by Ree *et al.* (1966).

[e] Numerals in the parentheses indicate the standard deviation in the last digits; i.e., $21.680(9) = 21.680 \pm 0.009$.

[f] Numerals in parentheses indicate possible errors in the last digit due to numerical integrations carried out to obtain these results.

significant improvement in the BGY2 $g^3(r)$ and B_5 over those from the BGY1 theory. Likewise, the PYII theory shows very good agreement with the Monte Carlo values. The PYII values of $g^3(r)$ and B_5 are evaluated from the exact data on $g^3(r)$ and B_5 subtracted by contributions of a modified graph integral which is absent in the PYII approximation (Wertheim, 1967). In view of this, a numerical solution of the PYII equation will be undoubtedly worthwhile, although an attempt in this direction will encounter considerable practical complications, arising from geometries involving as many as six particles. In view of this, the BGY2 equation discussed in Section IV,E has a practical advantage,

TABLE XI

COMPARISON OF THE FIFTH VIRIAL COEFFICIENT B_5 FOR HARD SPHERES CALCULATED FROM THE VARIOUS APPROXIMATE THEORIES OF $g(r)$[a]

Theory	$B_5/(B_2)^4$		Reference
	Virial	Compressibility	
Exact	0.1103(3)		Ree and Hoover (1964a); Katsura and Abe (1963); Kilpatrick and Katsura (1966); Ree *et al.* (1966)
BGY	0.0475	0.1335	Nijboer and Fieschi (1953)
BGY2	0.1090(8)	0.1112(5)	Lee *et al.* (1968)
PY	0.0859	0.1211	Rushbrooke (1963); Wertheim (1963); Thiele (1963)
PY2	0.1240	0.1074	Rushbrooke (1965); Oden *et al.* (1966)
PYII[b]	0.1103(4)	0.1105(3)	Rowlinson (1968); Wertheim (1967)
CHNC	0.1447	0.0493	Rushbrooke and Hutchinson (1961)
CHNC2	0.0657	0.1230	Rushbrooke (1965); Oden *et al.* (1966)
KS[c]	0.136(1)	0.082(1)	Chung and Espenscheid (1968)

[a] Numerals in parentheses indicate the possible errors in the last digit, i.e., 0.1103(3) $\equiv 0.1103 \pm 0.0003$.

[b] Standard deviations are estimated using the Monte Carlo values of a modified rooted graph integral [the fifth graph in Table II in Ree *et al.* (1966)] which is a measure of deviation of the PYII$g_3(r)$ from the exact values.

[c] Kirkwood–Salsburg equation with the Kirkwood superposition approximation.

although the latter equation, which is a set of differential equations, does not permit a convenient graph-theoretic interpretation of coefficients in the density series of $g(r)$ or g_{123}.

3. *Application—Attractive Potentials*

Aside from providing a useful criterion for judging the adequacy of various approximate theories of fluids, knowledge of B_n or $g^n(r)$ for attractive potentials helps us to understand several interesting theoretical questions involving the low-temperature behavior of individual virial coefficients and the convergence of virial series as a function of temperature. Neither question has been answered in a satisfactory way for a wide class of potentials. At present, however, the virial coefficients for a square-well potential have been rigorously shown to become negative when the temperature is lowered sufficiently (Hoover and Ree, 1965).

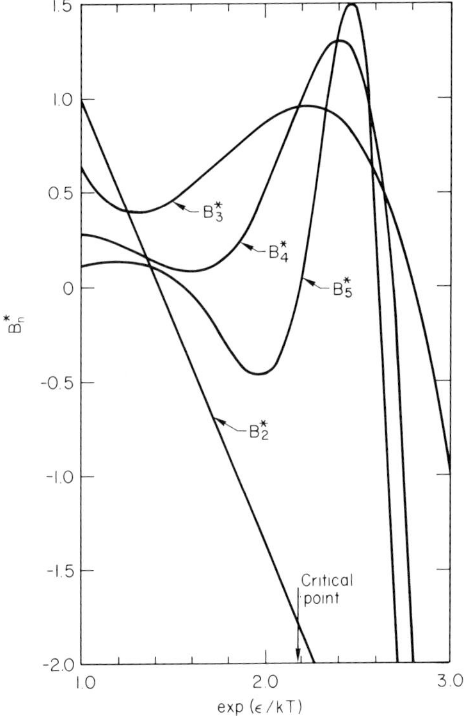

FIG. 16. The reduced virial coefficients B_n^* $[\equiv B_n(2\pi\sigma^3/3)^{-n+1}]$ for three-dimensional square-well particles, Eq. (3.32); $\lambda = 1.15$ (Barker and Henderson, 1967b).

Figure 16 shows the second, third, fourth, and fifth virial coefficients of the three-dimensional square-well potential [Eq. (3.32)] with a spherical hard core (Barker and Henderson, 1967b), in which the parameter λ, specifying the extent of the attractive range, is chosen to be 1.5. This figure demonstrates that the individual virial coefficients decrease rapidly to $-\infty$ as the temperature is decreased, and the rate of decrease is faster for the higher virial coefficients. This qualitative behavior of B_n can also be found in more realistic potentials, such as the Lennard-Jones potential (Barker *et al.*, 1966). If this feature is presumed to persist both for the more general class of potentials with an attractive tail and for higher virial coefficients for a given type of potential, the corresponding virial series will start to diverge at temperatures slightly below the critical temperature.

There is empirical evidence that the virial series could be used to predict some of the critical constants. This is demonstrated in Table XII

TABLE XII

CRITICAL CONSTANTS DERIVED FROM THE TRUNCATED VIRIAL SERIES FOR VARIOUS MODEL SYSTEMS

Potential	Constant	Truncated virial series—number of terms			Experiment
		(B_2, B_3)	(B_2, B_3, B_4)	(B_2, B_3, B_4, B_5)	
Lennard-Jones[a]	kT_c/ε	1.449	1.300	1.291	1.34 ± 0.02[e]
	$\varrho_c(2\pi\sigma^3/3)$	0.771	0.561	0.547	0.71 ± 0.04
	$P_c/kT_c\varrho_c$	0.339	0.352	0.352	0.33 ± 0.03
	$P_c/\varepsilon\varrho_c$	0.491	0.458	0.455	0.45
Square-well with spherical core (=1.5)[a,b,d]	kT_c/ε	1.33	1.16	1.13	1.26 ± 0.01[f]
	ϱ_c/ϱ_0	0.202	0.120	0.100	0.241 ± 0.01
	$P_c/kT_c\varrho_c$	0.333	0.366	0.416	0.28 ± 0.01
	$P_c/\varepsilon\varrho_c$	0.443	0.425	0.470	0.353
Cubic square-well with cubic core (=2)[c,d]	kT_c/ε	3.134	3.394	2.994	—
	ϱ_c/ϱ_0	0.1546	0.2175	0.1195	—
	$P_c/kT_c\varrho_0$	0.333	0.297	0.366	—
	$P_c/\varepsilon\varrho_c$	1.045	1.009	1.096	—

[a] Barker *et al.* (1966): $\phi(r) \equiv 4\varepsilon[(\sigma/r)^{12} - (\sigma/r)^6]$.
[b] Barker and Henderson (1967).
[c] Hoover and Ree (1965).
[d] ϱ_0 denotes the close-packing density.
[e] Verlet (1967).
[f] Alder and Hecht (1969).

by virial series truncated by the first three, four, and five virial coefficients for the Lennard-Jones and the three-dimensional, square-well particles having either spherical or cubic cores (Hoover and Ree, 1965). The latter potential, despite its unphysical shape, has the advantage in that the virial coefficients are analytically calculable. Comparison of the critical properties, in terms of the number of terms retained in the virial series, shows that PV at the critical point can be fixed with an accuracy of 10% or less. In the case of the Lennard-Jones potential, an apparently good convergence of the virial series is particularly notable. An apparent discrepancy between the "experimental" predictions and the calculations based on the truncated series may be reconciliable, provided that a proper correction is included to take care of the large density fluctuations, which are not properly accounted for by the numerical values given in the column "experiment" and which might appreciably alter the critical constants. Even if this problem is satisfactorily solved, further caution should be exercised in a comparison of this sort. The critical constants of the truncated virial series are determined analytically and might be shifted somewhat from the true values describing the critical point, in whose neighborhood the equilibrium properties exhibit a highly singular behavior.

Unlike the case for a hard-core potential (whose virial coefficients can be evaluated from unique geometric shapes of configurations fixed by their modified stars), the f-function for an attractive potential is not simply -1 or 0. Consequently, actual calculations for individual virial coefficients in this case require more complicated numerical procedures than the corresponding hard-core problem. One numerical procedure adaptable for this purpose was proposed by Barker and Monaghan (1962), who used a method of Legendre polynomials to expand the f-function and then calculated the fourth virial coefficients of the square-well and the Lennard-Jones potentials. Their method was later extended by Henderson and Oden (1966) to obtain $g^2(r)$ for the Lennard-Jones potential. Use of this method has resulted in a large amount of additional numerical data on B_n ($n \leq 5$) and $g^n(r)$ ($n \leq 3$) for the square-well potential (Barker and Henderson, 1967b) and the Lennard-Jones potential (Kim *et al.*, 1969; Barker *et al.*, 1966). In the following discussion, we explain the use of the Legendre polynomial expansion by evaluating an integral for the four-point complete star [the last graph in Eq. (3.80)].

Placing particle 1 at the origin and particle 2 along the z axis and using θ_{ijk} to denote the angle formed by straight lines r_{ij} and r_{jk} [see Eq. (3.55)], the four-point complete star integral can be rewritten as follows:

$$\begin{aligned} D_3(T) &\equiv \int\int\int d\mathbf{r}_2\, d\mathbf{r}_3\, d\mathbf{r}_4 (f_{12} f_{13} f_{14} f_{23} f_{24} f_{34}) \\ &= 8\pi^2 \int_0^\infty dr_{12}(r_{12}^2 f_{12}) \int_0^\infty dr_{13}(r_{13}^2 f_{13}) \int_0^\infty dr_{14}(r_{14}^2 f_{14}) \\ &\quad \times \int_0^\pi d\theta_{213}(\sin\theta_{213}) \int_0^\pi d\theta_{214}(\sin\theta_{214}) \int_0^{2\pi} d\phi(f_{23} f_{34} f_{24}), \end{aligned} \tag{3.97}$$

where ϕ denotes the angle formed between the planes 123 and 214. The quantities f_{23} and f_{24} occurring in Eq. (3.97) are expanded by Legendre polynomials in $P_l(\cos\theta_{213})$ and $P_l(\cos\theta_{214})$, respectively; i.e.,

$$f_{ij} = \sum_{l=0}^{\infty} [A_l(r_{1i}, r_{1j}) P_l(\cos\theta_{i1j})], \tag{3.98}$$

$$A_l(r_{1i}, r_{1j}) = [(2l+1)/2] \int_{-1}^{1} d(\cos\theta_{i1j}) f_{ij} P_l(\cos\theta_{i1j}). \tag{3.99}$$

Likewise, the quantity f_{34} in Eq. (3.97) is expanded by means of Eq. (3.98). Since the angle θ_{214} does not occur as an independent variable in the integrations in Eq. (3.97), $P_l(\cos\theta_{314})$ in the expansion of f_{34} has to be rewritten as functions of θ_{213}, θ_{214}, and ϕ. This can be done by using the addition theorem of the Legendre polynomials,

$$P_l(\cos\theta_{314}) = P_l(\cos\theta_{213})P_l(\cos\theta_{214}) + 2\sum_{m=1}^{l}\{[(l-m)!/(l+m)!] \times P_l^m(\cos\theta_{213})P_l^m(\cos\theta_{214})\cos\phi\}. \tag{3.100}$$

When the expansion (3.98) is used to replace f_{23}, f_{24}, and f_{34} in Eq. (3.97), and the factor $P_l(\cos\theta_{314})$ in the expansions is replaced by Eq. (3.100), the only contribution to integration over ϕ comes from the first term in the right-hand side of Eq. (3.100). Additional simplification occurs because of the orthogonality relation:

$$\begin{aligned}\int_{-1}^{1} d(\cos\theta)[P_l(\cos\theta)P_m(\cos\theta)] &= 2/(2l+1) \quad \text{if} \quad l=m, \\ &= 0 \quad \text{otherwise.}\end{aligned} \tag{3.101}$$

The resulting expression for $D_3(T)$ is

$$D_3(T) = 64\pi^2\left\{\sum_{l=0}^{\infty}(2l+1)^{-2}\int_0^\infty dr_{12}r_{12}^2\int_0^\infty dr_{13}r_{13}^2 \times \int_0^\infty dr_{14}[r_{14}^2 f_{12}f_{13}f_{14}A_l(r_{12},r_{14})A_l(r_{13},r_{14})A_l(r_{12},r_{13})]\right\}. \tag{3.102}$$

The above expression for $D_3(T)$, which is a single summation of a threefold integral, is a more convenient one to use for a numerical calculation than the original expression (3.97), which requires single sixfold integration. The integrand in Eq. (3.102) is a symmetric function in its independent variables r_{12}, r_{13}, and r_{14}. Since the limits of the integrations in Eq. (3.102) are the same for each variable, only one-sixth of the entire configuration (with restriction $r_{12} \leq r_{13} \leq r_{14}$) is required in the actual evaluation of $D_3(T)$. Introduction of the above symmetry property reduces further calculation time. In actual calculations, the summation in Eq. (3.102) is truncated beyond the first few terms, and the upper limit of the integrals in Eq. (3.102) is cut off at a finite distance, which is chosen large enough that negligible contribution to $D_3(T)$ is made by configurations beyond this distance. As temperature is progressively lowered, the series (3.102) for $D_3(T)$ converges more slowly and must include an increasingly higher number of terms in its truncated expression. In case of the Lennard-Jones potential at $kT/\varepsilon \approx 1.2$, the fourth term ($l = 3$) in the expansion (3.102) for $D_3(T)$ attains the same order of magnitude as $D_3(T)$ itself. Nevertheless, its contribution to the fourth virial coefficient is relatively small ($\approx$4%). At lower temperatures, however, an increasingly large numbers of terms must be retained in order to achieve proper convergence. These numbers are, for example, eight terms in calculations of $B_5(T)$ (Barker *et al.*, 1966) at $kT/\varepsilon \gtrsim 0.625$, and eleven terms for $g^3(r)$ (Kim *et al.*, 1969) at $kT/\varepsilon \gtrsim 0.7$.

For the sake of those who are interested in other approaches to calculating B_n ($n \leq 4$) and $g^n(r)$ ($n \leq 2$) for the square-well potential, we mention the work of Katsura and Nishihara (1969) [$g^2(r)$ and B_4 for $\lambda = 2$)], Hauge (1963) (some graph integrals in B_4 for general λ), and McQuarrie (1964) [some graph integrals in $g^2(r)$].

4. *Bounds on the Fugacity and Virial Coefficients*

To attain complete mathematical rigor, the virial series [Eq. (3.77) and (3.78)] should be shown to have a nonzero value of the radius of convergence R_ϱ. Similarly, the fugacity series of P and ϱ,

$$\beta P = z + b_2 z^2 + b_3 z^3 + b_4 z^4 + \dots, \tag{3.103}$$

$$\varrho = z + 2b_2 z^2 + 3b_3 z^3 + 4b_4 z^4 + \dots, \tag{3.104}$$

must have a nonzero radius of convergence R_z. Due to recent developments, it is felt that a detailed exposition of this important subject is both timely and desirable. However, it would require much more space than is available here. For this reason, plus the fact that this chapter is primarily intended to cover numerical work using a computer, the following discussion will be confined to a survey of available literature and brief comparisons and discussions of the work reported.

In the following discussion, the Mayer f-function of a pairwise additive potential is assumed to satisfy the following condition:

$$B \equiv \int d\mathbf{r} \, | f(\mathbf{r}) | < \infty. \tag{3.105}$$

Furthermore, the potential must belong to one of three possible classes: (a) the class of nonnegative potentials [$\phi(r) \geq 0$]; (b) the class of $\phi(r)$ having a hard-core with attraction satisfying the condition $\sum_{i=1, j\neq i}^{N} \phi_{ij} \geq$ negative constant; (c) the class of stable potentials, $\sum_{i,j=1, i\neq j}^{N} \phi_{ij} \geq N$

$\times$ negative constant. By choosing a zero width for the core, class (a) can be shown to belong to a subclass of class (b), which in turn belongs to one of the subclasses of class (c).

First significant progress on the bounds was made by Groeneveld (1962), who showed that for a class (a) potential the fugacity coefficient b_l in Eq. (3.103) is bounded by

$$l^{-1} \leq (-B)^{1-l} b_l \leq l^{l-2}/l!, \qquad l = 1, 2, 3, \ldots. \tag{3.106}$$

Three theoretically important consequences of Groeneveld's bound [Eq. (3.106)] are: first, b_l alternates in sign; second, the radius of convergence R_z of the fugacity series (3.103) is finite and bounded by

$$(eB)^{-1} \leq R_z \leq B^{-1}; \tag{3.107}$$

and third, the nearest singular point in z from the origin which prevents convergence of the series (3.103) lies on negative z axis; i.e., $z = -R_z$. This statement follows from the first statement on the sign of b_l. The lower bound on R_z, given by the left-hand inequality in Eq. (3.107), is shown to be also the lower bound for the class (c) potentials (Ruelle, 1963; Penrose, 1963).

For a potential belonging to class (c), Penrose obtained another upper bound for R_z:

$$R_z \leq | [l/(l-1)] b_l |^{1/(l-1)}, \qquad l = 2, 3, \ldots. \tag{3.108}$$

For a hard-core potential, the above bound can be replaced by a slightly tighter bound,

$$R_z \leq | \varrho_0 / l b_l |^{1/l}. \tag{3.109}$$

In practice, even for hard spheres, b_l are not known beyond $l = 7$. Therefore, use of the upper bounds (3.108) and (3.109) will probably not result in a significant improvement. For hard spheres, use of b_7 in the bound (3.109) yields poorer results than the one given by Groeneveld [Eq. (3.107)] (see Table XIII).

Groeneveld's approach to obtain a lower bound on R_z can be called the "Cayley tree" method, since Groeneveld's lower bound on R_z (therefore, an upper bound on b_l) can also be obtained (Penrose, 1967) by considering only the simplest subclass (i.e., Cayley trees) of all the connected graphs of l points appearing in the Mayer expression of b_l. An l-point Cayley tree is a connected Mayer graph of l points consisting only of lines of f-functions without triangles, squares, etc. (Uhlenbeck and Ford, 1962). The next improvement to the Cayley-tree method is the inclusion of all l-point Cayley trees, plus some l-point connected graphs composed of lines and triangles. Both upper and lower bounds on b_l and R_z can be obtained for a class (a) of nonnegative potentials (Ree, 1967). Using the superscripts $+$ and $-$ to represent the upper and lower bounds, respectively, these bounds are given by:

$$ll! \mid b_l^+ \mid = C^{(l-1/)2} \sum_{i=0}^{l-1} \left\{ \binom{l-1}{i} (-1)^i q_i[(l/2) - \alpha l] q_{l-i-1}[(l/2) + \alpha l] \right\}; \quad q_0 \equiv 1; \tag{3.110}$$

$$l(l-1) \mid b_l^- \mid = B \left\{ \sum_{j=1}^{l-1} j(l-j) \mid b_j^- \mid \mid b_{l-j}^- \mid \right\} + \tilde{C} \sum_{i=2}^{l-1} \sum_{i=1}^{j-1} \times \{ i(l-j)(j-i)(j-1)^{-1} \mid b_i^- \mid \mid b_{j-i}^- \mid \mid b_{l-j}^- \mid \}, \quad l > 2, \tag{3.111}$$

$$R_z^+ = B^{-1}(1-\lambda) \int_0^1 dt[(1+\lambda t)^{-1/(1+\lambda)}(1-t)^{-\lambda/(1+\lambda)}], \tag{3.112}$$

$$R_z^- = C^{-1/2}(2\alpha - 1)^{\alpha - 1/2}(2\alpha + 1)^{-\alpha - 1/2}, \tag{3.113}$$

where $b_1 \equiv 1$, and $b_2^- = b_2$. The following symbols are used in Eqs. (3.110)–(3.113): $C \equiv 3B_3$, $\tilde{C} = B^2 - C$, $\alpha = \frac{1}{2}BC^{-1/2}$, $\lambda = 1 + [B^2/(2\tilde{C})] + [(1 + B^2(2\tilde{C})^{-1})^2 - 1]^{1/2}$, and $q_i(x) = \prod_{j=1}^{i}(x - j + 1)$. As can be seen in Table XIII, the bounds R_z^+ and R_z^- improve Groeneveld's [Eq. (3.108)] and Penrose's [Eq. (3.109)] bounds for hard rods, disks, and spheres. For hard rods, R_z is exactly equal to $1/eB_2$ ($= 0.368/B_2$), in comparison to the bounds R_z^+ ($= 0.451/B_2$) and R_z^- ($= 0.219/B_2$).

TABLE XIII

UPPER (+) AND LOWER (−) BOUNDS OF THE RADII OF CONVERGENCE OF THE FUGACITY (z) AND VIRIAL (ϱ) SERIES OF PRESSURE FOR HARD RODS, DISKS, AND SPHERES[a]

	Lower bounds: R_z^-		Upper bounds: R_z^+			Lower bounds: R_ϱ^-		
	Eq.(3.113)	(3.107)	(3.112)	(3.107)	(3.109)	(3.118)	(3.116)	(3.115)
Rods	0.219	0.184	0.451	0.500	0.483	0.1392	0.0812	0.0724
Disks	0.208	0.184	0.428	0.500	0.492	0.1392	0.0788	0.0724
Spheres	0.202	0.184	0.414	0.500	0.515	0.1392	0.0773	0.0724

[a] Penrose's upper bound [Eq. (3.109)] was calculated by using b_7. The bounds are given in units of $B_2 \equiv 1$.

Penrose's bound for hard rods, for which b_l can be calculated to an arbitrary order, shows slow convergence. For hard spheres, the gap $R_z^+ - R_z^-$ is reduced to $0.2/B_2$ from $0.3/B_2$ for Groeneveld's bounds. Similar bounds on b_l for hard spheres are shown in Table XIV. Note that the bounds [Eqs. (3.110)–(3.113)] reduce to Groeneveld's bounds [Eqs. (3.106) and (3.107)] when C and $\tilde{C}$ are set to zero.

TABLE XIV

THE VALUES OF THE COEFFICIENT b_l AND ITS BOUNDS b_l^+ [Eq. (3.110)] AND b_l^- [Eq. (3.111)], AND THE BOUNDS GIVEN BY GROENEVELD [Eq. (3.106)] FOR A SYSTEM OF HARD SPHERES[a]

l =	3	4	5	6	7
b_l	1.688	−3.554	8.469	−21.838	59.490
b_l^+	1.688	−3.771	10.055	−30.178	98.369
b_l^-	1.688	−3.063	5.938	−11.978	24.842
Groeneveld:					
b_l^+	2.000	−5.333	16.667	−57.600	213.422
b_l^-	1.333	−2.000	3.200	− 5.333	9.143

[a] These bounds are expressed in units of $B_2 \equiv 1$.

Similar bounds on the radius of convergence R_ϱ of the virial series can be obtained. These are achieved using two different approaches. The first method, proposed by Lebowitz and Penrose (1964), is an indirect one which makes use of Groeneveld's upper bound (3.106). This procedure yields the following bounds for class (c) potentials:

$$|B_l| \leq (1/l)(R_\varrho^-)^{1-l}, \tag{3.114}$$

$$R_\varrho \geq R_\varrho^- \equiv 0.28952/(1+u)B, \tag{3.115}$$

where u is $\exp[-\min N^{-1} \sum_{i>j>1}^{N} (2\phi_{ij}/kT)]$. Slightly improved bounds for a class (c) potential can be obtained by using the tighter bound on R_ϱ^-, which can be derived by using b_l^- given by Eq. (3.111):

$$R_\varrho^- = \max\{C^{-1/2}x[2(1-x)^{\alpha-1/2}(1+x)^{-\alpha-1/2} - 1]\}, \qquad 0 \leq x < (2\alpha)^{-1}. \tag{3.116}$$

As shown in Table XIII, the bound (3.116) tightens the bound (3.115) only by a small amount ($0.005B_2^{-1}$).

Probably a more promising approach is the second method. This approach looks directly at star graphs in an expression of B_l, and then eliminates certain f-functions as well as subgraphs including points, provided that the resulting graphs are still star graphs of a possibly small number of points, This approach was extensively investigated by Groeneveld (1967), who obtained several bounds on B_l and R_ϱ. The following Groeneveld bounds are applicable for nonnegative potentials. These bounds are selected for the simplicity of the expressions and are by no means his best ones:

$$lB_l \leq B^{l-1}/(1 + C_0)C_0^{l-2}, \qquad l \geq 3, \tag{3.117}$$

$$R_\varrho \geq C_0/B = 0.27846/B, \tag{3.118}$$

where $C_0 = 0.27846$ is the positive root of $C_0 = \exp(-C_0 - 1)$.

From Table XIII, we see that the bound (3.118) considerably improves the lower bound obtained from Eq. (3.115) or (3.116). The bound (3.118) is still far off, either for hard rods (for which R_ϱ is known to be exactly $2/B$) or for hard spheres (for which the virial coefficients are empirically known to converge even at the freezing density $\varrho = 3.96/B$). Since the bounds (3.118) and (3.119) are obtained by considering only the relatively simple graphs, it is expected that inclusion of more complicated graphs in estimating R_ϱ will impose still tighter bounds than those currently available.

H. Additional Remarks

Besides the work described in Sections II and III, there are a great number of papers which deal with the Monte Carlo and the molecular-dynamic methods, and the literature is rapidly growing. In the following list, this literature is compiled without further comment. The list includes some references quoted earlier in the present chapter for their numerical techniques rather than their objectives. The variety of papers shows the applicability of the methods to a wide range of physically interesting problems.

(a) Velocity autocorrelation functions, diffusion coefficients, and liquid structures: Alder and Wainwright (1967, 1970), Rahman (1964, 1966);

(b) Liquid metals and ionic solutions: Paskin and Rahman (1966), Schiff (1969), Krogh-Moe *et al.* (1969), Vorontsov-Veliaminov *et al.* (1970), and the references quoted therein;

(c) Ground-state properties of quantum systems: McMillan (1965), Schiff and Verlet (1967), Hansen and Levesque (1968), Meissner and Hansen (1969), Hansen and Schiff (1969), Hansen (1969b);

(d) Elastic constants: Hoover *et al.* (1969), Squire *et al.* (1969);

(e) Vacancy motions and numbers: Benett and Alder (1968), Squire and Hoover (1969);

(f) Free-path distributions: Alder and Einwohner (1965), Einwohner and Alder (1968);

(g) Binary mixtures: Smith and Lea (1963), Alder (1964), Rotenberg (1965b), McDonald (1969), Singer (1969);

(h) Many-body (three or higher) forces and angle-dependent pair potentials: McDonald and Singer (1969), Barker *et al.* (1969), Beshinske and Lietzke (1969), Barker and Watts (1969);

(i) High-energy collision cascade and radiation damage in crystals: Gibson *et al.* (1960), for somewhat simplified models, see Beeler and Besco (1963), Robinson and Oen (1963);

(j) Magnetic systems: Binder (1969) and the references quoted therein, Watson *et al.* (1969);

(k) Electric dipoles: Bellemans *et al.* (1969) and the references quoted therein;

(l) Autocorrelations in diatomic liquids: Harp and Berne (1968);

(m) Time behavior of harmonic and anharmonic oscillators: Fermi *et al.* (1955), Hirooka and Saito (1969), Oogama *et al.* (1969).

Special References

Alder, B. J. (1964). *J. Chem. Phys.* **40**, 2724.

Alder, B. J., and Einwohner, T. (1965). *J. Chem. Phys.* **43**, 3399.

Alder, B. J., and Hecht, C. E. (1969). *J. Chem. Phys.* **50**, 2032.

Alder, B. J., and Hoover, W. G. (1968). *In* "Physics of Simple Liquids" (H. N. V. Temperley *et al.*, eds.), Chapter 4. North-Holland Publ., Amsterdam.

Alder, B. J., and Wainwright, T. E. (1957). *J. Chem. Phys.* **27**, 1208.

Alder, B. J., and Wainwright, T. E. (1959). *J. Chem. Phys.* **31**, 459.

Alder, B. J., and Wainwright, T. E. (1960). *J. Chem. Phys.* **33**, 1439.

Alder, B. J., and Wainwright, T. E. (1962). *Phys. Rev.* **127**, 359.

Alder, B. J., and Wainwright, T. E. (1967). *Phys. Rev. Lett.* **18**, 988.

Alder, B. J., and Wainwright, T. E. (1970). *Phys. Rev.* **1**, A18.

Alder, B. J., Frankel, S. P., and Lewinson, V. A. (1955). *J. Chem. Phys.* **23**, 417.

Alder, B. J., Hoover, W. G., and Wainwright, T. E. (1963). *Phys. Rev. Lett.* **11**, 241.

Alder, B. J., Hoover, W. G., and Young, D. A. (1968). *J. Chem. Phys.* **49**, 3688.

Alder, B. J., Carter, B. P., and Young, D. A. (1969). *Phys. Rev.* **183**, 831.

Baker, G. A. (1965). *In* "Advances in Theoretical Physics" (K. A. Bruckner, ed.), Vol. 1, Chapter 1. Academic Press, New York.

Barker, A. A. (1965). *Aust. J. Phys.* **18**, 119.

Barker, J. A., and Henderson, D. (1967a). *J. Chem. Phys.* **47**, 2856, 4714.

Barker, J. A., and Henderson, D. (1967b). *Can. J. Phys.* **45**, 3959.

Barker, J. A., and Monaghan, J. J. (1962). *J. Chem. Phys.* **36**, 2564.

Barker, J. A., and Watts, R. O. (1969). *Chem. Phys. Lett.* **3**, 144.

Barker, J. A., Leonard, P. J., and Pompe, A. (1966). *J. Chem. Phys.* **44**, 4206.

Barker, J. A., Henderson, D., and Smith, W. R. (1969). *Mol. Phys.* **17**, 579.

Beeler, J. R., Jr., and Besco, D. G. (1963). *J. Appl. Phys.* **34**, 2873.

Bellemans, A., and Nigam, R. K. (1966). *Phys. Rev. Lett.* **16**, 1038.

Bellemans, A., and Nigam, R. K. (1967). *J. Chem. Phys.* **46**, 2922.

Bellemans, A., Kohler, M., and Gancberg, M. (1969). *J. Chem. Phys.* **51**, 2578.

Benett, C. H., and Alder, B. J. (1968). *Solid State Commun.* **6**, 785.

Beshinske, R. J., and Lietzke, M. H. (1969). *J. Chem. Phys.* **51**, 2278.

Binder, K. (1969). *Phys. Lett.* **30**, A273.

Born, M., and Green, H. S. (1946). *Proc. Roy. Soc.* (*London*) **A188**, 10.

Brush, S. G., Sahlin, H. L., and Teller, E. (1966). *J. Chem. Phys.* **45**, 2102.

Buehler, R. J., Wentorf, R. H., and Hirschfelder, J. O. (1951). *J. Chem. Phys.* **19**, 61.

Byckling, E. (1961). *Physica* **27**, 1030.

Carley, D. D. (1963a). Radical Distribution Functions of an Electron Gas, Ph. D. dissertation. Univ. of Florida.

Carley, D. D. (1963b). *Phys. Rev.* **131**, 1406.

Chae, D. G., Ree, F. H., and Ree, T. (1969). *J. Chem. Phys.* **50**, 1581.

Chesnut, D. A. (1963). *J. Chem. Phys.* **39**, 2081.

Chesnut, D. A., and Salsburg, Z. W. (1963). *J. Chem. Phys.* **38**, 2861.

Chung, H. S., and Espenscheid, W. F. (1968). *Mol. Phys.* **14**, 317.

Cohen, E. G. D., and Murphy, T. J. (1969). *Phys. Fluids* **12**, 1404.

Debye, P., and Hückel, E. (1923). *Z. Phys.* **24**, 185.

del Rio, F., and DeWitt, H. E. (1969). *Phys. Fluids* **12**, 791.

Einwohner, T., and Alder, B. J. (1968). *J. Chem. Phys.* **49**, 1458.

Ewald, P. P. (1921). *Ann. Phys.* **64**, 253.

Fermi, E., Pas, J., and Ulam, S. (1955). Los Alamos Scientific Laboratory Rep. LA-1940.

Feller, W. (1950). "An Introduction to Probability Theory and Its Applications," Vol. I, Sec. 15. Wiley, New York.

Fickett, W., and Wood, W. W. (1960). *Phys. Fluids* **3**, 204.

Fisher, M. E. (1966). *In* "Critical Phenomena" (M. S. Green and J. V. Sengers, eds.), p. 108. Nat. Bur. Std. Misc. Publ. 273. U. S. Govt. Printing Office, Washington, D. C.

Fisher, I. Z., and Kopeliovich, B. L. (1960). *Dokl. Akad. Nauk USSR* **133**, 81.

Fixman, M. (1969). *J. Chem. Phys.* **51**, 3270.

Frenkel, J. (1955). "Kinetic Theory of Liquids." Dover, New York. See Sect. 4 in Chapter 3. This is a republication of English translation published by Oxford Univ. Press.

Gaunt, D. S., and Fisher, M. E. (1965). *J. Chem. Phys.* **43**, 2840.

Gibson, J. B., Goland, A. N., Milgram, M., and Vineyard, G. H. (1960). *Phys. Rev.* **120**, 1229.

GOSMAN, A. L., MCCARTY, R. D., and HUST, J. G. (1969). Thermodynamic Properties of Argon from the Triple Point to 300°K at Pressures to 1000 Atmospheres, Nat. Std. Ref. Data Service - NBC 27. U. S. Govt. Printing Office, Washington, D. C.
GROENEVELD, J. (1962). *Phys. Lett.* **3**, 50.
GROENEVELD, J. (1967). *Proc. Konikl Nederl. Acad. van Wetenschappen-Amsterdam Series B* **70**, 454, 468, 475, 490.
HANSEN, J.-P. (1969a). The Classical Lennard-Jones Solid above the Triple Point Temperature. Orsay Rep. 69/46.
HANSEN, J.-P. (1969b). *Phys. Lett.* **30**, A214.
HANSEN, J.-P. (1970). *Phys. Rev.* **2**, A221.
HANSEN, J.-P., and LEVESQUE, D. (1968). *Phys. Rev.* **165**, 293.
HANSEN, J.-P., and SCHIFF, D. (1969). *Phys. Rev. Lett.* **23**, 1488.
HANSEN, J.-P., and VERLET, L. (1969). *Phys. Rev.* **184**, 151.
HARP, G. D., and BERNE, B. J. (1968). *J. Chem. Phys.* **49**, 1249.
HAUGE, E. H. (1963). *J. Chem. Phys.* **39**, 389.
HENDERSON, D., and ODEN, L. (1966). *Mol. Phys.* **10**, 405.
HILL, T. L. (1956). "Statistical Mechanics," Chapter 6. McGraw-Hill, New York.
HIRSCHFELDER, J. O., STEVENSON, D., and EYRING, H. (1937). *J. Chem. Phys.* **5**, 896.
HIROOKA, H., and SAITO, N. (1969). *J. Phys. Soc. Japan* **26**, 624.
HIRT, C. W. (1965). *Phys. Fluids* **8**, 693.
HIRT, C. W. (1967). *Phys. Fluids* **10**, 565.
HOOVER, W. G. (1966). *J. Chem. Phys.* **44**, 221.
HOOVER, W. G., and ALDER, B. J. (1966). *J. Chem. Phys.* **45**, 2361.
HOOVER, W. G., and ALDER, B. J. (1967). *J. Chem. Phys.* **46**, 686.
HOOVER, W. G., and DEROCCO, A. G. (1962). *J. Chem. Phys.* **36**, 3141.
HOOVER, W. G., and REE, F. H. (1965). *J. Chem. Phys.* **43**, 375.
HOOVER, W. G., and REE, F. H. (1967). *J. Chem. Phys.* **47**, 4873.
HOOVER, W. G., and REE, F. H. (1968). *J. Chem. Phys.* **49**, 3609.
HOOVER, W. G., ALDER, B. J., and REE, F. H. (1964). *J. Chem. Phys.* **41**, 3258.
HOOVER, W. G., HOLT, A. C., and SQUIRE, D. R. (1969). *Physica* **42**, 123.
HOOVER, W. G., ROSS, M., JOHNSON, K. W., HENDERSON, D., BARKER, J. A., and BROWN, B. C. (1970). *J. Chem. Phys.* **52**, 4931.
KATSURA, S., and ABE, Y. (1963). *J. Chem. Phys.* **39**, 2068.
KATSURA, S., and NISHIHARA, K. (1969). *J. Chem. Phys.* **50**, 3579.
KILPATRICK, J. E., and KATSURA, S. (1966). *J. Chem. Phys.* **45**, 1866.
KIM, S., HENDERSON, D., and ODEN, L. (1969). *Trans. Faraday Soc.* **65**, 2308.
KIRKWOOD, J. G. (1935). *J. Chem. Phys.* **3**, 300.
KIRKWOOD, J. G. (1950). *J. Chem. Phys.* **18**, 380.
KRAMERS, H. A., and WANNIER, G. H. (1941). *Phys. Rev.* **60**, 252, 253.
KROGH-MOE, J., ØSTVOLD, T., and FØRLAND, T. (1969). *Acta Chem. Scand.* **23**, 2421.
LADO, F., and WOOD, W. W. (1968). *J. Chem. Phys.* **49**, 3092.
LEBOWITZ, J. L., and PENROSE, O. (1964). *J. Math. Phys.* **5**, 841.
LEBOWITZ, J. L., and PERCUS, J. K. (1961). *Phys. Rev.* **124**, 1673.
LEBOWITZ, J. L., PERCUS, J. K., and VERLET, L. (1967). *Phys. Rev.* **153**, 250.
LEE, Y.-T. (1970). The Born-Green-Yvon 2 Theory for Liquids. Ph. D. dissertation, Univ. of Utah.
LEE, Y.-T., REE, F. H., and REE, T. (1968). *J. Chem. Phys.* **48**, 3506.
LEVESQUE, D. (1966). *Physica* **32**, 1985.

Levesque, D., and Verlet, L. (1969). *Phys. Rev.* **182**, 307.
Lindemann, F. A. (1910). *Physik. Z.* **11**, 609.
Mayer, J. E., and Mayer, M. G. (1940). "Statistical Mechanics." Wiley, New York.
Mayer, J. E., and Montroll, E. W. (1941). *J. Chem. Phys.* **9**, 2.
Mayer, J. E., and Wood, W. W. (1965). *J. Chem. Phys.* **42**, 4268.
McDonald, I. R. (1969). *Chem. Phys. Lett.* **3**, 241.
McDonald, I. R., and Singer, K. (1967a). *Disc. Faraday Soc.* **43**, 40.
McDonald, I. R., and Singer, K. (1967b). *J. Chem. Phys.* **47**, 4766.
McDonald, I. R., and Singer, K. (1969). *J. Chem. Phys.* **50**, 2308.
McMillan, W. L. (1965). *Phys. Rev.* **138**, A442.
McQuarrie, D. A. (1964). *J. Chem. Phys.* **40**, 3455.
Meeron, E. (1958). *J. Chem. Phys.* **28**, 630.
Meissner, G., and Hansen, J.-P. (1969). *Phys. Lett.* **30**, A61.
Mermin, N. D. (1968). *Phys. Rev.* **176**, 250.
Metropolis, N., Rosenbluth, A. W., Rosenbluth, M. N., Teller, A. H., and Teller, E. (1953). *J. Chem. Phys.* **21**, 1087.
Michels, A., Levelt, J. M. H., and deGraaf, W. (1958). *Physica* **24**, 659.
Mitchell, D. J., and Ninham, B. W. (1968). *Phys. Rev.* **174**, 280.
Nijboer, B. R. A., and Fieschi, R. (1953). *Physica* **19**, 545.
Oden, L. Henderson, D., and Chen, R. (1966). *Phys. Lett.* **21**, 420.
Oogama, N., Hirooka, H., and Saito, N. (1969). *J. Phys. Soc. Japan* **27**, 815.
Oppenheim, I., and Mazur, P. (1957). *Physica* **23**, 197.
Orban, J. (1969). *Chem. Phys. Lett.* **3**, 702.
Orban, J., Van Craen, J., and Bellemans, A. (1968). *J. Chem. Phys.* **49**, 1778.
Paskin, A., and Rahman, A. (1966). *Phys. Rev. Lett.* **16**, 300.
Penrose, O. (1963). *J. Math. Phys.* **4**, 1312.
Penrose, O. (1967). *In* "Proceedings of the I.U.P.A.P. Conference on Statistical Mechanics and Thermodynamics, Copenhagen, July 1966." Benjamin, New York.
Poirier, J. C. (1966). *In* "Chemical Physics of Ionic Solutions" (B. E. Conway and R. G. Barradao, eds.), Chapter 2. Wiley, New York.
Powell, M. J. (1964). *Mol. Phys.* **7**, 591.
Rahman, A. (1964). *Phys. Rev.* **136**, A405.
Rahman, A. (1966). *J. Chem. Phys.* **45**, 2585.
Ree, F. H. (1967). *Phys. Rev.* **155**, 84.
Ree, F. H. (1970). *J. Chem. Phys.* **53**, 920.
Ree, F. H., and Chesnut, D. A. (1966). *J. Chem. Phys.* **45**, 3983.
Ree, F. H., and Chesnut, D. A. (1967). *Phys. Rev. Lett.* **18**, 5.
Ree, F. H., and Hoover, W. G. (1964a). *J. Chem. Phys.* **40**, 939.
Ree, F. H., and Hoover, W. G. (1964b). *J. Chem. Phys.* **41**, 1635.
Ree, F. H., and Hoover, W. G. (1967). *J. Chem. Phys.* **46**, 4181.
Ree, F. H., Keeler, R. N., and McCarthy, S. L. (1966). *J. Chem. Phys.* **44**, 3407.
Ree, F. H., Lee, Y.-T., and Ree, T. (1971). (to be published).
Rice, O. K. (1938). *J. Chem. Phys.* **6**, 476.
Robinson, M. T., and Oen, O. S. (1963). *Phys. Rev.* **132**, 2385.
Rosenbluth, M. N., and Rosenbluth, A. W. (1954). *J. Chem. Phys.* **22**, 881.
Ross, M., and Alder, B. J. (1966). *Phys. Rev. Lett.* **16**, 1077.
Rotenberg, A. (1965a). *J. Chem. Phys.* **43**, 1198.
Rotenberg, A. (1965b). *J. Chem. Phys.* **43**, 4377.

ROWLINSON, J. S. (1963). *Mol. Phys.* **6**, 517.

ROWLINSON, J. S. (1968). *In* "Physics of Simple Liquids" (H. N. V. Temperley, J. S. Rowlinson, and G. S. Rushbrooke, eds.), Chapter 3. North-Holland Publ., Amsterdam.

RUDD, W. G., SALSBURG, Z. W., YU, A. P., and STILLINGER, F. H., Jr. (1969). *J. Chem. Phys.* **49**, 4857.

RUELLE, D. (1963). *Ann. Phys.* **25**, 109.

RUNNELS, L. K. (1965). *Phys. Rev. Lett.* **15**, 581.

RUNNELS, L. K., COMBS, L. L., and SALVANT, J. P. (1967). *J. Chem. Phys.* **47**, 4015.

RUSHBROOKE, G. S. (1963). *J. Chem. Phys.* **38**, 1262.

RUSHBROOKE, G. S. (1965). *In* "Statistical Mechanics of Equilibrium and Non–Equilibrium" (J. Meixner, ed.), p. 222. North-Holland Publ., Amsterdam.

RUSHBROOKE, G. S. (1968). *In*"Physics of Simple Liquids" (H. N. V. Temperley *et al.*, eds.), Chapter 2. North-Holland Publ. Amsterdam.

RUSHBROOKE, G. S., and HUTCHINSON, P. (1961). *Physica* **27**, 647.

SACK, R. A. (1959). *Mol. Phys.* **2**, 8.

SALSBURG, Z. W., and WOOD, W. W. (1962). *J. Chem. Phys.* **37**, 798.

SALSBURG, Z. W., ZWANZIG, R. W., and KIRKWOOD, J. G. (1953). *J. Chem. Phys.* **21**, 1098.

SALSBURG, Z. W., JACOBSON, J. D., FICKETT, W., and WOOD, W. W. (1959). *J. Chem. Phys.* **30**, 65.

SALSBURG, Z. W., RUDD, W. G., and STILLINGER, F. H., Jr. (1967). *J. Chem. Phys.* **47**, 4534.

SCHIFF, D. (1969). *Phys. Rev.* **186**, 151.

SCHIFF, D., and VERLET, L. (1967). *Phys. Rev.* **160**, 208.

SHAW, J. N. (1963). Monte Carlo Calculations for a System of Hard-Sphere Ions. Ph. D. Dissertation, Duke Univ.

SIMON, F., and GLATZEL, G. (1929). *Z. Anorg. Allgem. Chem.* **178**, 309.

SINGER, K. (1966). *Nature* **212**, 1449.

SINGER, K. (1969). *Chem. Phys. Lett.* **3**, 164.

SMITH, E. B., and LEA, K. R. (1963). *Trans. Faraday Soc.* **59**, 1535.

SQUIRE, D. R., and HOOVER, W. G. (1969). *J. Chem. Phys.* **50**, 701.

SQUIRE, D. R., and SALSBURG, Z. W. (1961). *J. Chem. Phys.* **35**, 486.

SQUIRE, D. R., HOLT, A. C., and HOOVER, W. G. (1969). *Physica* **42**, 678.

STELL, G. (1964). *In* "The Equilibrium Theory of Classical Fluids" (H. F. Frisch and J. L. Lebowitz, eds.). Benjamin, New York.

STILLINGER, F. H., Jr., SALSBURG, Z. W., and KORNEGAY, R. K. (1965). *J. Chem. Phys.* **43**, 932.

THIELE, E. (1963). *J. Chem. Phys.* **39**, 474.

TONKS, L. (1936). *Phys. Rev.* **50**, 955.

VAN HORN, H. M. (1969). *Phys. Lett.* **28**, A706.

VERLET, L. (1966). *Physica* **32**, 304.

VERLET, L. (1967). *Phys. Rev.* **159**, 98.

VERLET, L., and LEVESQUE, D. (1967). *Physica* **36**, 254.

VORONTSOV-VELIAMINOV, P. N., ELIASHEVICH, A. M., RASAIAH, J. C., and FRIEDMAN, H. L. (1970). *J. Chem. Phys.* **52**, 1013.

WATSON, R. E., BLUME, M., and VINEYARD, G. H. (1969). *Phys. Rev.* **181**, 811.

WEISSBERG, H. L., and PRAGER, S. (1962). *Phys. Fluids* **5**, 1390.

WERTHEIM, M. S. (1963). *Phys. Rev. Lett.* **10**, 321.
WERTHEIM, M. S. (1967). *J. Math. Phys.* **8**, 927.
WOOD, W. W. (1952). *J. Chem. Phys.* **20**, 1334.
WOOD, W. W. (1963). Monte Carlo Calculations of the Equation of State of Systems of 12 and 48 Hard Circles. Los Alamos Scientific Laboratory Rep. LA-2827.
WOOD, W. W. (1968b). *J. Chem. Phys.* **48**, 415.
WOOD, W. W. (1970). *J. Chem. Phys.* **52**, 729.
WOOD, W. W., and JACOBSON, J. D. (1957). *J. Chem. Phys.* **27**, 1207.
WOOD, W. W., and PARKER, F. R. (1957). *J. Chem. Phys.* **27**, 720.
YOUNG, D. A., and ALDER, B. J. (1971). (to be published).
YVON, J. (1935). Actualités Sci. et Ind. No. 203.

GENERAL REFERENCES

ALDER, B. J. (1963). Experimental arithmetic, high speed computing and mathematics, *Proc. Symp. Appl. Math.* **15**, 335–349.
BEELER, J. R., Jr. (1966). *In* "Physics of Many-Particle Systems," (E. Meeron, ed.), Chapter 1. Gordon and Beach, New York.
HILL, T. L. (1956). "Statistical Mechanics." McGraw-Hill, New York.
NEECE, G. A., and WIDOM, B. (1969). *In* "Annual Review of Physical Chemistry" (H. Eyring, C. J. Christensen, and H. S. Johnston, eds.), p. 167. Annual Review, Palo Alto.
UHLENBECK, G. E., and FORD, G. W. (1962). *In* "Studies in Statistical Mechanics," (J. de Boer and G. E. Uhlenbeck, eds.), Vol. I. North-Holland Publ., Amsterdam.
WOOD, W. W. (1968a). *In* "Physics of Simple Liquids" (H. N. V. Temperley, J. S. Rowlinson, and G. S. Rushbrooke, eds.), Chapter 5. North-Holland Publ., Amsterdam.

Chapter 4

Distribution Functions

R. J. BAXTER

I. Introduction

One of the most important aims of statistical mechanics is to predict the equilibrium behavior of fluids and solids. Although this problem is mathematically well-defined, insofar as one can write down expressions for the thermodynamic quantities in terms of the partition function, these expressions are far too complicated to evaluate directly. For this reason a number of approximate theories have been evolved which involve calculating the distribution functions. These functions measure the probability of finding a number of particles in some particular configuration, and once they are known, certain thermodynamic properties can be deduced immediately.

In this chapter, some of these theories will be outlined and discussed. Throughout Sections II–VIII attention is focused on the case of a classical homogeneous one-component fluid of spherically symmetrical molecules. This is clearly a very idealized situation, and can only be regarded as applicable to the heavier inert gases. However, this is the case for which the equations become tractable and for which a reasonable amount of numerical data is available.

In Section IX it is shown how the equations can be formally extended to include inhomogeneous fluids, solids, and mixtures. The theory of functionals is used to define the various distribution and correlation functions, and to rederive two of the approximate theories previously discussed.

II. Statistical-Mechanical Properties in the Grand-Canonical Ensemble

A. Partition Function

Consider a sample of a single-component fluid of given volume V, free to exchange both particles and energy with its surroundings. Then the probability that it contains a total of N particles, located in the volume elements $d\mathbf{r}_1, \ldots, d\mathbf{r}_N$ centered on $\mathbf{r}_1, \ldots, \mathbf{r}_N$, with momenta in the elements $d\mathbf{p}_1, \ldots, d\mathbf{p}_N$ centered on $\mathbf{p}_1, \ldots, \mathbf{p}_N$, respectively, is

$$\mathscr{P}\, d\mathbf{r}_1 \cdots d\mathbf{r}_N\, d\mathbf{p}_1 \cdots d\mathbf{p}_N,$$

where

$$\mathscr{P} = (1/h^{3N}\Xi^*) \exp[(N\mu - E)/kT], \tag{2.1}$$

E being the energy of this configuration, h Planck's constant, k Boltzmann's constant, T the temperature, μ the chemical potential, and Ξ^* a normalization factor (Kittel, 1958).

If X is some quantity which is a function of the positions and momenta of the particles, its average value $\bar{X}$ is obtained by multiplying its value for a given configuration by $\mathscr{P}$ and summing over all possible configurations. The sum over configurations containing N particles is obtained by integrating with respect to $\mathbf{r}_1, \ldots, \mathbf{r}_N$ over the volume V, and with respect to $\mathbf{p}_1, \ldots, \mathbf{p}_N$ over all momentum space, except that this procedure counts separately configurations which differ only by interchanging the labels $1, \ldots, N$ of the particles. As such configurations are identical, the integral must be divided by $N!$. Summing over N then gives

$$\bar{X} = \sum_{N=0}^{\infty} \frac{1}{N!} \int_V \cdots \int d\mathbf{r}_1 \cdots d\mathbf{r}_N \int \cdots \int d\mathbf{p}_1 \cdots d\mathbf{p}_N\, X\mathscr{P}. \tag{2.2}$$

[Throughout this chapter we adopt the convention that

$$\int \cdots \int dx_{L+1} \cdots dx_N\, f = f \qquad \text{when} \quad N = L, \tag{2.3}$$

for any function f and any variables x_i. Thus, the $N = 0$ term in the series (2.2) is the value of $X\mathscr{P}$ when no particles are present.]

If X is a constant, then $\bar{X} = X$, and Eq. (2.2) simply expresses the fact that the sum of the probabilities of all possible configurations must be unity. Setting X and $\bar{X}$ equal to one and using the form (2.1) of $\mathscr{P}$, it follows that the normalization factor Ξ^* is given by

$$\Xi^* = \sum_{N=0}^{\infty} \frac{\exp(N\mu/kT)}{h^{3N}N!} \int_V \cdots \int d\mathbf{r}_1 \cdots d\mathbf{r}_N \int \cdots \int d\mathbf{p}_1 \cdots d\mathbf{p}_N\, e^{-E/kT}. \tag{2.4}$$

The Ξ^* can be regarded as a function of μ, T, and V, and in this sense is known as the *grand partition function*. For a homogeneous fluid it can be shown (Kittel, 1958) that

$$\log \Xi^* = PV/kT, \tag{2.5}$$

where P is the pressure.

If the particles interact only via central two-particle forces and there are no external fields, then the energy E of an N-particle configuration is

$$E = \sum_{i=1}^{N} (\mathbf{p}_i^2/2m) + \sum_{i<j}^{N} \phi(r_{ij}), \tag{2.6}$$

where m is the particle mass, $r_{ij} = | \mathbf{r}_i - \mathbf{r}_j |$ is the distance between particles i and j, $\phi(r)$ is the intermolecular potential, and the second summation is over all values of i and j such that $1 \leq i < j \leq N$. Substituting the form (2.6) of E into Eq. (2.4), the momentum integrations can be performed, and it is found that

$$\Xi^* = \sum_{N=0}^{\infty} \frac{z^N}{N!} \int_V \cdots \int d\mathbf{r}_1 \cdots d\mathbf{r}_N \exp\Big[- \sum_{i<j}^{N} u(r_{ij}) \Big], \tag{2.7}$$

where z is the fugacity, given by*

$$z = h^{-3}(2\pi mkT)^{3/2} e^{\mu/kT}, \tag{2.8}$$

and

$$u(r) = \phi(r)/kT. \tag{2.9}$$

B. Distribution Functions

The L-particle distribution function $n^{(L)}(\mathbf{r}_1, \ldots, \mathbf{r}_L)$ is defined so that $n^{(L)}(\mathbf{r}_1, \ldots, \mathbf{r}_L)\, d\mathbf{r}_1 \cdots d\mathbf{r}_L$ is the probability of finding a particle in each of the volume elements $d\mathbf{r}_1, \ldots, d\mathbf{r}_L$ centered on $\mathbf{r}_1, \ldots, \mathbf{r}_L$ (where $L = 1, 2, 3$, etc.). The contribution to this probability arising from configurations containing N particles (where $N \geq L$) is obtained by integrating $\mathscr{P}$ with respect to all momenta and with respect to the position coordinates $\mathbf{r}_{L+1}, \ldots, \mathbf{r}_N$, except that this procedure counts separately configurations which differ only by interchanging the labels $L+1, \ldots, N$. The integral must therefore be divided by $(N-L)!$, and summing over N then gives

$$n^{(L)}(\mathbf{r}_1, \ldots, \mathbf{r}_L) = \sum_{N=L}^{\infty} \frac{1}{(N-L)!} \int_V \cdots \int d\mathbf{r}_{L+1} \cdots d\mathbf{r}_N \int \cdots \int d\mathbf{p}_1 \cdots d\mathbf{p}_N\, \mathscr{P}. \tag{2.10}$$

Substituting the form of $\mathscr{P}$ given by Eqs. (2.1) and (2.6), the mo-

* Strictly speaking, the fugacity is $e^{\mu/kT}$, but it is convenient for the present purposes to define z by Eq. (2.8).

mentum integrations can be performed to give

$$n^{(L)}(\mathbf{r}_1, \ldots, \mathbf{r}_L) = \frac{1}{\Xi^*} \sum_{N=L}^{\infty} \frac{z^N}{(N-L)!} \int_V \cdots \int d\mathbf{r}_{L+1} \ldots d\mathbf{r}_N \times \exp\left[- \sum_{i<j}^{N} u(r_{ij}) \right]. \tag{2.11}$$

The one-particle distribution function $n^{(1)}(\mathbf{r})$ is simply the number density of particles at $\mathbf{r}$. For a homogeneous fluid this density has the uniform value ϱ, and one can write

$$n^{(1)}(\mathbf{r}) = \varrho = \bar{N}/V, \tag{2.12}$$

where $\bar{N}$ is the mean number of particles in the fluid.

Except at the critical point of a gas–liquid phase transition, a fluid exhibits only short-range order, i.e., there is no correlation between particles a large distance apart. Thus, when $\mathbf{r}_1$ and $\mathbf{r}_2$ are far apart the joint probability of finding a particle at $\mathbf{r}_1$ and a particle at $\mathbf{r}_2$ is simply the product of the individual probabilities of these events, so that

$$n^{(2)}(\mathbf{r}_1, \mathbf{r}_2) \sim n^{(1)}(\mathbf{r}_1) n^{(1)}(\mathbf{r}_2) = \varrho^2. \tag{2.13}$$

Further, for a homogeneous fluid of particles interacting *via* central forces $n^{(2)}(\mathbf{r}_1, \mathbf{r}_2)$ depends only on the distance r_{12} between the particles, and not on the position of the particle pair, or its orientation. Thus, one can define a function $g(r)$ by

$$n^{(2)}(\mathbf{r}_1, \mathbf{r}_2) = \varrho^2 g(r_{12}). \tag{2.14}$$

This $g(r)$ is known as the *radial distribution function*; from Eqs. (2.13) and (2.14) it has the property that

$$g(r) \to 1 \quad \text{as} \quad r \to \infty. \tag{2.15}$$

The $g(r)$ can be obtained experimentally by measuring the scattering of radiation through a fluid. Some typical radial distribution functions are shown in Figs. 4 and 7.

C. Thermodynamic Properties in Terms of $g(r)$

Once the radial distribution function $g(r)$ is known for given density, temperature, and interaction potential, certain thermodynamic properties can be evaluated immediately. In this section it will be shown that the internal energy, pressure, and compressibility can all be expressed in terms of $g(r)$.

1. *Internal Energy*

The internal energy U is simply the mean value $\bar{E}$ of the energy E. Replacing X by E in formula (2.2) and using Eqs. (2.1) and (2.6), the momentum integrations can be performed to give

$$U = \frac{1}{\Xi^*} \sum_{N=0}^{\infty} \frac{z^N}{N!} \int_V \cdots \int d\mathbf{r}_1 \ldots d\mathbf{r}_N \left[\frac{3}{2} NkT + \sum_{i<j}^{N} \phi(r_{ij}) \right] \times \exp\left[- \sum_{i<j}^{N} u(r_{ij}) \right]. \tag{2.16}$$

The contribution to U of the first term in the brackets is $3\bar{N}kT/2$, while each of the $N(N-1)/2$ terms $\phi(r_{ij})$ gives the same integral. Thus,

$$U = \frac{3}{2} \bar{N}kT + \frac{1}{2\Xi^*} \sum_{N=2}^{\infty} \frac{z^N}{(N-2)!} \int_V \cdots \int d\mathbf{r}_1 \cdots d\mathbf{r}_N \, \phi(r_{12}) \times \exp\left[- \sum_{i<j}^{N} u(r_{ij}) \right]. \tag{2.17}$$

Comparing the series on the right-hand side of Eq. (2.17) with the definition (2.11) of $n^{(2)}(\mathbf{r}_1, \mathbf{r}_2)$, it follows that

$$U = \tfrac{3}{2}\bar{N}kT + \tfrac{1}{2} \iint_V d\mathbf{r}_1 \, d\mathbf{r}_2 \, \phi(r_{12}) n^{(2)}(\mathbf{r}_1, \mathbf{r}_2). \tag{2.18}$$

Substituting the form (2.14) of $n^{(2)}(\mathbf{r}_1, \mathbf{r}_2)$, taking the limit of V large, and using Eq. (2.12), Eq. (2.18) can be cast into the form

$$U/\bar{N} = \tfrac{3}{2}kT + \tfrac{1}{2}\varrho \int d\mathbf{r} \, \phi(r) g(r), \tag{2.19}$$

where the integration is now over all space. Using spherical polar coordinates, this result can be written as

$$U/\bar{N} = \tfrac{3}{2}kT + 2\pi\varrho \int_0^\infty r^2 \, dr \, \phi(r) g(r). \tag{2.20}$$

2. *Virial Pressure*

An expression for the pressure can be derived either from the virial theorem of Clausius (Kittel, 1958; Henderson and Davison, 1967), or alternatively by differentiating the expression (2.7) for Ξ^* with respect to V. Adopting this latter approach, introduce the scaled coordinates

$$\mathbf{s}_i = V^{-1/3}\mathbf{r}_i; \tag{2.21}$$

the Eq. (2.7) becomes

$$\Xi^* = \sum_{N=0}^{\infty} \frac{z^N V^N}{N!} \int\cdots\int_D d\mathbf{s}_1\cdots d\mathbf{s}_N \exp\left[-\sum_{i<j}^{N} u(V^{1/3}s_{ij})\right], \tag{2.22}$$

where the domain D may be thought of as some fixed shape (e.g., a sphere centered on the origin) of volume unity. Thus, D is independent of V, and differentiating Eq. (2.22) with respect to V gives

$$V\frac{\partial \Xi^*}{\partial V}\bigg|_{\mu,\mathrm{T}} = \sum_{N=0}^{\infty} \frac{z^N V^N}{(N-1)!} \int\cdots\int_D d\mathbf{s}_1\cdots d\mathbf{s}_N \exp\left[-\sum_{i<j}^{N} u(V^{1/3}s_{ij})\right]$$
$$-\frac{1}{6}\sum_{N=0}^{\infty} \frac{z^N V^N}{(N-2)!} \int\cdots\int_D d\mathbf{s}_1\cdots d\mathbf{s}_N V^{1/3}s_{12}u'(V^{1/3}s_{12})$$
$$\times \exp\left[-\sum_{i<j}^{N} u(V^{1/3}s_{ij})\right], \tag{2.23}$$

where $u'(r)$ is the derivative of $u(r)$, and we have used the fact that differentiating the exponent gives $N(N-1)/2$ equal terms.

Returning to the original coordinates $\mathbf{r}_1,\ldots,\mathbf{r}_N$ and using Eq. (2.11) for $L = 1$ and 2, Eq. (2.23) becomes

$$\frac{V}{\Xi^*}\frac{\partial \Xi^*}{\partial V}\bigg|_{\mu,\mathrm{T}} = \int_V d\mathbf{r}_1\, n^{(1)}(\mathbf{r}_1) - \frac{1}{6}\iint_V d\mathbf{r}_1\, d\mathbf{r}_2\, r_{12}u'(r_{12})n^{(2)}(\mathbf{r}_1, \mathbf{r}_2). \tag{2.24}$$

Using Eqs. (2.12) and (2.14) and taking the limit of V large, it follows that

$$\frac{\partial}{\partial V}\log \Xi^*\bigg|_{\mu,\mathrm{T}} = \varrho - \frac{\varrho^2}{6}\int d\mathbf{r}\, ru'(r)g(r), \tag{2.25}$$

where the integration is now over all space. However, in this limit $\log \Xi^*$ is given by Eq. (2.5), where P depends on μ and T, but not on V. Using spherical polar coordinates for the integration, it therefore follows that

$$P/kT = \varrho - (2\pi\varrho^2/3)\int_0^{\infty} d\mathbf{r}\, r^3u'(r)g(r). \tag{2.26}$$

3. *Compressibility*

The isothermal compressibility K_T can be defined by

$$K_{\mathrm{T}} = \frac{1}{\varrho} \frac{\partial \varrho}{\partial P}\bigg|_{\mathrm{V,T}}. \tag{2.27}$$

To evaluate the derivative of ϱ, first note that from Eqs. (2.5) and (2.12)

$$kT \frac{\partial \varrho}{\partial P}\bigg|_{\mathrm{V,T}} = \frac{\partial \bar{N}}{\partial(\log \Xi^*)}\bigg|_{\mathrm{V,T}}. \tag{2.28}$$

Now, for given values of V and T, both Ξ^* and $\bar{N}$ can be thought of as functions of the fugacity z. Thus,

$$kT \frac{\partial \varrho}{\partial P}\bigg|_{\mathrm{V,T}} = z \frac{\partial \bar{N}}{\partial z}\bigg|_{\mathrm{V,T}} \bigg/ z \frac{\partial(\log \Xi^*)}{\partial z}\bigg|_{\mathrm{V,T}}. \tag{2.29}$$

When the variable X depends only on the number of particles in the fluid, and not on their coordinates or momenta, the formula (2.2) for the average value $\bar{X}$ of X becomes, using Eqs. (2.1), (2.6), (2.8), and (2.9),

$$\bar{X} = \frac{1}{\Xi^*} \sum_{N=0}^{\infty} \frac{z^N X}{N!} \int \cdots \int_V d\mathbf{r}_1 \ldots d\mathbf{r}_N \exp\Big[-\sum_{i<j}^{N} u(r_{ij})\Big]. \tag{2.30}$$

Thus, differentiating Eq. (2.7) with respect to z,

$$z \frac{\partial(\log \Xi^*)}{\partial z}\bigg|_{\mathrm{V,T}} = \bar{N}. \tag{2.31}$$

Also, replacing X by N in Eq. (2.30), differentiating with respect to z, and using Eq. (2.31), we get (writing $\bar{N}$ as $\langle N \rangle_{\mathrm{av}}$),

$$z \frac{\partial \langle N \rangle_{\mathrm{av}}}{\partial z}\bigg|_{\mathrm{V,T}} = \langle N^2 \rangle_{\mathrm{av}} - \langle N \rangle_{\mathrm{av}}^2. \tag{2.32}$$

Equation (2.29) can therefore be written as

$$kT(\partial \varrho/\partial P)\Big|_{\mathrm{V,T}} = (\langle N^2 \rangle_{\mathrm{av}} - \langle N \rangle_{\mathrm{av}}^2)/\langle N \rangle_{\mathrm{av}}. \tag{2.33}$$

It is of interest to note that this equation can be cast into the form

$$kT(\partial \varrho/\partial P)\Big|_{\mathrm{V,T}} = \langle N - \langle N \rangle_{\mathrm{av}} \rangle_{\mathrm{av}}^2/\langle N \rangle_{\mathrm{av}}, \tag{2.34}$$

which makes it clear that the compressibility is a measure of the magnitude of the fluctuations of N about its mean value. It is also apparent that the compressibility must be positive, since it is proportional to the average of a perfect square.

The right-hand side of Eq. (2.33) can be expressed in terms of the radial distribution function $g(r)$ by noting from Eqs. (2.11) and (2.30) that

$$\int_V \cdots \int d\mathbf{r}_1 \cdots d\mathbf{r}_L \, n^{(L)}(\mathbf{r}_1, \ldots, \mathbf{r}_L) = \langle N!/(N-L)! \rangle_{\mathrm{av}}. \qquad (2.35)$$

Writing

$$\langle N^2 \rangle_{\mathrm{av}} - \langle N \rangle_{\mathrm{av}}^2 = \langle N \rangle_{\mathrm{av}} + \langle N(N-1) \rangle_{\mathrm{av}} - \langle N \rangle_{\mathrm{av}}^2, \qquad (2.36)$$

it therefore follows that

$$\langle N^2 \rangle_{\mathrm{av}} - \langle N \rangle_{\mathrm{av}}^2 = \langle N \rangle_{\mathrm{av}} + \iint_V d\mathbf{r}_1 \, d\mathbf{r}_2 \, [n^{(2)}(\mathbf{r}_1, \mathbf{r}_2) - n^{(1)}(\mathbf{r}_1) n^{(2)}(\mathbf{r}_2)]. \qquad (2.37)$$

Using Eqs. (2.12) and (2.13) and taking the limit of V large, Eq. (2.37) becomes

$$\langle N^2 \rangle_{\mathrm{av}} - \langle N \rangle_{\mathrm{av}}^2 = \langle N \rangle_{\mathrm{av}} \{1 + \varrho \int d\mathbf{r} \, [g(r) - 1]\}. \qquad (2.38)$$

Substituting this result in Eq. (2.33) and using spherical polar coordinates in the integration gives

$$kT(\partial \varrho / \partial P)\Big|_{\mathrm{V,T}} = 1 + 4\pi\varrho \int_0^\infty r^2 \, dr \, h(r), \qquad (2.39)$$

where

$$h(r) = g(r) - 1. \qquad (2.40)$$

The function $h(r)$ occurs frequently in statistical mechanics and is known as the *indirect correlation function*. From Eq. (2.15) it has the property that it tends to zero with increasing r.

Rather than regarding the density ϱ as a function of the pressure P, it is usually more convenient theoretically to regard P as a function of ϱ. In this case Eq. (2.39) can be inverted to give

$$\frac{1}{kT} \frac{\partial P}{\partial \varrho}\Big|_{\mathrm{V,T}} = [1 + \varrho \int d\mathbf{r} \, h(r)]^{-1} \qquad (2.41)$$

(returning to the original three-dimensional integral).

4. *Other Thermodynamic Properties*

The internal energy, pressure, and compressibility are the only thermodynamic properties that can be evaluated if $g(r)$ is known at one particular density and temperature. However, if $g(r)$ is known for all densities in the range $(0, \varrho)$ at a given temperature, then the other thermodynamic variables can be obtained by integrating Maxwell's relations. In particular, the free energy F can be obtained from the relation

$$\left.\frac{\partial(F/\bar{N})}{\partial\varrho}\right|_{\mathrm{V,T}} = \varrho^{-2}P. \tag{2.42}$$

Alternatively, if $g(r)$ is known for a series of temperatures at a given density, then if F is known at one of these temperatures, it can be obtained at the others by integrating the relation

$$\left.\frac{\partial(F/T)}{\partial T}\right|_{\mathrm{V},\varrho} = -\,T^{-2}U. \tag{2.43}$$

Provided $g(r)$ is known exactly, the various methods of evaluating the thermodynamic properties must give the same results. However, if $g(r)$ is only known approximately (as is usually the case), then different answers may be obtained. In particular, the "compressibility pressure" obtained by integrating Eq. (2.41) with respect to ϱ may differ from the "virial pressure" given by Eq. (2.26).

III. Born–Green Approximation

In addition to obtaining relations between the thermodynamic properties and the one- and two-particle distribution functions, it is also possible to obtain relations between the higher distribution functions. For instance, differentiating Eq. (2.11) with respect to $\mathbf{r}_1$ gives

$$\frac{\partial}{\partial\mathbf{r}_1} n^{(L)}(\mathbf{r}_1,\ldots,\mathbf{r}_L) = -\frac{1}{\Xi^*}\sum_{N=L}^{\infty}\frac{z^N}{(N-L)!}\int\cdots\int_V d\mathbf{r}_{L+1}\cdots d\mathbf{r}_N$$
$$\times\left\{\sum_{i=2}^{N}\frac{\partial u(r_{1i})}{\partial\mathbf{r}_1}\right\}\exp\left[-\sum_{i<j}^{N} u(r_{ij})\right]. \tag{3.1}$$

The $i = 2,\ldots, L$ terms in the bracketed sum can be taken outside the integrals and the sum over N. The remaining $(N - L)$ terms all give the same integral, so that in these cases i can be replaced by $L + 1$. Using Eq. (2.11), it follows that

$$\frac{\partial}{\partial \mathbf{r}_1} n^{(L)}(\mathbf{r}_1, \ldots, \mathbf{r}_L) = - n^{(L)}(\mathbf{r}_1, \ldots, \mathbf{r}_L) \sum_{i=2}^{L} \frac{\partial u(r_{1i})}{\partial \mathbf{r}_1} - \int d\mathbf{r}_{L+1} \frac{\partial u(r_{1,L+1})}{\partial \mathbf{r}_1} n^{(L+1)}(\mathbf{r}_1, \ldots, \mathbf{r}_{L+1}), \quad (3.2)$$

where in the limit of V large the integration with respect to $\mathbf{r}_{L+1}$ can be extended over all space.

The set of Eqs. (3.2) with $L = 1, 2, 3$, etc., is known as the Born–Green–Yvon hierarchy. It is the equilibrium case of the Bogoliubov–Born–Green–Kirkwood–Yvon (BBGKY) hierarchy of equations for a time-dependent system (Cohen, 1962).

Although exact, the equations (3.2) are not of immediate use, since it is necessary to know the $(L + 1)$-particle distribution function in order to calculate the L-particle function. However, Kirkwood (1935) suggested that to a good degree of approximation the three-particle function is given by

$$n^{(3)}(\mathbf{r}_1, \mathbf{r}_2, \mathbf{r}_3) = \varrho^3 g(r_{23}) g(r_{31}) g(r_{12}), \quad (3.3)$$

i.e., that the probability of finding particles at $\mathbf{r}_1$, $\mathbf{r}_2$, and $\mathbf{r}_3$ is proportional to the product of the probabilities of finding two particles at $\mathbf{r}_2$ and $\mathbf{r}_3$, at $\mathbf{r}_3$ and $\mathbf{r}_1$, and at $\mathbf{r}_1$ and $\mathbf{r}_2$. This approximation is known as the Kirkwood superposition approximation.

Born and Green (1946) used this approximation to close the hierarchy (3.2). Setting $L = 2$ and using Eqs. (2.14) and (3.3) gives

$$\frac{\partial}{\partial \mathbf{r}_1} g(r_{12}) = - g(r_{12}) \frac{\partial u(r_{12})}{\partial \mathbf{r}_1} - \varrho \int d\mathbf{r}_3 \frac{\partial u(r_{13})}{\partial \mathbf{r}_1} g(r_{23}) g(r_{31}) g(r_{12}). \quad (3.4)$$

This equation can be simplified by noting that

$$\partial g(r_{12})/\partial \mathbf{r}_1 = (\partial r_{12}/\partial \mathbf{r}_1)\, g'(r_{12}), \quad (3.5)$$

where $g'(r)$ is the derivative of $g(r)$. Further, using Cartesian coordinates x, y, and z,

$$r_{12} = [(x_1 - x_2)^2 + (y_1 - y_2)^2 + (z_1 - z_2)^2]^{1/2}, \quad (3.6)$$

so that

$$\frac{\partial r_{12}}{\partial \mathbf{r}_1} = \left(\frac{\partial r_{12}}{\partial x_1}, \frac{\partial r_{12}}{\partial y_1}, \frac{\partial r_{12}}{\partial z_1}\right)$$

$$= \left(\frac{x_1 - x_2}{r_{12}}, \frac{y_1 - y_2}{r_{12}}, \frac{z_1 - z_2}{r_{12}}\right)$$

$$= \frac{\mathbf{r}_1 - \mathbf{r}_2}{r_{12}}. \tag{3.7}$$

Using Eqs. (3.5) and (3.7), together with the corresponding results for the derivatives of $u(r_{12})$ and $u(r_{13})$, on dividing by $g(r_{12})$ Eq. (3.4) becomes

$$\frac{\mathbf{r}_1 - \mathbf{r}_2}{r_{12}}\left[\frac{g'(r_{12})}{g(r_{12})} + u'(r_{12})\right] = -\varrho \int d\mathbf{r}_3 \, \frac{\mathbf{r}_1 - \mathbf{r}_3}{r_{13}} \, u'(r_{13})g(r_{23})g(r_{31}). \tag{3.8}$$

Both sides of Eq. (3.8) are vectors which are parallel to $\mathbf{r}_1 - \mathbf{r}_2$. Setting $\mathbf{r} = \mathbf{r}_1 - \mathbf{r}_2$, $\mathbf{s} = \mathbf{r}_1 - \mathbf{r}_3$, and performing a scalar multiplication by $\mathbf{r}/r$ gives

$$\frac{d}{dr}\{\log g(r) + u(r)\} = -\varrho \int d\mathbf{s} \, \frac{\mathbf{r} \cdot \mathbf{s}}{rs} \, u'(s)g(s)g(|\, \mathbf{r} - \mathbf{s} \,|), \tag{3.9}$$

where r $= |\, r \,|$, and s $= |\, s \,|$.

The three-dimensional integration with respect to $\mathbf{s}$ can conveniently be expressed in terms of spherical polar coordinates (s, θ, α) about an axis parallel to $\mathbf{r}$, i.e., θ is the angle between $\mathbf{s}$ and $\mathbf{r}$, and α is the angle that the plane through $\mathbf{r}$ and $\mathbf{s}$ makes with some fixed plane through $\mathbf{r}$ (see Fig. 1). The volume element in this coordinate system is

$$d\mathbf{s} = s^2 \sin\theta \, ds \, d\theta \, d\alpha \tag{3.10}$$

and

$$\mathbf{r} \cdot \mathbf{s} = rs \cos\theta, \tag{3.11}$$

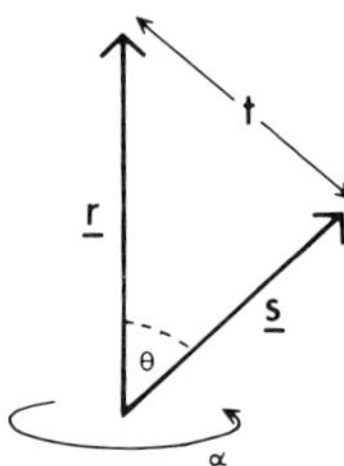

FIG. 1. Spherical polar coordinates (s, θ, α) of the vector $\mathbf{s}$ about an axis along the vector $\mathbf{r}$.

so the right-hand side of Eq. (3.9) becomes

$$-\varrho \int_0^\infty s^2\, ds \int_0^\pi \sin\theta\, d\theta \int_0^{2\pi} d\alpha\, (\cos\theta)\, u'(s)g(s)g(t), \tag{3.12}$$

where $t = |\mathbf{r} - \mathbf{s}|$ is the positive root of the equation

$$t^2 = r^2 + s^2 - 2rs\cos\theta. \tag{3.13}$$

Inspection of the expression (3.12) shows that it is unaltered by replacing the last factor $g(t)$ by $g(t) - 1$, i.e., $h(t)$. This facilitates the interchanging of orders of integration later in this discussion.

The integration with respect to α is trivial, giving a factor 2π. The θ-integration can be simplified by changing variables from θ to t (holding s fixed). From Eq. (3.13)

$$t\, dt = rs \sin\theta\, d\theta, \tag{3.14}$$

and as θ varies from 0 to π, t varies from $|r - s|$ to $r + s$. The right-hand side of Eq. (3.9) can therefore be written as

$$-2\pi\varrho \int_0^\infty s^2\, ds \int_{|r-s|}^{r+s} \frac{t\, dt}{rs} \frac{r^2 + s^2 - t^2}{2rs} u'(s)g(s)h(t). \tag{3.15}$$

Equation (3.9) can be integrated with respect to r. To do this, it is convenient to replace $h(t)$ by $h(|t|)$ in the expression (3.15). The integrand is then an odd function of t and the lower limit of integration $|r - s|$ can be replaced by $r - s$. Integrating with respect to r from r_0 to infinity and interchanging the orders of integration then gives

$$\log g(r_0) + u(r_0) = \pi\varrho \int_0^\infty ds\, u'(s)g(s) \int_{r_0-s}^\infty dt\, th(|t|)A(r_0, s, t), \tag{3.16}$$

where

$$A(r_0, s, t) = \int_{\max\{r_0, t-s\}}^{t+s} dr \left[1 + \frac{s^2 - t^2}{r^2}\right]. \tag{3.17}$$

The integration in Eq. (3.17) can be performed explicitly to give

$$\begin{aligned} A(r_0, s, t) &= [s^2 - (t - r_0)^2]/r_0 \qquad &\text{if } t < r_0 + s, \\ &= 0 &\text{if } t > r_0 + s. \end{aligned} \tag{3.18}$$

Substituting this result in Eq. (3.16) and replacing r_0 by r and $h(r)$ by

$g(r) - 1$, it therefore follows that

$$\log g(r) + u(r) = \pi\varrho \int_0^\infty ds\, u'(s)g(s) \int_{r-s}^{r+s} dt\, t\, \frac{s^2 - (t-r)^2}{r} [g(|t|) - 1]. \tag{3.19}$$

Equation (3.19) is the Born–Green (BG) approximation for $g(r)$ in terms of ϱ and $u(r)$.

IV. Kirkwood Approximation

Another hierarchy of relations between the distribution functions can be obtained by introducing an extra particle (particle O) which interacts with the others by a potential $\xi\phi(r)$, rather than $\phi(r)$. The partition function and distribution functions then depend on the parameter ξ, and, in analogy with Eqs. (2.7) and (2.11), are given by

$$\Xi^*(\xi) = \sum_{N=0}^{\infty} \frac{z^N}{N!} \int_V \cdots \int d\mathbf{r}_1 \cdots d\mathbf{r}_N\, e^{-\Phi(\xi)}, \tag{4.1}$$

$$n_\xi^{(L)}(\mathbf{r}_1, \ldots, \mathbf{r}_L) = \frac{1}{\Xi^*(\xi)} \sum_{N=L}^{\infty} \frac{z^N}{(N-L)!} \int_V \cdots \int d\mathbf{r}_{L+1} \cdots d\mathbf{r}_N\, e^{-\Phi(\xi)}, \tag{4.2}$$

where

$$\Phi(\xi) = \xi \sum_{i=1}^{N} u(r_{0i}) + \sum_{1 \le i < j \le N} u(r_{ij}). \tag{4.3}$$

When $\xi = 0$ the particle O does not interact with the rest of the system, and Eqs. (4.1) and (4.2) are identical with Eqs. (2.7) and (2.11), respectively. As ξ increases, the interaction grows in strength until, when $\xi = 1$, the particle O behaves as a typical particle. In this case the system can be regarded as containing $N + 1$ particles, instead of the original N, and it follows from the equations that

$$\Xi^*(1) = z^{-1}\Xi^* n^{(1)}(\mathbf{r}_0), \tag{4.4}$$

$$n_1^{(L)}(\mathbf{r}_1, \ldots, \mathbf{r}_L) = n^{(L+1)}(\mathbf{r}_0, \mathbf{r}_1, \ldots, \mathbf{r}_L)/n^{(1)}(\mathbf{r}_0) \tag{4.5}$$

[the functions on the right-hand side of Eqs. (4.4) and (4.5) being the normal functions defined by Eqs. (2.7) and (2.11)].

Consider the effect of differentiating with respect to ξ. From equations

(4.1) and (4.3),

$$\frac{\partial \Xi^*(\xi)}{\partial \xi} = -\sum_{N=1}^{\infty} \frac{z^N}{(N-1)!} \int \cdots \int_V d\mathbf{r}_1 \cdots d\mathbf{r}_N \, u(r_{01}) \, \mathrm{e}^{-\Phi(\xi)} \tag{4.6}$$

(using the fact that there are N terms $u(r_{0i})$, each corresponding to the same integral). From Eq. (4.2) it therefore follows that

$$\partial[\log \Xi^*(\xi)]/\partial \xi = -\int d\mathbf{r}_1 \, u(r_{01}) n_\xi^{(1)}(\mathbf{r}_1). \tag{4.7}$$

Now differentiate Eq. (4.2) with respect to ξ. As before, a set of terms $-u(r_{0i})$ arises from differentiating the exponent, but for $1 \leq i \leq L$ these are just multiplicative factors. The remaining $(N - L)$ terms all give the same integral, so that i can be replaced by $L + 1$. Using Eqs. (4.2) and (4.7), it is then found that

$$\begin{aligned} \frac{\partial}{\partial \xi} n_\xi^{(L)}(\mathbf{r}_1, \ldots, \mathbf{r}_L) \\ = - n_\xi^{(L)}(\mathbf{r}_1, \ldots, \mathbf{r}_L) \sum_{i=1}^{L} u(r_{0i}) - \int d\mathbf{r}_{L+1} \, u(r_{0,L+1}) \\ \times [n_\xi^{(L+1)}(\mathbf{r}_1, \ldots, \mathbf{r}_{L+1}) - n_\xi^{(1)}(\mathbf{r}_{L+1}) n_\xi^{(L)}(\mathbf{r}_1, \ldots, \mathbf{r}_L)]. \end{aligned} \tag{4.8}$$

In the limit of V large the integrations in Eqs. (4.7) and (4.8) can be extended over all space.

Like the hierarchy (3.2), the infinite set of equations (4.8) (for $L = 1$, 2, 3, etc.) is only useful if some closure of the equations can be made, such as the superposition approximation (3.3). In this case $n_\xi^{(2)}(\mathbf{r}_1, \mathbf{r}_2)$ depends also on $\mathbf{r}_0$, so it is effectively a three-particle function, and the appropriate analog of Eq. (3.3) is

$$n_\xi^{(2)}(\mathbf{r}_1, \mathbf{r}_2) = n_\xi^{(1)}(\mathbf{r}_1) n_\xi^{(1)}(\mathbf{r}_2) g(r_{12}), \tag{4.9}$$

where $g(r)$ is the normal radial distribution function defined in Section II,B. Equation (4.9) is exact to first order in an expansion in powers of the fugacity z, so that it is accurate for dilute systems. For $\xi = 0$ it is simply equivalent to the exact equation (2.14), and for $\xi = 1$ it becomes the approximation (3.3).

Setting $L = 1$ in Eq. (4.8) and using the approximation (4.9), it follows that

$$\partial[n_\xi^{(1)}(\mathbf{r}_1)]/\partial \xi = -u(r_{01}) n_\xi^{(1)}(\mathbf{r}_1) - \int d\mathbf{r}_2 \, u(r_{02}) n_\xi^{(1)}(\mathbf{r}_1) n_\xi^{(1)}(\mathbf{r}_2) h(r_{12}), \tag{4.10}$$

where $h(r) = g(r) - 1$ is the indirect correlation function.

The $n_\xi^{(1)}(\mathbf{r}_1)$ depends on $\mathbf{r}_0$ and $\mathbf{r}_1$ only insofar as it depends on their separation r_{01}. A ξ-dependent radial distribution function $g_\xi(r)$ can therefore be defined by

$$n_\xi^{(1)}(\mathbf{r}_1) = \varrho g_\xi(r_{01}). \tag{4.11}$$

When $\xi = 0$, $n_\xi^{(1)}(\mathbf{r})$ is simply the density ϱ, so that

$$g_0(r) = 1, \tag{4.12}$$

while when $\xi = 1$ it follows from Eqs. (2.12), (2.14), and (4.5) that

$$g_1(r) = g(r). \tag{4.13}$$

Setting $\mathbf{r} = \mathbf{r}_0 - \mathbf{r}_1$ and $\mathbf{s} = \mathbf{r}_0 - \mathbf{r}_2$, and dividing by $n_\xi^{(1)}(\mathbf{r}_1)$, Eq. (4.10) can now be written as

$$\partial[\log g_\xi(r)]/\partial\xi = -u(r) - \varrho \int ds\, u(s) g_\xi(s) h(|\, \mathbf{r} - \mathbf{s}\,|), \tag{4.14}$$

where $r = |\,\mathbf{r}\,|$ and $s = |\,\mathbf{s}\,|$.

The three-dimensional integration with respect to $\mathbf{s}$ can be simplified by the same techniques as were used with Eq. (3.9), the most significant difference being that the term $\cos\theta$ no longer occurs in the integrand. Thus,

$$\partial[\log g_\xi(r)]/\partial\xi = -\, u(r) - (2\pi\varrho/r) \int_0^\infty ds\, s u(s) g_\xi(s) \int_{|r-s|}^{r+s} dt\, t h(t). \tag{4.15}$$

Replacing ξ by ξ', integrating with respect to ξ' from 0 to ξ, and using Eqs. (2.40) and (4.12), it follows that

$$\log g_\xi(r) = -\, \xi u(r) - (2\pi\varrho/r) \int_0^\xi d\xi' \int_0^\infty ds\, s u(s) g_{\xi'}(s) \int_{|r-s|}^{r+s} dt\, t[g(t) - 1]. \tag{4.16}$$

Equation (4.16) is the Kirkwood approximation for $g_\xi(r)$, and hence $g(r) = g_1(r)$, in terms of ϱ and $u(r)$.

In addition to being able to evaluate the internal energy, pressure, and compressibility from Eqs. (2.20), (2.26), and (2.39), it is also possible in this approximation to obtain an expression for the fugacity z, and hence the chemical potential μ. Integrating Eq. (4.7) with respect to ξ from 0 to 1 and using Eqs. (4.4) and (4.11), it is found that

$$\log[n^{(1)}(\mathbf{r}_0)/z] = -\varrho \int_0^1 d\xi \int d\mathbf{r}_1\, u(r_{01}) g_\xi(r_{01}). \tag{4.17}$$

$n^{(1)}(\mathbf{r}_0)$ is simply the density ϱ, so, using spherical polar coordinates about $\mathbf{r}_0$, Eq. (4.17) becomes

$$\log(\varrho/z) = -4\pi\varrho \int_0^1 d\xi \int_0^\infty dr\, r^2 u(r) g_\xi(r). \tag{4.18}$$

V. Percus–Yevick and Convolution Hypernetted-Chain Approximations

Two quite successful approximations in classical statistical mechanics are the Percus–Yevick (PY) approximation (Percus and Yevick, 1958; Percus, 1962) and the convolution hypernetted chain (CHNC) approximation (Van Leeuwen *et al.*, 1959; Green, 1960; Meeron, 1960; Morita and Hiroike, 1960; Rusbrooke, 1960; Verlet, 1960). These can be derived in a number of ways—for instance, in Section IX it is shown that they can be obtained by truncating functional expansions. In this section they are obtained by diagrammatic techniques.

A. Virial Expansions

If

$$f(r) = e^{-u(r)} - 1, \tag{5.1}$$

then the expression (2.7) for the grand partition function can be written as

$$\Xi^* = 1 + \frac{z}{1!} \int_V d\mathbf{r}_1 + \frac{z^2}{2!} \iint_V d\mathbf{r}_1\, d\mathbf{r}_2\, [1 + f(r_{12})] + O(z^3). \tag{5.2}$$

Thus, Ξ^* is of the form of a power series in the fugacity z, with coefficients determined by the function $f(r)$.

Similarly, the expression (2.11) for the distribution functions is of the form of a ratio of two such power series. The division by Ξ^* can be systematically performed to give

$$\varrho = n^{(1)}(\mathbf{r}_1) = z + z^2 \int d\mathbf{r}_2 f(r_{12}) + O(z^3), \tag{5.3}$$

$$\begin{aligned} \varrho^2 g(r_{12}) &= n^{(2)}(\mathbf{r}_1, \mathbf{r}_2) \\ &= \exp[-u(r_{12})]\{z^2 + z^3 \int d\mathbf{r}_3\, [f(r_{13}) + f(r_{23}) + f(r_{13})f(r_{23})] \\ &\quad + O(z^4)\}, \end{aligned} \tag{5.4}$$

where in the limit of a large system the integrations can be extended over all space.

The power-series expansion (5.3) of ϱ in powers of z can be systematically inverted to give z in powers of ϱ; thus,

$$z = \varrho - \varrho^2 \int d\mathbf{r}_2 f(r_{12}) + O(\varrho^3). \tag{5.5}$$

Substituting this form for z into Eq. (5.4), it follows that $g(r_{12})$ can be expanded in powers of ϱ, giving

$$g(r_{12}) \exp[u(r_{12})] = 1 + \varrho \int d\mathbf{r}_3 f(r_{13}) f(r_{23}) + O(\varrho^2). \tag{5.6}$$

Expansions in powers of the density are used frequently in statistical mechanics and are known generally as virial expansions. In particular, if the form (5.6) of $g(r)$ is substituted into the inverse compressibility relation (2.41) [using Eq. (2.40)], then on integrating with respect to ϱ and using the condition that $P = 0$ when $\varrho = 0$, it is found that

$$P/kT = \sum_{m=1}^{\infty} B_m \varrho^m, \tag{5.7}$$

where

$$\begin{aligned} B_1 = 1, \qquad B_2 &= -\tfrac{1}{2} \int d\mathbf{r}_2 f(r_{12}), \\ B_3 = -\tfrac{1}{3} \iint d\mathbf{r}_2\, d\mathbf{r}_3 &f(r_{23}) f(r_{31}) f(r_{12}), \qquad \text{etc.} \end{aligned} \tag{5.8}$$

The coefficients B_m are known as the virial coefficients.

B. Diagrammatic Representations

The terms occurring in the expansion (5.6) of $g(r_{12})$ in powers of ϱ can conveniently be represented by diagrams. These diagrams consist of two labeled white dots (usually labeled 1 and 2), possibly together with black dots and lines linking the dots. There can be at most one line linking any pair of dots, and the diagrams are to be interpreted according to the following rules:

(a) with each black dot associate a label i (these labels are arbitrary, except that no two dots, black or white, can have the same label).
(b) with each black dot associate a factor ϱ.
(c) with each line linking dots i and j associate a factor $f(r_{ij})$.
(d) multiply together all the ϱ and $f(r_{ij})$ factors and integrate over all the position coordinates $\mathbf{r}_i$ corresponding to black dots.

From Eq. (2.11) for $L = 2$ it can be seen that an integral with respect to $\mathbf{r}_3, \ldots, \mathbf{r}_N$ occurs in conjunction with a weighting factor $1/(N-2)!$, and it can be shown that this is also true of the integrals in the expansion (5.6). Further, if there are J different integrands that can be obtained by permuting $\mathbf{r}_3, \ldots, \mathbf{r}_N$, then each of the corresponding J integrals occurs in the expansion. If there are M permutations of $\mathbf{r}_3, \ldots, \mathbf{r}_N$ that leave the integrand unchanged, then $(N-2)! = JM$.

Since any two integrals which can be obtained from one another by interchanging the variables of integration are identical, it is convenient to adopt the convention that we sum only over integrals corresponding to distinguishable diagrams, i.e., diagrams that cannot be obtained from one another by rearranging the black dots. The contribution of such an integral is then multiplied by a factor $J/(N-2)! = M^{-1}$. This can be built into the interpretation of a diagram by the rule:

(e) Multiply the integral by a factor M^{-1}, where M is the number of permutations of the black dots that leave the diagram unchanged (the lines are to be regarded here as pieces of elastic fixed to the dots at each end).

The diagrammatic representations of the terms of order ϱ and ϱ^2 in the expansion (5.6) are shown in Fig. 2, together with their associated weighting factors M^{-1}.

The usefulness of such diagrams lies in the fact that they can be classified according to their structure, and that the expansions that occur in statistical mechanics can often be expressed as a sum over a certain

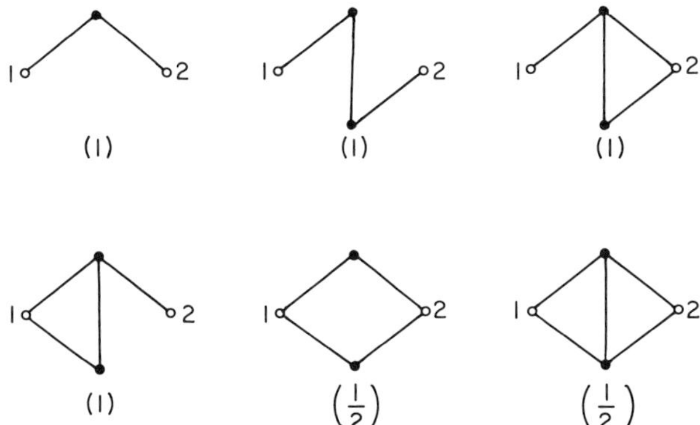

FIG. 2. Diagrammatic representations of the terms of order ϱ and ϱ^2 in the expansion (5.6) of $g(r_{12}) \exp[u(r_{12})]$. The associated weighting factors of the integrals are shown in parentheses.

class of diagrams. In order to discuss such classifications of the diagrams, it is convenient to define certain terms concerning their properties. Rather than speak of white dots and black dots, we shall refer to them as *external points* and *internal points*, respectively.

A *route* between two points i and j is a sequence of points $k_1, \ldots, k_m$ such that there are lines between all the point pairs (i, k_1), (k_1, k_2), (k_2, k_3), ..., (k_{m-1}, k_m), (k_m, j).

An *external link* is a line directly connecting the two external points.

A diagram is *connected* if there exists at least one route between any pair of points.

A connected diagram is *simply connected* if there exists at least one route between any pair of internal points which does not pass through an external point, and if there is no external link.

An *articulation point* is such that the diagram can be cut into two or more parts at this point (i.e., all routes between the parts pass through this point), and at least one such part contains no external points, except perhaps the articulation point itself.

A diagram is *reducible* if it contains an articulation point, and *irreducible* if it does not.

A *node* is an internal point such that all routes between the two exterior points pass through this point.

A diagram is *nodal* if it contains a node, and *nonnodal* if it does not.

It should be noticed that these definitions are not all independent. For instance, a connected irreducible nodal diagram is necessarily simply connected. Also, a simply connected diagram is necessarily connected.

Some typical diagrams are shown in Fig. 3. Diagram (a) is not connected, since there are no routes linking points 4 and 5 to 1, 2, and 3. Diagram (b) is connected, but is reducible, since both 1 and 3 are articulation points. It also contains an external link. Diagram (c) is connected and irreducible, but not simply connected. Diagrams (d) and (e) are both simply connected and irreducible. The point 3 in diagram (d) is

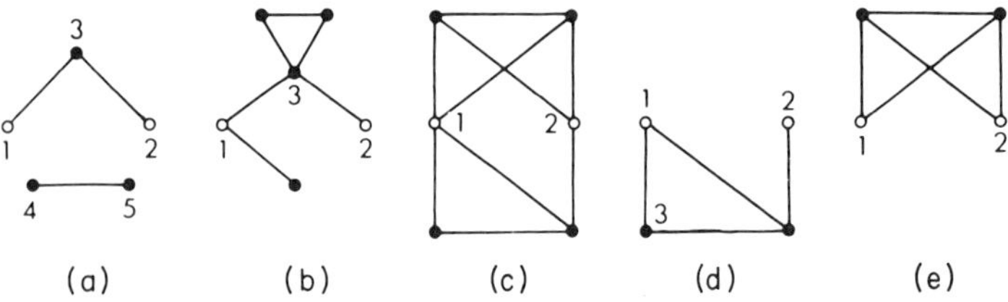

FIG. 3. Typical diagrams with two external points.

a node, while diagram (e) is nonnodal. It should be noticed that the integral corresponding to diagram (c) factorizes into a product of two integrals corresponding to diagrams (d) and (e).

In terms of these definitions it can be established (Stell, 1964) that the indirect correlation function $h(r)$ is given by

$$\begin{aligned} h(r_{12}) &= g(r_{12}) - 1 \\ &= \text{sum of all connected irreducible diagrams.} \end{aligned} \tag{5.9}$$

All the diagrams occurring in the following reasoning are connected and irreducible, so for the sake of brevity it is convenient to adopt the convention that this be understood.

The first diagram in the expansion (5.7) is simply two external points connected by an external link, i.e., $f(r_{12})$. To each of the remaining diagrams with an external link there corresponds one without, and vice versa. The diagrams of each such pair must have the same weighting factor M^{-1}, so that

$$h(r_{12}) = f(r_{12}) + [1 + f(r_{12})] \times \text{sum of diagrams with no external link.} \tag{5.10}$$

Using Eqs. (2.40) and (5.1), it follows that

$$g(r_{12}) \exp[u(r_{12})] = 1 + \text{sum of diagrams with no external link,} \tag{5.11}$$

which is the diagrammatic equivalent of the expansion (5.6).

A typical diagram contributing to the expansion (5.11) can be decomposed into a number of simply connected subdiagrams linking points 1 and 2 not otherwise connected to each other [e.g., Fig. 3(*c*)]. The corresponding integral factorizes into a product of integrals corresponding to the simply connected subdiagrams.

Suppose the simply connected diagrams are ordered in some way so that $w_1(r_{12})$, $w_2(r_{12})$, $w_3(r_{12})$, etc., are the corresponding weighted integrals. Then a typical term in the expansion (5.11) is proportional to

$$[w_1(r_{12})]^{m_1}[w_2(r_{12})]^{m_2}\cdots, \tag{5.12}$$

where m_λ is the number of times the simply connected subdiagram corresponding to $w_\lambda(r_{12})$ occurs in the main diagram.

To obtain the correct weighting factor given by rule (5) of such a term, note that the weighting factors due to possible interchanges of internal points within a particular subdiagram are already included in the expression (5.12). The only way internal points corresponding to different subdiagrams can be interchanged so as to leave the main diagram unaltered is to interchange *all* the internal points of a subdiagram corresponding to $w_\lambda(r_{12})$ with those of another subdiagram also corresponding to $w_\lambda(r_{12})$. Thus, there are $m_1!$ times as many ways of permuting the internal points within the subdiagrams of type $w_1(r_{12})$, $m_2!$ times as many ways of permuting the internal points within the subdiagrams of type $w_2(r_{12})$, etc. The correctly weighted contribution of the diagram corresponding to the expression (5.12) is therefore

$$\frac{[w_1(r_{12})]^{m_1}}{m_1!} \, \frac{[w_2(r_{12})]^{m_2}}{m_2!} \cdots . \tag{5.13}$$

For nonnegative integral values of $m_1, m_2, \ldots$, any term in the expansion (5.11) corresponds to one and only one expression of the type (5.13), and vice versa (the term unity corresponds to all the m_λ being zero). Thus, the right-hand side of Eq. (5.11) is obtained by summing the expression (5.13) over $m_1, m_2, \ldots$ from 0 to ∞. This gives

$$g(r_{12}) \exp[u(r_{12})] = \exp[w_1(r_{12}) + w_2(r_{12}) + \ldots]. \tag{5.14}$$

Taking the logarithm of both sides of Eq. (5.14), it therefore follows that

$$\log g(r_{12}) + u(r_{12}) = \text{sum of simply connected diagrams.} \tag{5.15}$$

C. Nodal and Nonnodal Diagrams—The Direct Correlation Function

The diagrams occurring in the above expansions can be further classified according to whether they are nodal or nonnodal. Let

$$c(r_{12}) = \text{sum of nonnodal diagrams,} \tag{5.16a}$$

$$d(r_{12}) = \text{sum of nodal diagrams.} \tag{5.16b}$$

Comparing these definitions with Eq. (5.9), it is apparent that

$$h(r_{12}) = c(r_{12}) + d(r_{12}). \tag{5.16c}$$

Another relation between $h(r_{12})$, $c(r_{12})$, and $d(r_{12})$ can be obtained. Suppose the terms in the expansions (5.9) and (5.16a) are ordered in some way so that

$$h(r_{12}) = \sum_\lambda h_\lambda(r_{12}), \tag{5.17}$$

$$c(r_{12}) = \sum_\lambda c_\lambda(r_{12}), \tag{5.18}$$

where each term $h_\lambda(r_{12})$ corresponds to one diagram, and each term $c_\lambda(r_{12})$ to one nonnodal diagram. Then

$$d_{\lambda\mu}(r_{12}) = \varrho \int d\mathbf{r}_3\, c_\lambda(r_{13}) h_\mu(r_{32}) \tag{5.19}$$

corresponds to a diagram consisting of a nonnodal subdiagram linking points 1 and 3, and a subdiagram linking points 3 and 2. The point 3 is clearly a node, so that this diagram occurs in the expansion (5.16b) of $d(r_{12})$.

Further, the expression (5.19) has the correct weighting factor given by rule 5, since the weighting factors due to possible interchanges of internal points within each subdiagram are already included in $c_\lambda(r_{13})$ and $h_\mu(r_{32})$, and no point in the 1–3 subdiagram can be interchanged with one in the 3–2 subdiagram without altering the main diagram.

Not only does any term of the type (5.19) contribute to the expansion (5.16b), but also any term of the expansion can be made to correspond to one and only one term of the type (5.19) by choosing the point 3 to be the node closest to the point 1. Thus,

$$d(r_{12}) = \sum_\lambda \sum_\mu d_{\lambda\mu}(r_{12}). \tag{5.20}$$

Substituting the form (5.19) of $d_{\lambda\mu}(r_{12})$ and using equations (5.17) and (5.18), it follows that

$$d(r_{12}) = \varrho \int d\mathbf{r}_3\, c(r_{13}) h(r_{32}). \tag{5.21}$$

The function $d(r_{12})$ can be eliminated from Eqs. (5.16c) and (5.21) to give

$$h(r_{12}) = c(r_{12}) + \varrho \int d\mathbf{r}_3\, c(r_{13}) h(r_{32}). \tag{5.22}$$

This relation can be regarded as defining $c(r_{12})$ in terms of ϱ and $h(r_{12})$. It was first given by Ornstein and Zernike (1914), and is therefore known as the Ornstein–Zernike (OZ) relation.

The function $c(r)$ is known as the direct correlation function. Like $h(r)$, it has the property that it tends to zero with increasing r, but it seems that it does so much more rapidly. For instance, in the case of a one-dimensional gas of hard rods it can be shown that $c(r)$ vanishes exactly outside the range of the potential, while $h(r)$ oscillates to zero comparatively slowly.

D. PY Approximation

No nodal diagram can have an external link, so Eq. (5.11) can be written as

$$g(r_{12}) \exp[u(r_{12})] = 1 + d(r_{12}) + B(r_{12}), \tag{5.23}$$

where

$$B(r_{12}) = \text{sum of nonnodal diagrams with no external link.} \tag{5.24}$$

Expending the function $B(r_{12})$, it is found that

$$B(r_{12}) = \tfrac{1}{2}\varrho^2 \iint d\mathbf{r}_3\, d\mathbf{r}_4\, f(r_{13})f(r_{23})f(r_{14})f(r_{24}) \exp[-u(r_{34})] + O(\varrho^3). \tag{5.25}$$

For short-range potentials with a strong repulsive core the first term in this expansion must be small, since when $\mathbf{r}_1$, $\mathbf{r}_2$, $\mathbf{r}_3$, and $\mathbf{r}_4$ are close enough for each of the f-functions to be significant r_{34} and $\exp[-u(r_{34})]$ are in general small. Similar arguments appear to apply to the higher terms, suggesting the approximation

$$B(r_{12}) = 0. \tag{5.26}$$

Using Eqs. (5.16c) and (2.40) to eliminate $d(r_{12})$ and $h(r_{12})$, when $B(r_{12})$ vanishes Eq. (5.23) can be written as

$$c(r_{12}) = \{1 - \exp[u(r_{12})]\}\, g(r_{12}). \tag{5.27}$$

The exact equations (2.40) and (5.22), together with the approximate equation (5.27), form a closed set of equations for $g(r)$, $h(r)$, and $c(r)$ in terms of ϱ and $u(r)$. Eliminating $h(r)$ and $c(r)$, and setting $\mathbf{r} = \mathbf{r}_1 - \mathbf{r}_2$ and $\mathbf{s} = \mathbf{r}_1 - \mathbf{r}_3$, it is found that

$$g(r)e^{u(r)} = 1 + \varrho \int d\mathbf{s}\, [1 - e^{u(s)}]g(s)[g|\mathbf{r} - \mathbf{s}| - 1]. \tag{5.28}$$

Using spherical polar coordinates as in Section III, the three-dimensional integration with respect to $\mathbf{s}$ can be simplified to give

$$g(r)e^{u(r)} = 1 + (2\pi\varrho/r) \int_0^\infty ds\, s[1 - e^{u(s)}]g(s) \int_{|r-s|}^{r+s} dt\, t[g(t) - 1]. \tag{5.29}$$

Equation (5.29) is the PY approximation for $g(r)$ in terms of ϱ and $u(r)$.

E. CHNC Approximation

Another approximation can be obtained by noting that any nodal diagram is necessarily simply connected. Thus Eq. (5.15) can be written as

$$\log g(r_{12}) + u(r_{12}) = d(r_{12}) + E(r_{12}), \tag{5.30}$$

where

$$E(r_{12}) = \text{sum of nonnodal simply connected diagrams.} \tag{5.31}$$

Neglecting $E(r_{12})$ in Eq. (5.30) and using Eqs. (5.16c) and (2.40) to eliminate $d(r_{12})$ and $h(r_{12})$, it follows that

$$c(r_{12}) = g(r_{12}) - 1 - u(r_{12}) - \log g(r_{12}). \tag{5.32}$$

Equations (2.40), (5.22), and (5.32) form a closed set of equations for $g(r)$, $h(r)$, and $c(r)$ in terms of ϱ and $u(r)$. Eliminating $h(r)$ and $c(r)$ and simplifying the three-dimensional integral gives

$$\log g(r) + u(r) = (2\pi\varrho/r) \int_0^\infty ds\, s[g(s) - 1 - u(s) - \log g(s)] \int_{|r-s|}^{r+s} dt\, t[g(t) - 1]. \tag{5.33}$$

This is the CHNC approximation for $g(r)$ in terms of ϱ and $u(r)$.

As any diagram contributing to $E(r_{12})$ necessarily contributes to $B(r_{12})$, but not vice versa, the CHNC approximation neglects less diagrams than the PY. However, it does not follow that the CHNC approximation is more accurate; rather, it seems that the diagrams neglected by the PY approximation tend to cancel one another out.

Attempts have been made to extend the PY and CHNC approximations to include larger classes of diagrams (Verlet, 1964 and 1965; Wertheim, 1967; Baxter, 1968a). However, the resulting equations are very difficult to handle numerically, even on fast computing machines.

VI. Numerical Results

A. Potentials

A number of forms of the potential function $\phi(r)$ have been used by various workers. The most popular semirealistic potential is the Lennard-Jones 12–6 potential, defined by

$$\phi(r) = 4\varepsilon[(\sigma/r)^{12} - (\sigma/r)^6]. \tag{6.1}$$

This potential has a short-range strongly repulsive core, together with a long-range attraction proportional to r^{-6}, which are both properties possessed by real molecules. The parameters σ and ε measure the range of the potential and the depth of the attractive well, and must be chosen to fit experimental data.

Another potential which is often used in theoretical calculations is the hard-sphere potential:

$$\begin{aligned} \phi(r) &= +\infty, \qquad && r < a, \\ &= 0, && r > a. \end{aligned} \tag{6.2}$$

This consists solely of an infinitely repulsive core, corresponding to hard spheres of diameter a.

An attractive well is sometimes added to the potential of Eq. (6.2) to make it the square-well potential:

$$\begin{aligned} \phi(r) &= +\infty, \qquad && r < \sigma, \\ &= -\varepsilon, && \sigma < r < a, \\ &= 0, && r > a. \end{aligned} \tag{6.3}$$

The advantage of the hard-sphere and square-well potentials is that they are simpler to use in calculations than more realistic forms, while they do embody some of the properties of real potentials.

B. Monte Carlo Calculations

In the previous sections, four approximate equations (BG, Kirkwood, PY and CHNC) for the radial distribution function $g(r)$ have been obtained. Once $g(r)$ is known, the thermodynamic properties can be obtained using the equations of Section II,C. Although the numerical solution of these approximations can be quite lengthy, the amount of work involved is still very small compared with that involved in calculating the partition function and distribution functions from their definitions (2.7) and (2.11).

However, with the advent of high-speed computers, it has been possible to make some progress in the exact calculations. The methods used have been based on sampling the configurations available to a large, but finite, number of particles and forming statistical-mechanical averages over the sample. Such techniques are known as "Monte Carlo" or "molecular-dynamics" calculations.

Only a very limited number of such calculations have been made, due to the fact that they take many hours on even the fastest machines. One useful purpose they serve is to test the adequacy of the potential models used to represent real molecules. For instance, the Lennard-Jones potential, Eq. (6.1), with $\varepsilon/k = 119.76°\text{K}$ and $\sigma = 3.405$ Å, is often regarded as appropriate to an argon atom (Michels *et al.*, 1949). In Fig. 4

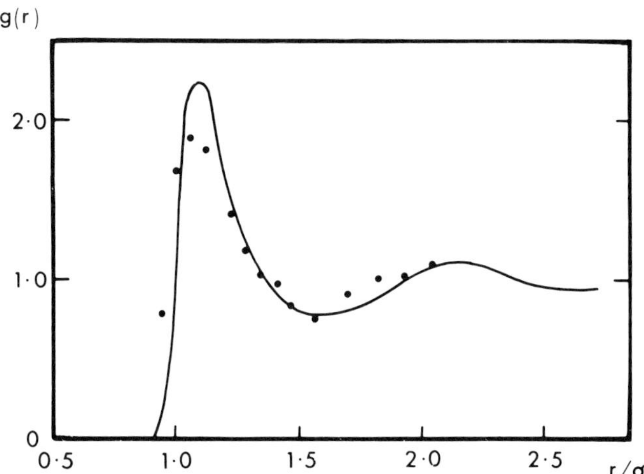

FIG. 4. Comparison of radial distribution functions $g(r)$ obtained from Monte Carlo calculations and experiment; the smooth curve is that obtained by Wood *et al.* (1958) for the Lennard-Jones potential, Eq. (6.1), with $kT/\varepsilon = 1.0579$ and $\varrho\sigma^3 = 0.654$, while the dots are the experimental results for argon obtained by Eisenstein and Gingrich (1942).

the radial distribution function obtained for this potential by Monte Carlo (MC) techniques (Wood *et al.*, 1958) is compared with that obtained by X-ray diffraction measurements on argon (Eisenstein and Gingrich, 1942) for $kT/\varepsilon = 1.0579$ and $\varrho\sigma^3 = 0.654$. It can be seen that the two exhibit qualitatively the same behavior, but there are differences, due presumably to the inadequacy of the potential model and the neglect of three-particle forces.

Although the ultimate aim of statistical mechanics is of course to describe real systems, from the point of view of testing the approximations outlined above it is preferable to compare them with the MC calculations, rather than with experiment. In this way errors due to an inadequate potential model can be avoided and the approximations tested for any potential that one may care to assume.

C. Hard-Sphere Potential

Kirkwood *et al.* (1950) solved the Kirkwood and BG approximations for hard spheres. The resulting values of the virial pressure are plotted in Fig. 5 and compared with the MC results of Alder and Wainwright (1960). It can be seen that the Kirkwood approximation is less accurate in this case than the BG, and this conclusion is also borne out by data on the virial coefficients (Henderson and Davison, 1967). Further, the Kirkwood approximation requires the calculation of $g_\xi(r)$, which is a function of the two variables ξ and r, while the other approximations

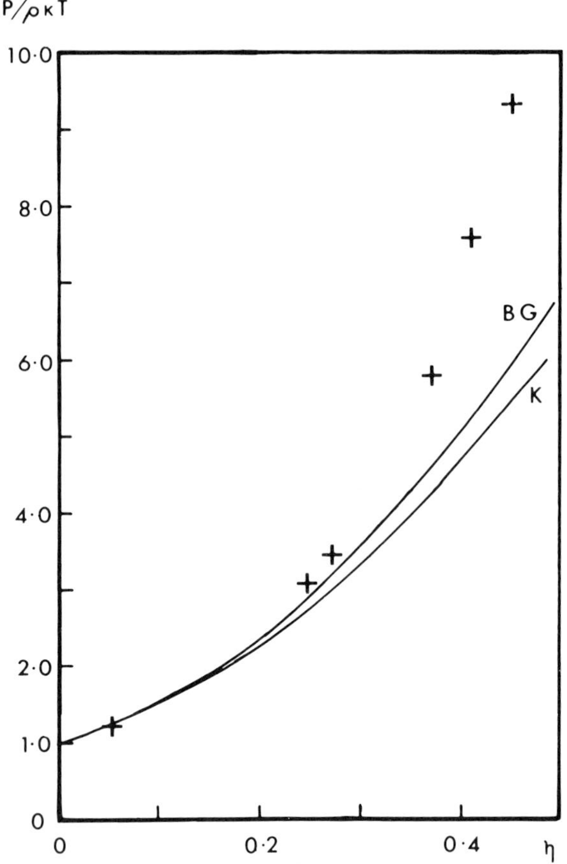

Fig. 5. The virial equation of state for hard spheres; the smooth curves are those obtained by Kirkwood *et al.* (1950) using the Born–Green (BG) and Kirkwood (K) approximations, while the crosses are the Monte Carlo results of Alder and Wainwright (1960).

only require the calculation of $g(r)$. This makes the numerical solution of the Kirkwood approximation much more lengthy than the others. For these reasons very few calculations of the Kirkwood approximation have been made.

Klein and Green (1963b) have solved the CHNC approximation for hard spheres. Levesque (1966) has systematically performed calculations for the BG, PY, and CHNC approximations applied to the Lennard-Jones, hard-sphere, and square-well potentials. (It will be shown in the next section that the PY approximation can be solved analytically for the

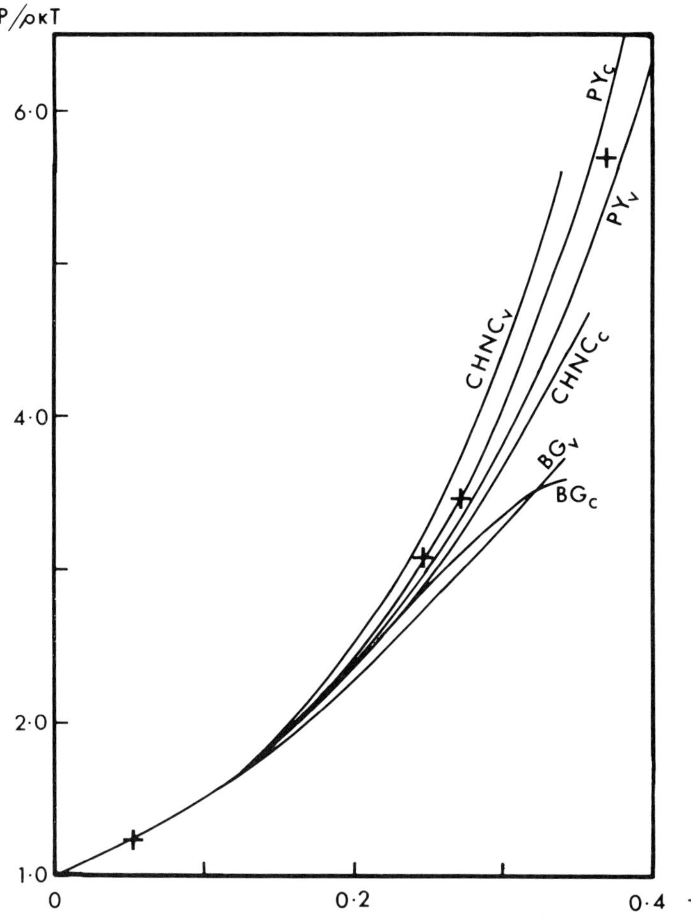

FIG. 6. Compressibility and virial equations of state for hard spheres obtained from the BG, CHNC, and PY approximations (Levesque, 1966), together with the MC results (+) of Alder and Wainwright (1960).

hard-sphere potential.) The resulting values of the virial and compressibility pressures for hard spheres are plotted in Fig. 6, together with the MC values.

The results given in Figs. 5 and 6 apply to the fluid phase of the hard-sphere system. The MC results show that at a density $\eta \approx 0.48$ (where $\eta = \pi \varrho R^3/6$) the system undergoes a phase transition to a crystalline solid state. There has been some speculation as to whether the approximate theories predict this transition; for instance, Kirkwood *et al.* (1950) found that the Kirkwood and BG approximations had no solutions for $\eta > 0.6$ and 0.5, respectively (the close-packed density is $\eta = \pi/3\sqrt{2} \approx 0.74$). Although the CHNC and PY theories have continuous solutions for all permissible values of η, for $\eta > 0.62$ the PY radial distribution function ceases to be everywhere positive, which is physically inadmissible (Throop and Bearman, 1965).

Temperley (1964) suggested that there were other solutions to the PY approximation, but Hutchinson (1967) has shown that these are also not allowed on physical grounds.

In the author's opinion it is asking too much of the approximations as formulated above to describe a solid state. Unlike a fluid, a solid has a long-range crystalline structure, which must be incorporated into the equations by abandoning the requirement that the system be homogeneous. The distribution functions $n^{(1)}(\mathbf{r}_1)$ and $n^{(2)}(\mathbf{r}_1, \mathbf{r}_2)$ then no longer satisfy Eqs. (2.12) and (2.14).

D. Lennard-Jones Potential

Broyles *et al.* (1962) solved the BG, PY, and CHNC approximations for the Lennard-Jones potential, Eq. (6.1), for the reduced temperature $kT/\varepsilon \times 2.74$, and compared their results with the MC calculations of Wood and Parker (1957). The resulting radial distribution functions for a reduced density $\varrho\sigma^3 = 10/9$ are plotted in Fig. 7, while the pressures over a range of densities are plotted in Fig. 8.

The attractive well in the Lennard-Jones potential encourages the particles to cluster together, and in three dimensions it is found that the system undergoes a first-order gas–liquid phase transition. Unlike the hard-sphere system, this transition is between disordered homogeneous phases and is therefore observed in the BG, CHNC, and PY approximations as formulated.

At the critical point of a gas–liquid phase transition the compressibility becomes infinite. From Eq. (2.39) this implies that the integral $\int_0^\infty dr\, r^2 h(r)$ is divergent, i.e., the indirect correlation function $h(r)$ tends only slowly to zero with increasing r. In some numerical calculations this long-range behavior of $h(r)$ creates problems, for it becomes necessary to calculate $h(r)$ over a very large range of values.

Watts (1968a, 1969) avoided this difficulty by using a transformation (Baxter, 1967a) of the OZ relation (5.22), together with a truncated Lennard-Jones potential given by

$$\begin{aligned} \phi(r) &= 4\varepsilon\{(\sigma/r)^{12} - (\sigma/r)^6\}, & r \leq a, \\ &= 0, & r > a. \end{aligned} \tag{6.4}$$

In this way he was able to solve the PY approximation without needing to calculate $h(r)$ beyond the range a. He chose a to be sufficiently large ($\geq 6\sigma$) for the potential (6.4) to be a reasonable approximation to the untruncated potential (6.1).

Watts found that for $kT/\varepsilon > 1.32$ the PY approximation had two solutions for all densities, while for $kT/\varepsilon < 1.32$ there was an inter-

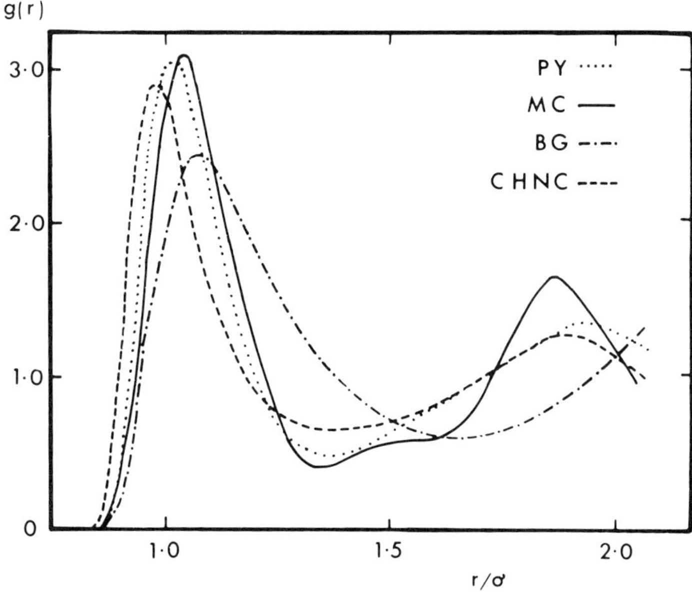

FIG. 7. Radial distribution functions $g(r)$ for the Lennard-Jones potential (6.1) with $kT/\varepsilon = 2.74$ and $\varrho\sigma^3 = 10/9$. The BG, CHNC, and PY results are those of Broyles *et al.* (1962), while the MC results are those of Wood and Parker (1957).

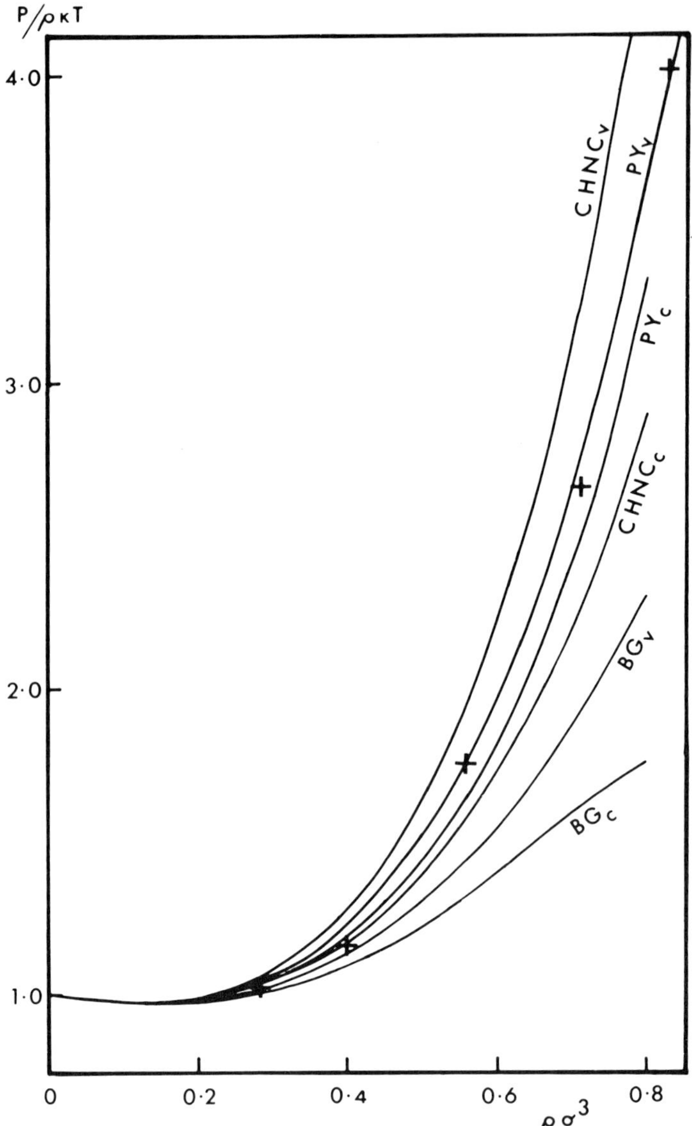

FIG. 8. Compressibility and virial equations of state for the Lennard-Jones potential, Eq. (6.1), with $kT/\varepsilon = 2.74$. The BG, CHNC, and PY results are those of Levesque (1966) and Broyles *et al.* (1962), while the MC results (+) are those of Wood and Parker (1957).

mediate range of densities within which the equation had no solutions. This behavior is illustrated in Fig. 9, where $-c(0)$ is plotted against density for various temperatures [$c(r)$ being the direct correlation function].

At zero density $c(r)$ is given by

$$c(r) = \mathrm{e}^{-\phi(r)/kT} - 1, \tag{6.5}$$

so that $-c(0) = 1$. On continuity grounds it follows that the physical states of the system correspond to the lower branches of the isotherms

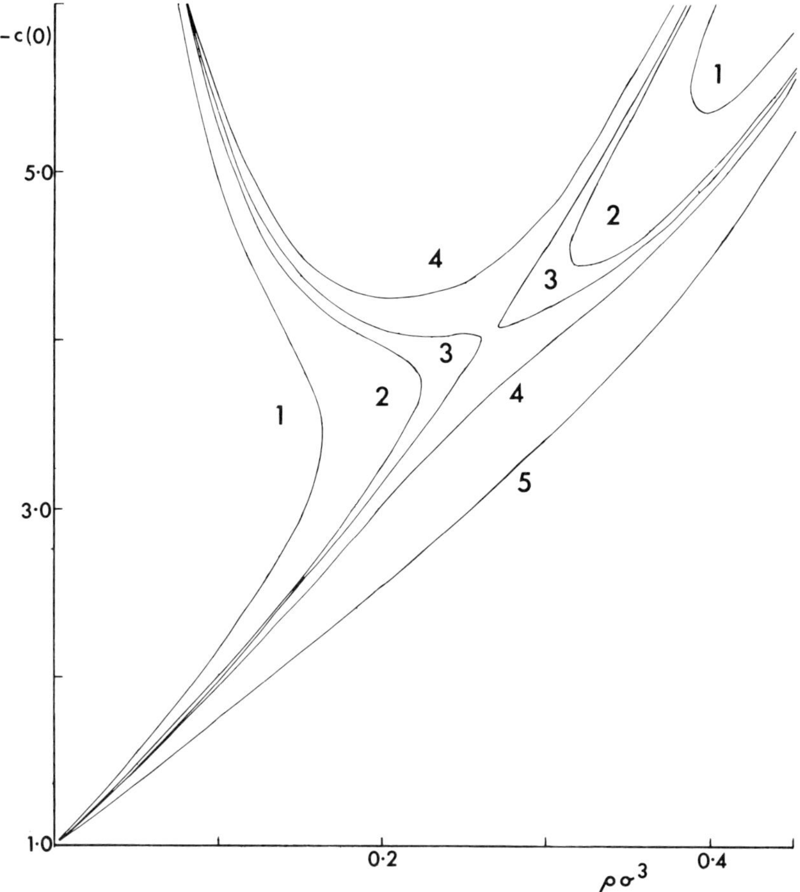

FIG. 9. Behavior of the solutions of the PY approximation for the Lennard-Jones potential, Eq. (6.1), as found by Watts (1968a); $-c(0)$ is plotted against the dimensionless density $\varrho\sigma^3$ for (1) $kT/\varepsilon = 1.20$; (2) 1.30; (3) 1.32; (4) 1.33; and (5) 1.5.

shown in Fig. 9, i.e., to the points lying below the broken line. Thus, for temperatures greater than $T_c \approx 1.32\varepsilon/k$ the properties of the fluid vary smoothly with density, while for $T < T_c$ the fluid undergoes a discontinuous transition between states of different density. Such behavior is qualitatively typical of a first-order phase transition, the critical point being the "saddle point" shown in Fig. 9.

In Fig. 10 the compressibility pressure for the physical solutions is plotted against the volume per particle for various temperatures. The compressibility is found to be infinite at the critical point, and the be-

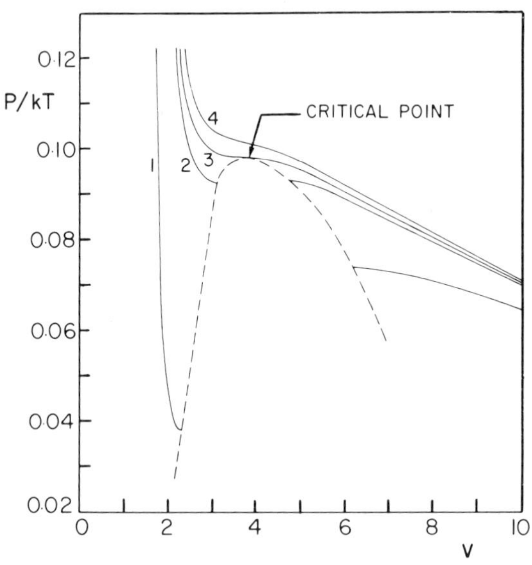

FIG. 10. The compressibility equation of state for the Lennard-Jones potential, Eq. (6.1), as obtained by Watts (1968a). P/kT (in units of σ^{-3}) is plotted against the dimensionless volume per particle $v = \varrho^{-1}\sigma^{-3}$ for the temperatures $kT/\varepsilon =$ (1) 1.20; (2) 1.30; (3) 1.32; and (4) 1.33. The broken line represents the boundary of the region within which the PY approximation has no solutions.

havior is qualitatively similar to a van der Waal gas. Unlike the van der Waal theory, however, no solutions are obtained corresponding to a negative compressibility, i.e., to a positive slope of the isotherms.

The virial pressure obtained from Eq. (2.26) for the physical solutions is plotted in Fig. 11. For a given temperature below T_c there exists a range of pressures within which the isotherm does not exist. The compressibility (obtained by differentiating this expression with respect to ϱ and inverting) does not appear to be infinite anywhere. Thus, in the virial

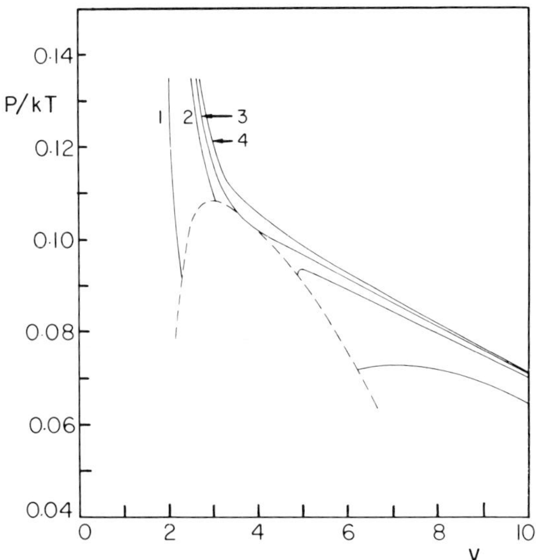

FIG. 11. The virial equation of state for the Lennard-Jones potential, Eq. (6.1), as obtained by Watts (1968a). The parameters and curves have the same interpretation as in Fig. 10.

equation of state the T_c isotherm does not have all the properties of a critical isotherm. Such behavior is nonphysical.

In Section VII it will be shown that the PY approximation can be solved analytically for a hard-sphere potential with an infinitely deep and narrow attractive well. The results obtained for this model suggest that at least some of the nonphysical alternative solutions found by Watts correspond to radial distribution functions $g(r)$ which diverge to infinity as r becomes large.

Similar results have been obtained by Levesque (1966) and Throop and Bearman (1966) for the PY approximation and Verlet and Levesque (1962), Klein and Green (1963a), de Boer *et al.* (1964), Levesque (1966), and Watts (1969, 1970) for the CHNC approximation.

E. Discussion of Approximations

From Figs. 5–8 it appears that the approximations can be classified in decreasing order of accuracy as: PY, CHNC, BG, Kirkwood. It should be emphasized, however, that this is a generalization. There are some indications that at moderate to low densities the CHNC approximation is more accurate than the PY for potentials with a significant attractive well (Khan, 1964).

It is disappointing that the virial and compressibility pressures in the PY approximation for the Lennard-Jones potential disagree so markedly around the critical point. Certainly, it seems that if the PY approximation is used, more-consistent values of the pressure are obtained from the compressibility relation than from the virial theorem.

VII. Ornstein–Zernike Relation

A. Fourier-Transform Formulation

Setting $\mathbf{r} = \mathbf{r}_1 - \mathbf{r}_2$ and $\mathbf{s} = \mathbf{r}_1 - \mathbf{r}_3$, the OZ relation (5.22) can be written as

$$h(r) = c(r) + \varrho \int d\mathbf{s}\, c(s) h(|\mathbf{r} - \mathbf{s}|). \tag{7.1}$$

This equation can be conveniently expressed in terms of Fourier transforms. Multiply both sides by $\exp(i\mathbf{k} \cdot \mathbf{r})$ and integrate with respect to $\mathbf{r}$ over all space. Then, interchanging the orders of integration in the second term on the right-hand side and replacing $\mathbf{r}$ by $\mathbf{s} + \mathbf{t}$ gives

$$\tilde{h}(k) = \tilde{c}(k) + \varrho \tilde{c}(k) \tilde{h}(k), \tag{7.2}$$

where $\tilde{h}(k)$ and $\tilde{c}(k)$ are the three-dimensional Fourier transforms of $h(r)$ and $c(r)$, i.e.,

$$\tilde{h}(k) = \int d\mathbf{r} \exp(i\mathbf{k} \cdot \mathbf{r})\, h(r) \tag{7.3}$$

$$\tilde{c}(k) = \int d\mathbf{r} \exp(i\mathbf{k} \cdot \mathbf{r})\, c(r). \tag{7.4}$$

$\tilde{h}(k)$ and $\tilde{c}(k)$ depend only on the magnitude k of the vector $\mathbf{k}$.

The integrations in Eqs. (7.3) and (7.4) can be simplified by using spherical polar coordinates (r, θ, α) about an axis along the vector $\mathbf{k}$. Equation (7.3) then becomes

$$\tilde{h}(k) = \int_0^\infty dr\, r^2 \int_0^\pi d\theta \sin\theta \int_0^{2\pi} d\alpha\, e^{ikr\cos\theta} h(r). \tag{7.5}$$

The angle integrations can be performed, giving

$$\tilde{h}(k) = (4\pi/k) \int_0^\infty dr \sin kr\, r h(r). \tag{7.6}$$

Integrating by parts, Eq. (7.6) can be cast into the form

$$\tilde{h}(k) = 4\pi \int_0^\infty dr\,(\cos kr)H(r), \tag{7.7}$$

where

$$H(r) = \int_r^\infty dt\, th(t). \tag{7.8}$$

Similarly, Eq. (7.4) can be reduced to

$$\tilde{c}(k) = 4\pi \int_0^\infty dr\,(\cos kr)C(r), \tag{7.9}$$

where

$$C(r) = \int_r^\infty dt\, tc(t). \tag{7.10}$$

B. Compressibility in Terms of $c(r)$

From Eq. (7.2), it can be deduced that

$$1 + \varrho\tilde{h}(k) = \{1 - \varrho\tilde{c}(k)\}^{-1}. \tag{7.11}$$

Now, from the definition (7.3) of $\tilde{h}(k)$ it can be seen that the inverse compressibility relation (2.41) can be written as

$$\frac{1}{kT}\left.\frac{\partial P}{\partial \varrho}\right|_{\mathrm{V,T}} = \{1 + \varrho\tilde{h}(0)\}^{-1}. \tag{7.12}$$

Using Eq. (7.11), it follows that

$$\frac{1}{kT}\left.\frac{\partial P}{\partial \varrho}\right|_{\mathrm{V,T}} = 1 - \varrho\tilde{c}(0), \tag{7.13}$$

i.e.,

$$\frac{1}{kT}\left.\frac{\partial P}{\partial \varrho}\right|_{\mathrm{V,T}} = 1 - \varrho \int d\mathbf{r}\, c(r). \tag{7.14}$$

Thus, the inverse compressibility can be expressed quite naturally in terms of $c(r)$. From a numerical viewpoint Eq. (7.14) is easier to use than (2.41), since $c(r)$ tends quite rapidly to zero with increasing r.

C. Finite-Range Transformation

If $c(r)$ can be assumed to vanish beyond some range a, then the OZ relation can be transformed into a form suitable for computation (Baxter, 1967a, 1968b). In view of the known short-range behavior of $c(r)$, this does not seem a very drastic assumption. Further, for a potential of finite range the PY approximation (5.27) predicts that beyond the range of the potential $c(r)$ should vanish exactly. The transformation is therefore particularly appropriate to this case.

Set

$$c(r) = 0, \qquad r > a; \tag{7.15}$$

then from Eqs. (7.9) and (7.10)

$$1 - \varrho\tilde{c}(k) = 1 - 4\pi\varrho \int_0^a dr \cos kr \, C(r). \tag{7.16}$$

For a fluid with no long-range order the integral (7.3) must be convergent for all real values of k, so that $\tilde{h}(k)$ is finite. It therefore follows from Eq. (7.11) that $1 - \varrho\tilde{c}(k)$, and hence the right-hand side of Eq. (7.16), cannot be zero for any real k. When this is so it is a theorem of Fourier transforms (Krein, 1962; Baxter, 1968b) that the right-hand side of Eq. (7.16) can be factorized into the form

$$1 - 4\pi\varrho \int_0^a dr \cos kr \, C(r) = \tilde{Q}(k)\tilde{Q}(-k), \tag{7.17}$$

where $\tilde{Q}(k)$ is a regular function of k and cannot be zero unless k has a negative imaginary part. (In complex-variable language, $\tilde{Q}(k)$ has zeros only in the lower half-plane.) Further, $\tilde{Q}(k)$ can be written as

$$\tilde{Q}(k) = 1 - 2\pi\varrho \int_0^a dr \, e^{ikr} Q(r), \tag{7.18}$$

where $Q(r)$ is a real function.

An explicit relation between $Q(r)$ and $C(r)$ can be obtained by replacing $\tilde{Q}(k)$ and $\tilde{Q}(-k)$ in Eq. (7.17) by their forms as given by Eq. (7.18) and inverting the Fourier transforms. This gives

$$C(r) = Q(r) - 2\pi\varrho \int_r^a dt \, Q(t)Q(t-r) \tag{7.19}$$

for $0 < r < a$.

It is also possible to obtain a relation between $Q(r)$ and $H(r)$. To do this, note from Eqs. (7.11) and (7.17) that

$$\tilde{Q}(k)[1 + \varrho\tilde{h}(k)] = [\tilde{Q}(-k)]^{-1}. \tag{7.20}$$

Now invert the Fourier transforms by multiplying Eq. (7.20) by $\exp(-ikr)$ and integrating with respect to k from $-\infty$ to ∞. When r is positive the integration on the right-hand side can be closed around the lower half k-plane, where $\tilde{Q}(-k)$ is regular, has no zeros, and can be seen from Eq. (7.18) to tend to 1 at infinity. From Cauchy's theorem in complex-variable theory it follows that this integral vanishes.

Substituting the forms (7.18) and (7.7) of $\tilde{Q}(k)$ and $\tilde{h}(k)$ into the left-hand side of Eq. (7.20) and performing the integration, it follows that

$$-Q(r) + H(r) - 2\pi\varrho \int_0^a dt\, Q(t)H(|r-t|) = 0, \tag{7.21}$$

for $r > 0$, where the range of $Q(r)$ is extended by defining

$$Q(r) = 0 \qquad \text{for} \quad r > a. \tag{7.22}$$

From Eqs. (7.10) and (7.19) it can be seen that $Q(r) \to 0$ as $r \to a$ from below, so the definition (7.22) ensures that $Q(r)$ is continuous at $r = a$.

Equations (7.19) and (7.21) can be expressed in terms of $c(r)$ and $h(r)$, rather than $C(r)$ and $H(r)$, by differentiating them with respect to r. Using Eqs. (7.8) and (7.10), and the fact that $Q(a) = 0$, it is found that

$$rc(r) = -Q'(r) + 2\pi\varrho \int_r^a dt\, Q'(t)Q(t-r) \tag{7.23}$$

for $0 < r < a$, and

$$rh(r) = -Q'(r) + 2\pi\varrho \int_0^a dt\, (r-t)h(|r-t|)Q(t) \tag{7.24}$$

for $r > 0$, where $Q'(r)$ is the derivative of $Q(r)$.

Equations (7.23) and (7.24) are the desired transformation of the OZ relation. Their advantage lies in the fact that if r is restricted to lie in the range $(0, a)$, then they only involve the functions $c(r)$, $h(r)$, and $Q(r)$ over this range, even though $h(r)$ may be quite a long-ranged function.

D. PY Approximation for Adhesive Hard Spheres

Consider the square-well potential defined by

$$\begin{aligned} \phi(r)/kT = u(r) &= +\infty, && r < \sigma, \\ &= -\log[a/12\tau(a-\sigma)], && \sigma < r < a, \\ &= 0, && r > a. \end{aligned} \tag{7.25}$$

The parameters in this potential are chosen so that τ is a dimensionless measure of the temperature, being zero at zero temperature and large at high temperatures.

It turns out that the PY approximation can be solved analytically for this potential provided one takes the limit of $\sigma \to a$, τ and a being held fixed (Baxter, 1968c). In this limit the potential consists of a hard core together with an infinitely deep and narrow attractive well. Such a potential can be regarded as applicable to hard spheres of diameter a with surface adhesion. Although it is highly idealized, it does at least incorporate the two most significant features of a real potential.

The following discussion can be made quite rigorous by working directly with the potential (7.25), exhibiting the first-order terms in $a - \sigma$ in the various equations, and deriving bounds on the higher-order remainder terms. However, for the sake of clarity the limit $\sigma \to a$ will be taken explicitly where possible. Thus, the function $f(r)$ given by Eqs. (5.1) and (7.25) becomes:

$$\begin{aligned} f(r) &= -1 + (a/12\tau)\,\delta(r - a_-), \qquad & 0 < r < a, \\ &= 0, & r > a, \end{aligned} \tag{7.26}$$

where the notation a_- is used to denote the fact that the delta-function lies just to the left of a.

When $r > a$, it follows from Eq. (7.25) and the PY approximation (5.27) that $c(r) = 0$, so the transformation (7.23)–(7.24) of the OZ relation can be used. Also, when $r < \sigma$ it follows from Eqs. (7.25) and (5.27) that $g(r) = 0$. Hence, $h(r) = -1$ is a known function in the range $(0, \sigma)$.

The behavior in the range (σ, a) is more complicated, but when $\sigma \to a$ it can be seen from the virial expansion that $h(r)$ has a delta-function singularity similar to that in $f(r)$. Thus, in this limit

$$h(r) = -1 + (\lambda a/12)\,\delta(r - a_-) \qquad \text{for} \quad 0 < r < a, \tag{7.27}$$

where λ is some dimensionless parameter, as yet unknown. (The factor 12 is introduced purely for convenience in the later equations.)

If Eq. (7.24) is restricted to the range $0 < r < a$, then $h(r)$ can be replaced by its form (7.27), giving

$$Q'(r) = \alpha r + \beta a - (\lambda a^2/12)\,\delta(r - a_-) \tag{7.28}$$

for $0 < r < a$, where

$$\alpha = 1 - 2\pi\varrho \int_0^a dt\, Q(t), \qquad \beta = (2\pi\varrho/a) \int_0^a dt\, tQ(t). \tag{7.29}$$

$Q'(r)$ also has finite discontinuities at 0 and a, but these do not affect the following working.

Integrating Eq. (7.28) with respect to r, using the condition that $Q(a) = 0$, gives

$$Q(r) = (\alpha/2)(r^2 - a^2) + \beta a(r - a) + (\lambda a^2/12) \qquad (7.30)$$

for $0 < r < a_-$. Substituting this form for $Q(r)$ into the equations (7.29) gives two linear equations for α and β, which can be solved to give

$$\alpha = (1 + 2\eta - \mu)/(1 - \eta)^2, \qquad \beta = \tfrac{1}{2}(-3\eta + \mu)/(1 - \eta)^2, \quad (7.31)$$

where η is the dimensionless density, given by

$$\eta = \tfrac{1}{6}\pi\varrho a^3, \qquad (7.32)$$

and

$$\mu = \lambda\eta(1 - \eta). \qquad (7.33)$$

Now that $Q(r)$ is known in terms of the parameter λ, $c(r)$ can be evaluated from Eq. (7.23), and the PY approximation (5.27) for $\sigma < r < a$ used to give an equation for λ in terms of τ. To do this, some care has to be taken in handling the delta-functions, and it is best to use Eqs. (2.40) and (5.1) to write the PY approximation as

$$1 + h(r) = [1 + f(r)]p(r), \qquad (7.34)$$

where

$$p(r) = 1 + h(r) - c(r). \qquad (7.35)$$

From the original OZ relation, Eq. (7.1), it can be seen that in the limit $\sigma \to a$, $p(r)$ is continuous at $r = a$. Substituting the limiting forms (7.26) and (7.27) of $f(r)$ and $h(r)$ into Eq. (7.34), it follows that for $\sigma < r < a$ the PY approximation is

$$p(a) = \lambda\tau. \qquad (7.36)$$

The $p(a)$ can be obtained in terms of λ by adding r to both sides of Eq. (7.24), subtracting Eq. (7.23), and setting $r = a$. Using Eq. (7.35), this gives

$$ap(a) = a + 2\pi\varrho \int_0^a dt\,(a - t)h(a - t)Q(t). \qquad (7.37)$$

Substituting the form (7.27) of $h(r)$, using the equations (7.29), and noting that $Q(r)$ is continuous at $r = 0$, it follows that

$$p(a) = \alpha + \beta + (\eta\lambda/a^2)Q(0), \qquad (7.38)$$

or, using Eqs. (7.30) and (7.31),

$$p(a) = \frac{1 + (\eta/2)}{(1-\eta)^2} - \frac{\eta\lambda}{1-\eta} + \frac{\eta\lambda^2}{12}. \tag{7.39}$$

Eliminating $p(a)$ between Eqs. (7.36) and (7.39) gives a quadratic equation for λ in terms of the "temperature" τ and the dimensionless density η. The solutions obtained are plotted in Fig. 12. When τ is greater than a value

$$\tau_c = (2 - \sqrt{2})/6 \approx 0.0976 \tag{7.40}$$

there are two real solutions for λ throughout the permissible density range $0 < \eta < 1$.* Below this temperature there exists an intermediate range of densities within which there are no real solutions for λ, so that the system has to undergo a discontinuous transition between states of different density. The PY approximation therefore predicts a first-order phase transition for this system, with critical temperature τ_c and critical density

$$\eta_c = (3\sqrt{2} - 4)/2 \approx 0.1213. \tag{7.41}$$

For the values of τ and η, for which there exist two real solutions, it is clearly necessary to decide which solution (if either) corresponds to an allowed physical state of the system. On thermodynamic grounds the larger solutions for λ must be rejected, as λ must be a continuous function of η and τ in the regions where it is real, and must have the known zero-density value τ^{-1}.

There is, however, a further condition that must be satisfied, for the Fourier transform $\tilde{Q}(k)$ introduced by Eq. (7.17) was required to have zeros only in the lower half k-plane. If this condition is violated, then it can be shown that the indirect correlation function $h(r)$ given by Eq. (7.24) in general diverges to infinity when r becomes large.

Now it can be seen from Eq. (7.18) that $\tilde{Q}(k)$ is real and continuous when k lies on the positive imaginary axis (i.e., $k = iy$, where $y > 0$), and tends to 1 as k tends to infinity. Thus, $\tilde{Q}(0)$ must be positive, for otherwise there would be a zero on this axis.

Comparing Eqs. (7.18), (7.29), and (7.31), it is apparent that

$$\tilde{Q}(0) = \alpha = (1 + 2\eta - \mu)/(1-\eta)^2. \tag{7.42}$$

* The density $\eta = 1$ is that which would be obtained if the spheres could be deformed and packed so as to fill all the available space, while the true close-packed density is $\eta = \pi/3\sqrt{2} \approx 0.74$.

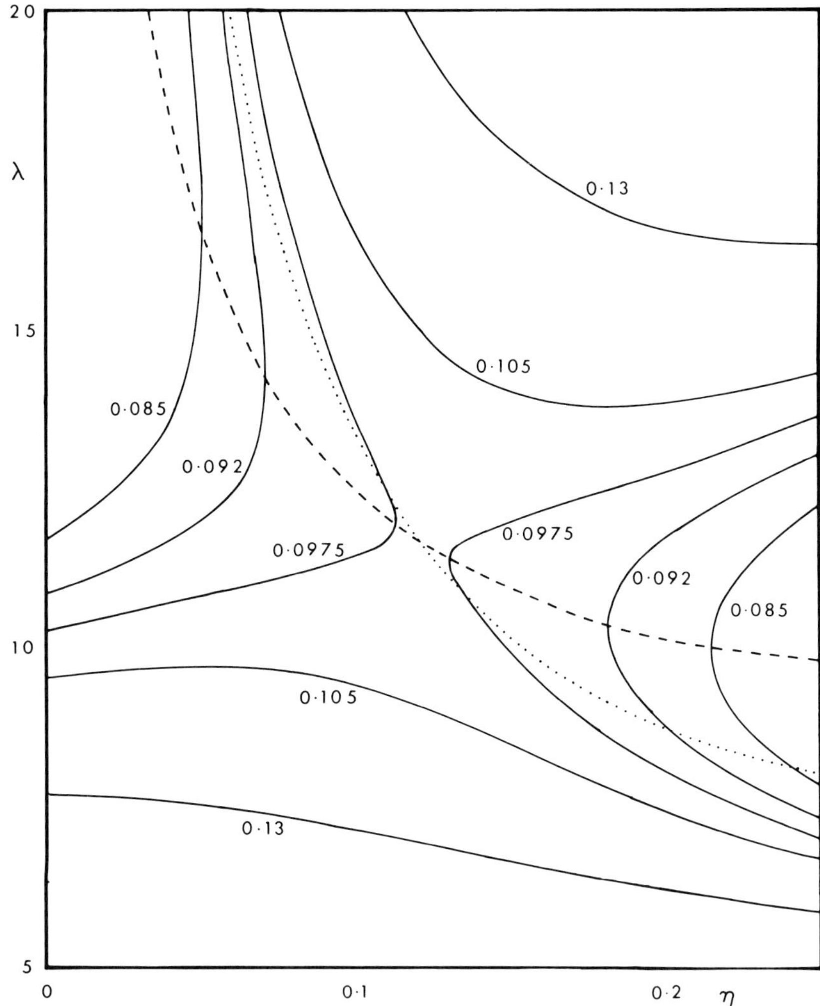

FIG. 12. Behavior of the solutions of the PY approximation for adhesive hard spheres; the parameter λ obtained from Eqs. (7.36) and (7.39) plotted against the dimensionless density η for $\tau = 0.085$, 0.092, 0.0975, 0.105, and 0.13. All points above the broken curve correspond to the nonphysical alternative solutions of the quadratic equation. The dotted curve is the locus where the compressibility is infinite, and all points above this correspond to divergent radial distribution functions. These two curves cross at the critical point.

Thus, any solution for λ such that $\mu \geq 1 + 2\eta$ corresponds to a non-positive value of $\tilde{Q}(0)$ and must be discarded.

The compressibility equation of state is very easily obtained, for, from Eqs. (7.13), (7.16), and (7.17)

$$\frac{1}{kT}\,\frac{\partial P}{\partial \varrho}\bigg|_{\mathrm{V,T}} = [\tilde{Q}(0)]^2. \tag{7.43}$$

Thus, the locus of values of τ and η for which $\tilde{Q}(0)$ vanishes is also the locus for which the compressibility is infinite. This locus is shown in Fig. 12 and is found to pass through the critical point. All points above it correspond to divergent indirect correlation functions.

Using Eqs. (7.33), (7.36), (7.39), and (7.42) to express the right-hand side of Eq. (7.43) in terms of τ and η, and then integrating with respect to η (or ϱ), it is found that

$$\frac{P}{\varrho kT} = \frac{1 + \eta + \eta^2 - \mu(1 + \frac{1}{2}\eta) + (\eta^{-1}\mu^3/36)}{(1 - \eta)^3}. \tag{7.44}$$

Thus, Eq. (7.44) gives the compressibility pressure.

The virial pressure can be obtained from Eq. (2.26). Noting from Eqs. (2.40), (5.1), and (7.34) that

$$g(r) = \mathrm{e}^{-u(r)}p(r), \tag{7.45}$$

Eq. (2.26) can be written as

$$\frac{P}{\varrho kT} = 1 + \frac{2\pi\varrho}{3}\int_0^\infty dr\, r^3 f'(r)p(r). \tag{7.46}$$

Integrating by parts and using the form of $f(r)$ given by Eqs. (5.1) and (7.25), it follows that

$$\frac{P}{\varrho kT} = 1 + 4\eta\, p(a) - \frac{\eta[a^3p(a) - \sigma^3p(\sigma)]}{3\tau a^2(a - \sigma)}. \tag{7.47}$$

In this case rather more care has to be taken in evaluating the limit $\sigma \to a$, for in this limit $p(r)$ is not differentiable at $r = a$. It is found that

$$\frac{P}{\varrho kT} = \frac{1}{(1 - \eta)^2}\left\{1 + 2\eta + 3\eta^2 - \frac{1}{3}(2 + 10\eta)\mu + \frac{1}{3}\mu^2\right\} - \frac{\eta}{3\tau(1 - \eta)^3}\left\{1 - 5\eta - 5\eta^2 + 6\eta\mu - \mu^2 + \frac{\mu^3}{24\eta}\right\}. \tag{7.48}$$

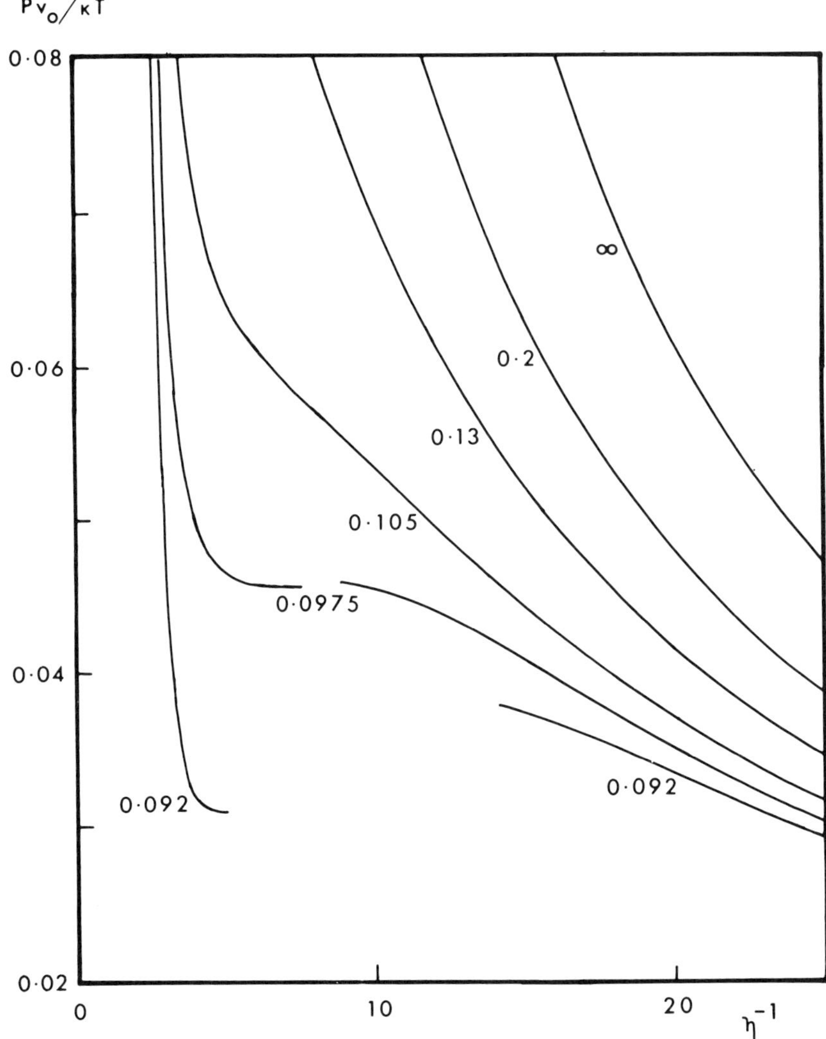

FIG. 13. The compressibility equation of state for adhesive hard spheres given by Eq. (7.44). Pv_0/kT (where $v_0 = \frac{1}{6}\pi a^3$) is plotted against the dimensionless volume per particle η^{-1} for $\tau = 0.092$, 0.0975, 0.105, 0.13, 0.2, and ∞. Only the physical branches of the isotherms are shown.

The compressibility and virial pressures are plotted in Figs. 13 and 14. It can be seen that they exhibit qualitatively the same behavior as Watts found for the Lennard-Jones potential (Figs. 10 and 11). In particular, the compressibility equation of state is similar to that of a van der Waal gas, while the virial equation of state is physically inconsistent.

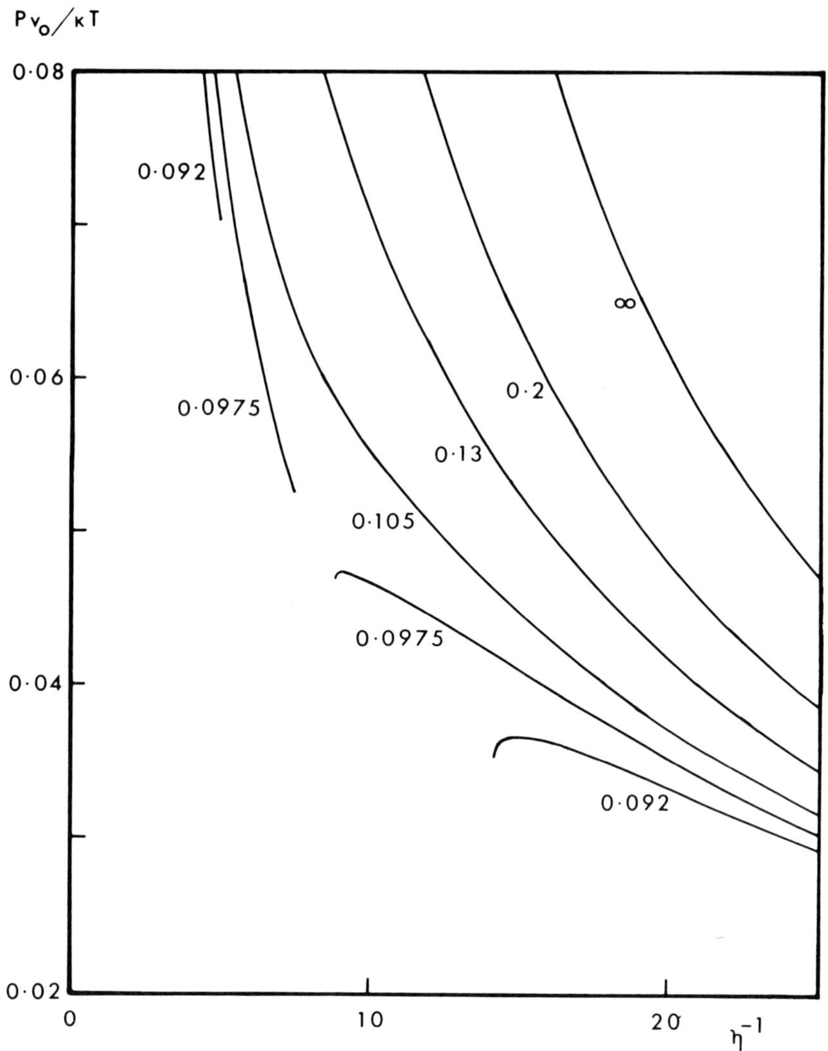

FIG. 14. The virial equation of state for adhesive hard spheres given by Eq. (7.48). The parameters and curves have the same interpretation as in Fig. 13.

When $\tau \to \infty$ the potential (7.25) becomes that of a simple hard-sphere gas and both λ and μ become zero. The PY approximation for this case was solved by Wertheim (1963) and Thiele (1963). Their results for the compressibility and virial pressures can be obtained by setting $\mu = 0$ in Eqs. (7.44) and (7.48) and allowing τ to tend to infinity.

VIII. Scaled-Particle Theory

An interesting and simple treatment of the simple hard-sphere fluid is provided by scaled-particle theory (Reiss *et al.*, 1959; Reiss, 1965). For hard-sphere molecules of diameter a the virial equation of state (2.26) can be transformed by the same techniques that led to Eq. (7.47), giving

$$P/\varrho kT = 1 + \tfrac{2}{3}\pi\varrho a^3 g(a), \tag{8.1}$$

where $g(a)$ is the value of the radial distribution function $g(r)$ for r just greater than a.

It is apparent from Eq. (8.1) that in this case a complete knowledge of $g(r)$ is unnecessary for obtaining the pressure; rather, it is sufficient to obtain its value for $r = a$. It was this observation that led to the development of scaled-particle theory.

Consider a spherical region of radius r inside the fluid, and let $p_0(r)$ be the probability that no molecule center lies inside this region. Further, let $\varrho G(r)$ be the concentration of centers just outside the sphere when there are no centers lying inside it.

From a microscopic viewpoint an empty spherical cavity of radius r inside the fluid plays exactly the same role as a hard-sphere molecule of diameter $2r - a$ (it is this concept which gives the theory its name). When $r = a$ the cavity therefore behaves as a typical molecule and $\varrho G(a)$ can be regarded as the concentration of molecule centers at a distance a from some specified particle, i.e.,

$$G(a) = g(a). \tag{8.2}$$

Thus, the equation of state (8.1) can be written as

$$P/\varrho kT = 1 + \tfrac{2}{3}\pi\varrho a^3 G(a). \tag{8.3}$$

Instead of calculating the radial distribution function $g(r)$, scaled-particle theory is concerned with evaluating the function $G(r)$ (in particular, its value at $r = a$).

One can relate $p_0(r)$ and $G(r)$ by noting that $p_0(r + dr)$ is the joint probability of finding no particles in the sphere of radius r, and no particles in the spherical shell between r and $r + dr$. Now, the probability of finding a particle in the shell is $4\pi r^2\varrho G(r)\,dr$, so that

$$p_0(r + dr) = p_0(r)[1 - 4\pi r^2\varrho G(r)\,dr]. \tag{8.4}$$

Taking the limit of dr infinitesimally small, it follows that

$$d[\log p_0(r)]/dr = -4\pi r^2 \varrho G(r). \tag{8.5}$$

When $r < \frac{1}{2}a$, $p_0(r)$ and $G(r)$ can be calculated exactly, for then at most one molecule center can lie in the sphere of radius r. The probability that the sphere contains a center is $4\pi r^3\varrho/3$, so that

$$p_0(r) = 1 - \tfrac{4}{3}\pi r^3 \varrho, \tag{8.6}$$

and hence, from Eq. (8.5)

$$G(r) = \{1 - \tfrac{4}{3}\pi r^3 \varrho\}^{-1} \tag{8.7}$$

for $0 < r < \frac{1}{2}a$.

It can be shown that $G(r)$ and its first derivative are continuous at $r = \frac{1}{2}a$, but that the second derivative has a finite discontinuity, due to the possible introduction of a second molecule center. At $r = a/\sqrt{3}$ it becomes possible to include three centers, and a discontinuity occurs in the fourth derivative of $G(r)$. Similarly, as r increases, more and more centers can be accommodated, and it appears that at the transition from accommodating L centers to $L + 1$ centers a discontinuity arises in the $(2L)$th derivative of $G(r)$.

It is also possible to obtain information concerning the behavior of $G(r)$ when r is large. To do this, note that on statistical-mechanical grounds (Tolman, 1938)

$$p_0(r) = e^{-W(r)/kT}, \tag{8.8}$$

where $W(r)$ is the reversible work required to produce a spherical cavity of radius r in the fluid. Comparing Eqs. (8.5) and (8.8), it is apparent that

$$dW(r) = 4\pi kTr^2 \varrho G(r)\, dr. \tag{8.9}$$

When r is large it follows from thermodynamic reasoning that

$$dW(r) = P\, dV + \sigma[1 - (2\delta/r)]\, dA, \tag{8.10}$$

where $dV = 4\pi r^2\, dr$ is the volume increment, $dA = 8\pi r\, dr$ is the surface increment, P is the pressure, and $\sigma[1 - (2\delta/r)]$ is the surface tension of a bubble of radius r. Substituting the explicit forms of dV and dA, and comparing Eqs. (8.9) and (8.10), it follows that

$$G(r) = \frac{P + (2\sigma/r) - (4\sigma\delta/r^2)}{\varrho kT}. \tag{8.11}$$

It has been pointed out that $G(r)$ is not an analytic function for $r > \frac{1}{2}a$. Nevertheless, it is a smooth function because the discontinuities occur in high-order derivatives. This suggests that it should be a good approximation to interpolate for $G(r)$ between $\frac{1}{2}a$ and ∞ by choosing it to be of the form (8.11) throughout this range. The quantities P, σ, and δ can then be evaluated from Eq. (8.3) and the requirement that $G(r)$ and its first derivative be continuous at $r = \frac{1}{2}a$. This gives

$$P/\varrho kT = (1 + \eta + \eta^2)/(1 - \eta)^3, \tag{8.12}$$

$$\sigma/a\varrho kT = -\tfrac{3}{4}(\eta + \eta^2)/(1 - \eta)^3, \tag{8.13}$$

$$\delta/a = \tfrac{1}{4}\,\eta/(1 + \eta), \tag{8.14}$$

where η, the dimensionless density, is given by

$$\eta = \tfrac{1}{6}\,\pi\varrho a^3. \tag{8.15}$$

It is interesting to note that the form (8.12) of the pressure is identical with the compressibility pressure of the PY approximation for hard spheres, given by Eq. (7.44) for $\mu = 0$.

In view of the simplicity of these results and the information they provide on the surface tension of a fluid, attempts have been made to apply them to real fluids by approximating the intermolecular potential by a simple hard core with a temperature-dependent diameter (Reiss,

TABLE I

Comparison of Experimental Surface Tensions of Some Simple Liquids with Those Calculated from Scaled-Particle Theory[a]

Substance	T (°K)	σ_{exp} (dynes/cm)	σ_{calc} (dynes/cm)
N	27.2	4.8	6.09
Ar	85.1	13.2	16.40
He	4.2	0.098	0.223
H_2	20.4	1.80	1.91
N_2	70.1	10.5	14.9
O_2	70.1	18.3	23.6
C_6H_6	273.1	29.02	34.3

[a] Reiss (1965).

1965). In Table I the surface tensions σ of some simple liquids are given, together with the values obtained from Eq. (8.13). When allowance is made for the gross simplifications that are necessary to apply the theory the agreement can be regarded as satisfactory.

IX. Functional Techniques Applied to Inhomogeneous Fluids, Solids, and Mixtures

A. Grand Partition Function and Distribution Functions

In the previous sections of this chapter attention has been focused on a one-component homogeneous fluid with central forces. In this section these restrictions will be relaxed and the general case of an inhomogeneous fluid or solid consisting of molecules of different types considered.* In general, it will be assumed that the particles of type α move in a spatially varying external field of potential $\phi_\alpha^{(1)}(\mathbf{r})$, and that a particle of type α at $\mathbf{r}$ interacts with one of type α' at $\mathbf{r}'$ *via* an interaction potential $\phi_{\alpha\alpha'}^{(2)}(\mathbf{r}, \mathbf{r}')$. The interactions may depend on the positions of the particles in the system, as well as on their relative separation and orientation.

Consider a configuration of the system in which it contains N particles, of types $\alpha_1, \ldots, \alpha_N$, located in the volume elements $d\mathbf{r}_1, \ldots, d\mathbf{r}_N$ centered on $\mathbf{r}_1, \ldots, \mathbf{r}_N$, with momenta in the elements $d\mathbf{p}_1, \ldots, d\mathbf{p}_N$ centered on $\mathbf{p}_1, \ldots, \mathbf{p}_N$, respectively. If m_α is the mass of each particle of type α, then the total energy of this configuration is

$$E = \sum_{i=1}^{N} \frac{\mathbf{p}_i^2}{2m_{\alpha_i}} + \sum_{i=1}^{N} \phi_{\alpha_i}^{(1)}(\mathbf{r}_i) + \sum_{1 \le i < j \le N}\sum \phi_{\alpha_i\alpha_j}^{(2)}(\mathbf{r}_i, \mathbf{r}_j), \tag{9.1}$$

and if μ_α is the chemical potential of the particles of type α, then the probability that the system is in this configuration is $\mathscr{P}\, d\mathbf{r}_1 \cdots d\mathbf{r}_N\, d\mathbf{p}_1 \cdots d\mathbf{p}_N$, where

$$\mathscr{P} = \frac{1}{h^{3N}\,\Xi^*} \exp\left\{\left[\sum_{i=1}^{N} \mu_{\alpha_i} - E\right] \Big/ kT\right\}. \tag{9.2}$$

The factor Ξ^* must be chosen so as to ensure that the sum of the probabilities of all possible configurations is unity. The "sum over con-

* Any two physically distinguishable molecules are regarded here as being of different types, e.g., chemically different molecules, polymers of different lengths, or asymmetrical molecules with different orientations.

figurations" is obtained in just the same way as in Section II, except that it is also necessary to sum over the possible types of each particle. The momentum integrations can then be performed, giving

$$\Xi^* = \sum_{N=0}^{\infty} \frac{1}{N!} \sum_{\alpha_1} \cdots \sum_{\alpha_N} \int \cdots \int_V d\mathbf{r}_1 \cdots d\mathbf{r}_N \exp\Big[-\sum_i u_i - \sum_{i<j} u_{ij}\Big], \tag{9.3}$$

where each of the summations over $\alpha_1, \ldots, \alpha_N$ denotes a summation over the types of particles in the system, the spatial integrations are over the volume V within which the system is confined, the two summations in the exponent are over all values of i such that $1 \le i \le N$, and over all values of i and j such that $1 \le i < j \le N$ and

$$\begin{aligned} u_i &\equiv u_{\alpha_i}(\mathbf{r}_i) = \frac{\phi_{\alpha_i}^{(1)}(\mathbf{r}_i) - \mu_{\alpha_i}}{kT} + \frac{3}{2} \log \frac{h^2}{2\pi m_{\alpha_i} kT}, \\ u_{ij} &\equiv u_{\alpha_i \alpha_j}(\mathbf{r}_i, \mathbf{r}_j) = \phi_{\alpha_i \alpha_j}^{(2)}(\mathbf{r}_i, \mathbf{r}_j)/kT. \end{aligned} \tag{9.4}$$

In Eqs. (9.3) and (9.4) the single suffix i has been used to denote dependence on the type α_i and position $\mathbf{r}_i$ of the particle i. Also, the number of suffixes has been regarded as sufficient to distinguish between one-particle and two-particle functions. These conventions provide a useful short-hand notation that will be used frequently in this section.

The equations can be further simplified by noting that an integration over a particle's coordinates is usually accompanied by a summation over possible types, and *vice versa*. Thus, it is convenient to introduce the notation

$$\int di = \sum_{\alpha_i} \int_V d\mathbf{r}_i. \tag{9.5}$$

The expression (9.3) for Ξ^*, which is the grand partition function of the system, then becomes

$$\Xi^* = \sum_{N=0}^{\infty} \frac{1}{N!} \int \cdots \int d1 \ldots dN \exp\Big[-\sum_i u_i - \sum_{i<j} u_{ij}\Big]. \tag{9.6}$$

These conventions can be used to simplify the notation for the distribution functions of a multicomponent system. Suppose $n_{\alpha_1 \cdots \alpha_L}^{(L)}(\mathbf{r}_1, \ldots, \mathbf{r}_L)\, d\mathbf{r}_1 \cdots d\mathbf{r}_L$ is the probability of finding particles of type $\alpha_1, \ldots, \alpha_L$ in the volume elements $d\mathbf{r}_1, \ldots, d\mathbf{r}_L$ centered on $\mathbf{r}_1, \ldots, \mathbf{r}_L$. Then one can use the short-hand notation

$$n_i = n_{\alpha_i}^{(1)}(\mathbf{r}_i), \qquad n_{ij} = n_{\alpha_i \alpha_j}^{(2)}(\mathbf{r}_i, \mathbf{r}_j), \qquad n_{ijk} = n_{\alpha_i \alpha_j \alpha_k}^{(3)}(\mathbf{r}_i, \mathbf{r}_j, \mathbf{r}_k), \quad \text{etc.} \tag{9.7}$$

Then $n_{1\cdots L}$ is the L-particle distribution function, and can be obtained by summing the probabilities of the appropriate configurations as in Section II, except that it is now also necessary to sum over the types of the particles $L+1,\ldots, N$. This gives

$$n_{1\cdots L} = \frac{1}{\Xi^*} \sum_{N=L}^{\infty} \frac{1}{(N-L)!} \int\cdots\int d(L+1)\cdots dN \exp\Big[-\sum_i u_i - \sum_{i<j} u_{ij}\Big] \tag{9.8}$$

where the two summations in the exponent are again over all values of i such that $1 \leq i \leq N$, and over all values of i and j such that $1 \leq i < j \leq N$, respectively.

B. Thermodynamic Properties

The mean value $\bar{X}$ of a quantity X is obtained by multiplying its value for a particular configuration by the corresponding probability $\mathscr{P}$ and summing over configurations. Proceeding in this way, it can be established from Eqs. (9.1)–(9.8) that

$$\bar{N}_\alpha = \int_V d\mathbf{r}\, n_\alpha(\mathbf{r}), \tag{9.9}$$

$$\bar{N} = \sum_\alpha \bar{N}_\alpha = \int d1\, n_1, \tag{9.10}$$

$$U = \tfrac{3}{2}\bar{N}kT + \int d1\, \phi_1 n_1 + \tfrac{1}{2} \iint d1\, d2\, \phi_{12} n_{12}, \tag{9.11}$$

$$TS = kT \log \Xi^* - \sum_\alpha \mu_\alpha \bar{N}_\alpha + U, \tag{9.12}$$

where $\bar{N}_\alpha$ is the mean number of particles of type α, $\bar{N}$ is the mean total number of particles, $U = \bar{E}$ is the internal energy, and $S = -k\langle \log \mathscr{P}\rangle_{\text{av}}$ is the entropy.

The Gibbs function G for a multicomponent system is

$$G = \sum_\alpha \mu_\alpha \bar{N}_\alpha, \tag{9.13}$$

so it follows from Eq. (9.12) that

$$kT \log \Xi^* = TS + G - U. \tag{9.14}$$

Comparing this result with the usual thermodynamic relations (Pippard, 1957), it follows that for a homogeneous system

$$\log \Xi^* = PV/kT, \tag{9.15}$$

where P is the pressure. The free energy F can now be obtained from the relations

$$F = U - TS = G - PV. \tag{9.16}$$

C. Functional Derivatives

One of the most-useful techniques in statistical mechanics is to consider the effect of varying the state of the system—for instance, by varying the external fields and/or the interaction potentials. The statistical-mechanical properties are then *functionals* (i.e., functions of a function) of the functions which are being varied. In order to fully exploit these functional techniques, it is necessary to define what is meant by a functional derivative.

Suppose one has a function $y_{1\cdots L}$ of particle coordinates[†] $1,\ldots, L$, which also depends on some one-particle function $x_\alpha(\mathbf{r})$. Then $y_{1\cdots L}$ is a functional of $x_\alpha(\mathbf{r})$, and this dependence can be exhibited explicitly by writing $y_{1\cdots L}$ as $y_{1\cdots L}\{x\}$.

The $y_{1\cdots L}\{x\}$ can be thought of as a function of many variables—namely, the values of $x_\alpha(\mathbf{r})$ at all points in space for all α. Like a function of many variables, it can in general be Taylor expanded about a particular set of values of its arguments, i.e., $y_{1\cdots L}\{x + \xi\}$ can be expanded in powers of the function $\xi_\alpha(\mathbf{r})$, giving

$$\begin{aligned} y_{1\cdots L}\{x+\xi\} = y_{1\cdots L}\{x\} &+ \frac{1}{1!}\int d(L+1)\, y^{(1)}_{1\cdots L|L+1}\{x\}\xi_{L+1} \\ &+ \frac{1}{2!}\iint d(L+1)\, d(L+2)\, y^{(2)}_{1\cdots L|L+1,L+2}\{x\} \\ &\times \xi_{L+1}\xi_{L+2} + \cdots. \end{aligned} \tag{9.17}$$

For $M \geq 2$ the "coefficients" $y^{(M)}_{1\cdots L|L+1,\ldots,L+M}\{x\}$ in this expansion are not uniquely defined, since the equation is unaltered by adding to them any antisymmetrical function of the particle coordinates $L+1,\ldots,$ $L+M$. This difficulty can be overcome by requiring them to be sym-

[†] The word "coordinates" is generally used here to denote both a particle's type and position.

metrical functions of the coordinates $L+1,\ldots, L+M$. With this restriction, $y^{(M)}_{1\cdots L|L+1,\ldots,L+M}\{x\}$ is known as the Mth functional derivative of $y_{1\cdots L}\{x\}$, and is written as

$$y^{(M)}_{1\cdots L|L+1,\ldots,L+M}\{x\} = \frac{\delta^{(M)} y_{1\cdots L}\{x\}}{\delta x_{L+1}\cdots\delta x_{L+M}}. \tag{9.18}$$

(This definition is analogous to that of partial derivatives of a function of many variables.)

Like partial differentiation, functional differentiation has the significant property that it is in general commutative, so that the $(M+N)$th derivative is the Mth derivative of the Nth derivative. In particular, this implies that

$$y^{(M)}_{1\cdots L|L+1,\ldots,L+M}\{x\} = \frac{\delta y^{(M-1)}_{1\cdots L|L+1,\ldots,L+M-1}\{x\}}{\delta x_{L+M}}, \tag{9.19}$$

or, Taylor expanding the $(M-1)$th derivative,

$$\begin{aligned} y^{(M-1)}_{1\cdots L|L+1,\ldots,L+M-1}\{x+\xi\} = {} & y^{(M-1)}_{1\cdots L|L+1,\ldots,L+M-1}\{x\} \\ & + \int d(L+M)\, y^{(M)}_{1\cdots L|L+1,\ldots,L+M}\{x\}\xi_{L+M} \\ & + O(\xi^2). \end{aligned} \tag{9.20}$$

D. Ξ^* As a Functional of u_i and u_{ij}

1. *Distribution and Indirect Correlation Functions*

From Eq. (9.6) it is apparent that Ξ^* depends on the two functions u_i and u_{ij}. Suppose u_{ij} is held fixed; then Ξ^* can be regarded as a functional of u_i, or, more conveniently, of the function

$$t_i = \exp(-u_i). \tag{9.21}$$

Working in terms of this function, and explicitly displaying the functional dependence of Ξ^* Eq. (9.6) can be written as

$$\Xi^*\{t\} = \sum_{N=0}^{\infty} \frac{1}{N!} \int\cdots\int d1\cdots dN \left[\exp\left(-\sum_{i<j} u_{ij}\right)\right] \prod_{i=1}^{N} t_i. \tag{9.22}$$

Now, replace t_i in Eq. (9.22) by $t_i + \delta t_i$, where δt_i is some arbitrary function. Expanding the right-hand side in powers of δt_i and using

Eqs. (9.6) and (9.8), it is found that

$$\Xi^*\{t + \delta t\} = \Xi^*\{t\}\left\{1 + \sum_{M=1}^{\infty} \frac{1}{M!} \int \cdots \int d1 \cdots dM \, \delta t_1 \cdots \delta t_M \frac{n_{1\cdots M}}{t_1 \cdots t_M}\right\}. \tag{9.23}$$

Equation (9.23) is a functional Taylor expansion of the general type (9.17), so the "coefficients" are the functional derivatives of Ξ^*, i.e.,

$$\frac{\Xi^* n_{1\cdots M}}{t_1 \cdots t_M} = \frac{\delta^M \, \Xi^*}{\delta t_1 \cdots \delta t_M} \tag{9.24}$$

(dropping the explicit dependence of Ξ^* on t).

Similarly, one can in principle Taylor expand $\log(\Xi^*\{t + \delta t\})$ in powers of δt, giving

$$\log \frac{\Xi^*\{t + \delta t\}}{\Xi^*\{t\}} = \sum_{M=1}^{\infty} \frac{1}{M!} \int \cdots \int d1 \cdots dM \, \delta t_1 \cdots \delta t_M \frac{p_{1\cdots M}}{t_1 \cdots t_M}, \tag{9.25}$$

where $p_{1\cdots M}/t_1 \ldots t_M$ is the Mth functional derivative of $\log \Xi^*$, i.e.,

$$\frac{p_{1\cdots M}}{t_1 \cdots t_M} = \frac{\delta^M \log \Xi^*}{\delta t_1 \cdots \delta t_M}. \tag{9.26}$$

The functions $p_{1\cdots M}$ can be expressed in terms of the distribution functions $n_{1\cdots M}$ by dividing Eq. (9.23) by $\Xi^*\{t\}$, taking logarithms of both sides, and expanding the right-hand side in powers of δt (taking care to choose the "coefficients" to be symmetrical with respect to the particle coordinates). Comparing the resulting expansion with Eq. (9.25) and equating "coefficients" then gives

$$\begin{gathered} p_1 = n_1, \qquad p_{12} = n_{12} - n_1 n_2, \\ p_{123} = n_{123} - n_1 n_{23} - n_2 n_{31} - n_3 n_{12} + 2 n_1 n_2 n_3, \qquad \text{etc.} \end{gathered} \tag{9.27}$$

The functions $p_{1\cdots M}$ are known as the Ursell–Mayer cluster functions, or indirect correlation functions. They have the property that they tend to zero when any group of the particles $1, \ldots, M$ becomes greatly separated from the remainder.

It is frequently convenient to work with normalized distribution and indirect correlation functions, defined by

$$g_{1\cdots L} = n_{1\cdots L}/n_1 n_2 \cdots n_L \tag{9.28}$$

$$h_{1\cdots L} = p_{1\cdots L}/n_1 n_2 \cdots n_L. \tag{9.29}$$

The functions $h_{1\cdots L}$ can then be expressed in terms of the functions $g_{1\cdots L}$ by using the relations (9.27). In particular,

$$h_1 = g_1 = 1, \tag{9.30}$$

and

$$h_{12} = h_{12} - 1. \tag{9.31}$$

For a single-component homogeneous fluid g_{12} and h_{12} reduce to the two-particle radial distribution and indirect correlation functions $g(r_{12})$ and $h(r_{12})$ defined by Eqs. (2.14) and (2.40).

2. *Derivatives with Respect to Chemical Potential*

The commutative property of functional differentiation can be used to obtain expressions for the derivatives of the distribution and indirect correlation functions with respect to the chemical potentials. In particular, from Eq. (9.26) it follows that

$$\frac{p_{1\cdots M}}{t_1\cdots t_M} = \frac{\delta}{\delta t_M}\left\{\frac{p_{1\cdots M-1}}{t_1\cdots t_{M-1}}\right\}, \tag{9.32}$$

so the increment induced in $p_{1\cdots M-1}/t_1\cdots t_{M-1}$ by a small variation δt_i in the function t_i is

$$\delta\left\{\frac{p_{1\cdots M-1}}{t_1\cdots t_{M-1}}\right\} = \int dM\,\delta t_M\,\frac{p_{1\cdots M}}{t_1\cdots t_M} + O\{(\delta t)^2\}. \tag{9.33}$$

The increment on the left-hand side of Eq. (9.33) can be expressed in terms of $\delta t_1, \ldots, \delta t_{M-1}$ and the increment $\delta p_{1\cdots M-1}$ in terms of $p_{1\cdots M-1}$, giving

$$\delta p_{1\cdots M-1} - p_{1\cdots M-1}\sum_{i=1}^{M-1}\frac{\delta t_i}{t_i} = \int dM\,\frac{\delta t_M}{t_M}\,p_{1\cdots M} + O\{(\delta t)^2\}. \tag{9.34}$$

For simplicity, consider a one-component system, and suppose that the temperature and interaction potentials are held fixed while the chemical potential μ is incremented by some small amount $\delta\mu$ (the suffixes α can be dropped in this case). Then from Eqs. (9.4) and (9.21) the corresponding increment in t_i is

$$\delta t_i = (\delta\mu/kT)t_i + O\{(\delta\mu)^2\}. \tag{9.35}$$

Substituting this form for δt_i in Eq. (9.34) gives

$$kT\,\delta p_{1\cdots M-1} - (M-1)\,\delta\mu\,p_{1\cdots M-1} = \delta\mu\int dM\,p_{1\cdots M} + O\{(\delta\mu)^2\}, \tag{9.36}$$

and taking the limit of $\delta\mu$ small, it follows that

$$\left\{kT\frac{\partial}{\partial\mu} - M + 1\right\}p_{1\cdots M-1} = \int dM\, p_{1\cdots M} \tag{9.37}$$

for $M \geq 2$.

A similar hierarchy of equations for the distribution functions can be obtained by using the commutative property of the functional derivatives in Eq. (9.24). It is found that

$$\left\{kT\frac{\partial}{\partial\mu} - M + 1\right\}n_{1\cdots M-1} = \int dM\, [n_{1\cdots M} - n_M n_{1\cdots M-1}]. \tag{9.38}$$

3. *First-Order Variations in* log Ξ^*

As yet we have not considered the effect on Ξ^* of varying the two-particle potential or the temperature. Suppose u_i and u_{ij} in Eq. (9.6) are replaced by $u_i + \delta u_i$ and $u_{ij} + \delta u_{ij}$, respectively, where δu_i and δu_{ij} are arbitrary functions. Then, using Eq. (9.8), it can be seen that the increment induced in Ξ^* is

$$\delta\Xi^* = \Xi^*\{-\int d1\, \delta u_1\, n_1 - \tfrac{1}{2}\iint d1\, d2\, \delta u_{12}\, n_{12}\} + O\{(\delta u)^2\}. \tag{9.39}$$

Dividing by Ξ^* and neglecting the second-order terms in δu_i and δu_{ij}, it follows that

$$\delta(\log \Xi^*) = -\int d1\, \delta u_1\, n_1 - \tfrac{1}{2}\iint d1\, d2\, \delta u_{12}\, n_{12}. \tag{9.40}$$

Thus, $-n_1$ and $-\frac{1}{2}n_{12}$ are the rates of change of log Ξ^* with respect to variations in u_1 and u_{12}, respectively. With a slight extension of the concept of a functional derivative, one can write

$$n_1 = -\delta(\log \Xi^*)/\delta u_i, \tag{9.41}$$

$$n_{12} = -2\, \delta(\log \Xi^*)/\delta u_{12}. \tag{9.42}$$

E. Free Energy As a Functional of n_i and u_{ij}

1. *First Functional Derivatives*

For a homogeneous one-component system it was shown in Eq. (5.5) that the fugacity z can in principle be expressed in terms of the density ϱ and the two-particle function $f(r)$, or, equivalently, $u(r)$. Similarly, in the general case the one-particle "potential" u_i can be expressed in terms of the one-particle density n_i and the two-particle "potential" u_{ij}. Substituting this expression for u_i in the various equations, the statistical-mechanical properties become functionals of n_i and u_{ij}.

It turns out that in this formulation the most natural quantity to consider is not the grand partition function Ξ^*, but rather a quantity $\mathscr{F}$ defined by

$$\mathscr{F} = \log \Xi^* + \int d1\, n_1[u_1 + (\log n_1) - 1]. \tag{9.43}$$

For a homogeneous system in the absence of external forces it can be established from Eqs. (9.4) and (9.9)–(9.16) that

$$\mathscr{F} = -\frac{F}{kT} + \sum_\alpha \bar{N}_\alpha\left\{\log \frac{\bar{N}_\alpha h^3}{V(2\pi m_\alpha kT)^{3/2}} - 1\right\}. \tag{9.44}$$

For a perfect gas F/kT is equal to the second term on the right-hand side of this equation, so $-kT\mathscr{F}$ is the free energy of the system due to interactions between the molecules.

Consider the effect on $\mathscr{F}$ of incrementing u_i and u_{ij} by small amounts δu_i and δu_{ij}. If the corresponding increment in n_i is δn_i, then it follows from Eq. (9.43) that to first order the increment induced in $\mathscr{F}$ is

$$\delta\mathscr{F} = \delta \log \Xi^* + \int d1\, \delta u_1\, n_1 + \int d1\, \delta n_1\, [u_1 + \log n_1]. \tag{9.45}$$

The increment $\delta(\log \Xi^*)$ is given by Eq. (9.40). Substituting this form, the terms explicitly involving δu_1 cancel, leaving

$$\delta\mathscr{F} = \int d1\, \delta n_1\, [u_1 + \log n_1] - \tfrac{1}{2} \iint d1\, d2\, \delta u_{12}\, n_{12}. \tag{9.46}$$

Thus, if $\mathscr{F}$ is regarded as a functional of n_1 and u_{12}, it follows from Eq. (9.46) that

$$u_1 + \log n_1 = \delta\mathscr{F}/\delta n_1, \tag{9.47}$$

$$n_{12} = -2\, \delta\mathscr{F}/\delta u_{12}. \tag{9.48}$$

2. *Direct Correlation Functions*

Just as the indirect correlation functions are obtained in terms of the functional derivatives of $\log \Xi^*$ with respect to u_i, so one can define another set of functions in terms of the derivatives of $\mathscr{F}$ with respect to n_i. Suppose the two-particle function u_{ij} is held fixed, while the density n_i is incremented by some small amount δn_i. Then the increment induced in $\mathscr{F}$ can be formally expanded to all orders in δn_1 as

$$\delta\mathscr{F} = \sum_{M=1}^{\infty} \frac{1}{M!} \int \cdots \int d1 \cdots dM\, \delta n_1 \cdots \delta n_M\, c_{1\cdots M}\,, \tag{9.49}$$

the "coefficients" $c_{1\cdots M}$ in this expansion being the functional derivatives of $\mathscr{F}$, i.e.,

$$c_{1\cdots M} = \delta^M \mathscr{F} / \delta n_1 \cdots \delta n_M\,. \tag{9.50}$$

Comparing Eqs. (9.47) and (9.50), it is apparent that

$$c_1 = u_1 + \log n_1\,, \tag{9.51}$$

and the higher coefficients c_{12}, c_{123}, etc., can be expressed in terms of the distribution functions by successive functional differentiation. In particular, from Eq. (9.50)

$$c_{12} = \delta c_1 / \delta n_2\,, \tag{9.52}$$

so that

$$\delta c_1 = \int d2\, \delta n_2\, c_{12} + O\{(\delta n)^2\}\,, \tag{9.53}$$

or, using the expression (9.51) for c_1,

$$\delta u_1 + (\delta n_1/n_1) = \int d2\, \delta n_2\, c_{12} + O\{(\delta n)^2\}\,. \tag{9.54}$$

However, using Eqs. (9.21) and (9.29), it can be established from the $M = 2$ equation of the set (9.34) that

$$\delta u_1 + (\delta n_1/n_1) = -\int d2\, \delta u_2\, n_2 h_{12} + O\{(\delta u)^2\}\,. \tag{9.55}$$

Replacing δn_i in Eq. (9.54) by its form as given by Eq. (9.55), and rearranging the integrations gives

$$\int d2\, \delta u_2\, n_2 \{h_{12} - c_{12} - \int d3\, n_3 c_{13} h_{32}\} + O\{(\delta u)^2\} = 0\,. \tag{9.56}$$

As the increment δu_i is an arbitrary function, Eq. (9.56) can only be satisfied if

$$h_{12} = c_{12} + \int d3\, n_3 c_{13} h_{32}\,. \tag{9.57}$$

Equation (9.57) can be regarded as defining the function c_{ij} in terms of n_i and h_{ij}. Comparing it with Eq. (5.22), it is apparent that it is the generalization to inhomogeneous systems and mixtures of the Ornstein–Zernike (OZ) relation, so that c_{ij} is the two-particle direct correlation function. Similarly, the higher functions c_{123}, c_{1234}, ... are known as the three-, four-, etc., particle direct correlation functions. Like the indirect correlation functions, they tend to zero when the separation between any two of the position coordinates in their arguments becomes large, but in general they do this much more rapidly.

A very simple expression for the derivatives of the direct correlation functions with respect to density can be obtained from their definition (9.50), for, using the commutative property of functional differentiation, it follows that the increment induced in $c_{1\cdots M-1}$ by a small increment δn_i in n_i is

$$\delta c_{1\cdots M-1} = \int dM\, \delta n_M\, c_{1\cdots M} + O\{(\delta n)^2\}. \tag{9.58}$$

For simplicity, consider a one-component homogeneous system, so that $\delta n_\alpha(\mathbf{r}) = \delta \varrho$, where $\delta\varrho$ is the change in density. Taking the limit of $\delta\varrho$ small, it follows that

$$\frac{\partial}{\partial \varrho} c_{1\cdots M-1} = \int dM\, c_{1\cdots M}, \tag{9.59}$$

the temperature and interaction potentials being regarded as constants.

Regarding the two-particle potential u_{12} as fixed, the functional derivative of $\log \Xi^*$ with respect to n_i can be obtained by setting $\delta u_{12} = 0$ in Eq. (9.40) and substituting the form of δu_1 given by Eq. (9.54). Rearranging the integrals, to first order this gives

$$\delta(\log \Xi^*) = \int d1\, \delta n_1\, [1 - \int d2\, n_2 c_{21}], \tag{9.60}$$

i.e.,

$$\delta(\log \Xi^*)/\delta n_1 = 1 - \int d2\, n_2 c_{21}. \tag{9.61}$$

For a one-component homogeneous fluid $\log \Xi^* = PV/kT$, and Eq. (9.61) reduces to the inverse compressibility equation (7.14).

3. *PY and CHNC Approximations*

From Eq. (9.53) it can be established that the increment in $\exp(c_1)$ is

$$\delta(\exp c_1) = (\exp c_1) \int d2\, \delta n_2\, c_{12} + O\{(\delta n)^2\}. \tag{9.62}$$

Percus (1962) suggested that $\exp(c_1)$ should be an almost linear functional of the density, so that it should be possible to neglect the second- and higher-order terms in δn in Eq. (9.62). He then chose the increment in

u_i to be

$$\delta u_\alpha(\mathbf{r}) = u_{\alpha_0\alpha}(\mathbf{r}_0, \mathbf{r}), \tag{9.63}$$

where α_0 and r_0 are regarded as parameters. Incrementing u_i by this amount is equivalent to introducing an additional external field of potential $\phi^{(2)}_{\alpha_0\alpha}(\mathbf{r}_0, \mathbf{r})$, and this is in turn equivalent to introducing an extra particle into the system, of type α_0 and position $\mathbf{r}_0$.

Replacing u_i by $u_i + \delta u_i$ in Eqs. (9.6) and (9.8), where δu_i is given by Eq. (9.63), it is found that the one-particle density n_1 becomes

$$[n_i]_{\text{incr}} = n_{0i}/n_0 = n_i g_{0i} \tag{9.64}$$

[using the definition (9.28) of the normalized distribution functions]. Hence, from Eq. (9.51) $\exp(c_i)$ becomes

$$[\exp c_i]_{\text{incr}} = n_i g_{0i} \exp(u_i + u_{0i}) = (\exp c_i) g_{0i} \exp u_{0i}. \tag{9.65}$$

Using Eq. (9.31), the increment in n_i is therefore

$$\delta n_i = [n_i]_{\text{incr}} - n_i = n_i h_{0i}, \tag{9.66}$$

while the increment in $\exp(c_i)$ is

$$\delta[\exp c_i] = [\exp c_i]_{\text{incr}} - \exp c_i = (\exp c_i)\{g_{0i}(\exp u_{0i}) - 1\}. \tag{9.67}$$

Substituting the forms (9.66) and (9.67) of δn_i and $\delta[\exp(c_i)]$ in Eq. (9.62) and neglecting the second-order terms in δn, it follows that

$$g_{01}(\exp u_{01}) - 1 = \int d2\, n_2 h_{02} c_{12}. \tag{9.68}$$

Remembering that $h_{02} = h_{20}$, it follows from the OZ relation (9.57) that the right-hand side of Eq. (9.68) is $h_{10} - c_{10}$, or $h_{01} - c_{01}$. Using Eq. (9.31), it then follows that

$$c_{01} = \{1 - \exp u_{01}\} g_{01}, \tag{9.69}$$

and this is the PY approximation (5.27) generalized to inhomogeneous systems and mixtures. Equations (9.31), (9.57), and (9.69) form a closed set of equations for the functions g_{ij}, h_{ij}, and c_{ij} in terms of n_i and u_{ij}.

The CHNC approximation can also be obtained by this method by assuming that c_1, rather than $\exp(c_1)$, should be a linear functional of the density. Replacing Eq. (9.62) by (9.53) in the above working gives

$$c_{01} = g_{01} - 1 - \log g_{01} - u_{01}, \tag{9.70}$$

which is the generalization of the CHNC approximation (5.32).

F. Entropy As a Functional of n_i and n_{ij}

1. *First Functional Derivatives*

Yet another possible approach to statistical mechanics is to regard the properties (including the "potentials" u_i and u_{ij}) as determined by the one- and two-particle distribution functions n_i and n_{ij}, or, equivalently, by n_i and the two-particle indirect correlation function h_{ij}. In this case the most natural quantity to consider is not the grand partition function or the free energy, but a quantity $\mathscr{S}$ defined by

$$\mathscr{S} = \log \Xi^* + \int d1\, n_1[u_1 + \log n_1 - 1] + \tfrac{1}{2} \iint d1\, d2\, n_1 n_2[g_{12}u_{12} + g_{12}(\log g_{12}) - g_{12} + 1]. \qquad (9.71)$$

$\mathscr{S}$ vanishes for a perfect gas, and from Eqs. (9.4) and (9.9)–(9.12) it can be established that

$$\mathscr{S} = k^{-1}S - \text{terms determined by } n_i \text{ and } h_{ij}, \qquad (9.72)$$

so $k\mathscr{S}$ can be regarded as the "interaction entropy" of the system.

Suppose u_i and u_{ij} are incremented by small amounts δu_i and δu_{ij}, thereby inducing increments δn_i and δn_{ij} in n_i and n_{ij}. Replacing g_{12} in Eq. (9.71) by $n_{12}/n_1 n_2$ and using the form (9.40) of $\delta(\log \Xi^*)$, the increment induced in $\mathscr{S}$ is found to first order to be

$$\delta\mathscr{S} = \int d1\, \delta n_1\, [\log n_1 + u_1 - \int d2\, n_2 h_{12}] + \tfrac{1}{2} \iint d1\, d2\, \delta n_{12}\, [u_{12} + \log g_{12}]. \qquad (9.73)$$

Thus, if $\mathscr{S}$ is regarded as a functional of n_1 and n_{12},

$$\log n_1 + u_1 - \int d2\, n_2 h_{12} = \delta\mathscr{S}/\delta n_1, \qquad (9.74)$$

$$u_{12} + \log g_{12} = 2\, \delta\mathscr{S}/\delta n_{12}. \qquad (9.75)$$

2. *Expansion of* $\mathscr{S}$

Using techniques similar to those employed in Section V,A, the equations (9.6) and (9.8) for n_i and n_{ij} in terms of u_i and u_{ij} can be systematically inverted to give u_i and u_{ij} as expansions in powers of n_i, with "coefficients" determined by the two-particle indirect correlation function h_{ij}. Substituting these expansions into Eq. (9.71), using the form (9.6) of Ξ^*, it is found that

$$\begin{aligned}\mathscr{S} = &\tfrac{1}{6} \iiint d1\, d2\, d3\, n_1 n_2 n_3 h_{12} h_{23} h_{31} \\ &- \tfrac{1}{8} \iiiint d1\, d2\, d3\, d4\, n_1 n_2 n_3 n_4 h_{12} h_{23} h_{34} h_{41} \\ &+ \tfrac{1}{24} \iiiint d1\, d2\, d3\, d4\, n_1 n_2 n_3 n_4 h_{12} h_{23} h_{34} h_{41} h_{13} h_{24} + O\{n_i^5\}. \quad (9.76)\end{aligned}$$

The terms occurring in this expansion can be represented by diagrams similar to those introduced in Section V, except that they contain only black dots (and lines linking the dots), and are to be interpreted according to the rules:

(a) with each black dot associate a label i and a factor n_i (the labels are arbitrary, except that no two dots can have the same label).
(b) With each line linking dots i and j associate a factor h_{ij}.
(c) Multiply together all the n_i and h_{ij} factors, sum over all the particle types α_i, and integrate with respect to each particle position $\mathbf{r}_i$ over the volume V.

Since we shall only use these diagrams as a short-hand notation for the corresponding integrals, we can write the weighting factors explicitly and do not need to include a rule to determine them.

The diagrammatic representations of the terms of order n_i^3, n_i^4, and n_i^5 in the expansion (9.76) are given in Fig. 15. Morita and Hiroike (1960) and De Dominicis (1962) have shown that the diagrams consist of two types, one set being m-sided polygons with weight factor $(-)^{m-1}/2m$, the other being diagrams which have no unconnected portions and which cannot be cut into unconnected portions by snipping at a point or pair of points (a line joining the pair of points does not count as a portion in this context). Referring to these latter diagrams as 2-

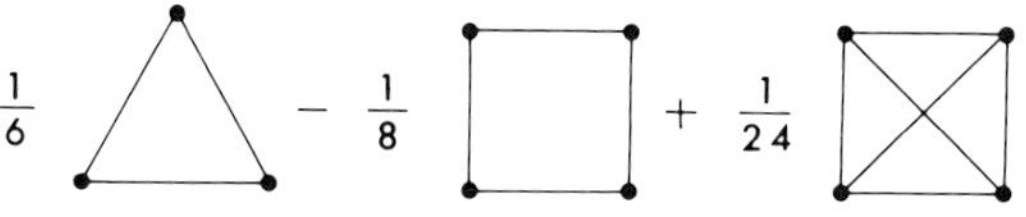

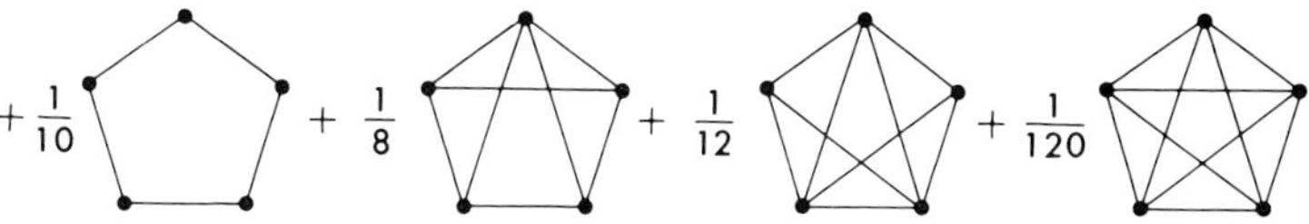

FIG. 15. Diagrammatic representations of the terms of order n_i^3, n_i^4, and n_i^5 in the expansion (9.76) of the "interaction entropy" $\mathscr{S}$.

irreducible, it follows that

$$\mathscr{S} = \mathscr{S}' + \mathscr{S}'', \tag{9.77}$$

where $\mathscr{S}'$ is the sum over polygon diagrams,

$$\mathscr{S}' = \sum_{m=3}^{\infty} \frac{(-)^{m-1}}{2m} \int \cdots \int d1 \cdots dm \; n_1 \cdots n_m h_{12} h_{23} \cdots h_{m-1,m} h_{m1} \tag{9.78a}$$

and $\mathscr{S}''$ is the sum over 2-irreducible diagrams. (9.78b)

3. *CHNC Approximation*

The decomposition (9.76) of $\mathscr{S}$ suggests the approximation of neglecting the 2-irreducible diagrams, i.e., replacing $\mathscr{S}$ by $\mathscr{S}'$ in Eqs. (9.71), (9.74), and (9.75). The functional derivatives of $\mathscr{S}'$ can be obtained by replacing the indirect correlation function h_{ij} in Eq. (9.78a) by $(n_{ij}/n_i n_j) - 1$ and considering the increments induced by varying n_i and n_{ij}. Proceeding in this way, it is found that

$$\delta\mathscr{S}'/\delta n_1 = \iint d2\, d3\; n_2 n_3 (\tfrac{1}{2} h_{12} - g_{12}) c_{13} h_{32}, \tag{9.79}$$

$$\delta\mathscr{S}'/\delta n_{12} = \tfrac{1}{2} \int d3\; n_3 c_{13} h_{32}, \tag{9.80}$$

where the function c_{ij} is defined by

$$c_{12} = h_{12} - \sum_{m=3}^{\infty} (-)^{m-1} \int \cdots \int d3 \cdots dm \; n_3 \cdots n_m h_{13} h_{34} \cdots h_{m-1,m} h_{m2}. \tag{9.81}$$

From Eq. (9.81) it can be seen that

$$c_{12} = h_{12} - \int d3\; n_3 h_{13} c_{32}, \tag{9.82}$$

so, remembering that c_{ij} and h_{ij} are symmetrical functions and comparing equations (9.57) and (9.82), it is apparent that c_{ij} is the two-particle direct correlation function.

Equations (9.79) and (9.80) can be simplified by using the OZ relation (9.57) or (9.82). Replacing $\mathscr{S}$ by $\mathscr{S}'$ in Eqs. (9.71), (9.74), and (9.75), it then follows that

$$\begin{aligned} \log \Xi^* = &- \int d1\; n_1[u_1 + \log n_1 - 1] \\ &- \tfrac{1}{2} \iint d1\, d2\; n_1 n_2 [g_{12} u_{12} + g_{12} \log g_{12} - h_{12}] + \mathscr{S}', \end{aligned} \tag{9.83}$$

$$\log n_1 + u_1 = \int d2\; n_2 [c_{12} - \tfrac{1}{2} h_{12}(h_{12} - c_{12})], \tag{9.84}$$

$$u_{12} + \log g_{12} = h_{12} - c_{12}. \tag{9.85}$$

Equation (9.85) is precisely the CHNC approximation (9.70). The advantage of this derivation is that it manifests an important property of the CHNC approximation, namely, that it occurs naturally in conjunction with the two equations (9.83) and (9.84). Equation (9.85) (coupled with the OZ relation) enables the two-particle correlation functions to be evaluated in terms of u_{ij} and n_i, and Eq. (9.84) then enables n_i to be evaluated in terms of u_i. This is particularly important for inhomogeneous systems, where the density n_i is not a known function.

Further, Eq. (9.83), together with the relations (9.9)–(9.16), enables all the thermodynamic properties of the system to be calculated for any particular solution of the equations. The properties calculated in this way are self-consistent, inasmuch as their first derivatives satisfy the second law of thermodynamics and Maxwell's relations. However, since the results are only approximate, it is still possible for other inconsistencies to arise; for instance, the pressure obtained in this way will not in general agree with that obtained by integrating the inverse compressibility relation (7.14).

In view of these comments, it seems worthwhile to write the equations (9.83)–(9.85) explicitly for a one-component homogeneous system. In this case $n_i = \varrho$ and $h_{ij} = h(r_{ij})$, so the convolution integrals in the definition (9.78a) of $\mathscr{S}'$ are conveniently expressed in terms of the Fourier transform $h(k)$ of $h(r)$ given by Eqs. (7.3) and (7.6). It is then found that in the limit of V large

$$\mathscr{S}' = \frac{V}{(2\pi)^3} \sum_{m=3}^{\infty} \frac{(-)^{m-1}\varrho^m}{2m} \int d\mathbf{k}\, [h(k)]^m \tag{9.86}$$

(the factor V occurs as the integrand of the last integration is constant when V is large).

The summation over m in Eq. (9.86) can be performed explicitly. Using Eqs. (9.4) and (9.15), Eqs. (9.83)–(9.85) become

$$\begin{aligned}\beta P = \varrho - \varrho \log(\varrho/z) - \tfrac{1}{2}\varrho^2 \int d\mathbf{r}\, \{&g(r)[\log g(r)] \\ &+ g(r)u(r) - h(r)\} \\ + (1/16\pi^3) \int d\mathbf{k}\, \{&\log[1 + \varrho h(k)] - \varrho h(k) + \tfrac{1}{2}\varrho^2 h^2(k)\},\end{aligned} \tag{9.87}$$

$$\log(\varrho/z) = \varrho \int d\mathbf{r}\, \{c(r) - \tfrac{1}{2}h(r)[h(r) - c(r)]\}, \tag{9.88}$$

$$u(r) + \log g(r) = h(r) - c(r), \tag{9.89}$$

where the fugacity z is given by Eq. (2.8), and $\beta = 1/kT$.

4. *PY Approximation*

Unfortunately, no such self-consistent formulation of the PY approximation is as yet known. However, some progress in this direction can be made (Baxter, 1967b) by considering the quantity

$$\Psi = \int d1\, n_1 + \tfrac{1}{2} \iint d1\, d2\, n_1 n_2 [(\exp u_{12}) g_{12}^2 - 2g_{12} + 1] - 2\mathscr{S}'. \quad (9.90)$$

Holding the functions n_i and u_{ij} fixed, the functional derivative of Ψ with respect to n_{ij} is [using Eq. (9.80)]

$$\delta\Psi/\delta n_{12} = g_{12}(\exp u_{12}) - 1 - \int d3\, n_3 c_{13} h_{32}. \quad (9.91)$$

From Eq. (9.68) the right-hand side of Eq. (9.91) vanishes when the PY approximation is satisfied, so Ψ is stationary with respect to small variations of n_{ij}.

When n_{ij} is given by the PY equation, Ψ can be regarded as a functional of n_i and u_{ij}. Holding u_{ij} fixed, varying n_i, and using the PY approximation (9.68) or (9.69), it is found that

$$\delta\Psi/\delta n_1 = 1 - \int d2\, n_2 c_{12} \quad (9.92)$$

(the terms arising from the increment induced in n_{ij} must cancel, as Ψ is stationary with respect to variations in n_{ij}).

Remembering that $c_{12} = c_{21}$ and comparing Eqs. (9.61) and (9.92), it is apparent that Ψ and $\log \Xi^*$ have the same functional derivative with respect to n_1. Further, they both vanish when n_i is identically zero, so that in the PY approximation

$$\Psi = \log \Xi^*. \quad (9.93)$$

Specializing to one-component homogeneous fluids, it follows from Eq. (9.15) that $\Psi = PV/kT$, and Eq. (9.90) becomes

$$\beta P = \varrho + \tfrac{1}{2}\varrho^2 \int d\mathbf{r}\, \{g^2(r)\, e^{u(r)} - 2g(r) + 1\} - (1/8\pi^3) \int d\mathbf{k}\, \{\log[1 + \varrho h(k)] - \varrho h(k) + \tfrac{1}{2}\varrho^2 h^2(k)\}. \quad (9.94)$$

Since Eq. (9.61) reduced to the inverse compressibility relation (7.14) in this case, the pressure given by Eq. (9.94) is the "compressibility pressure" obtained by integrating Eq. (7.14) with respect to the density ϱ.

General References

Cole, G. H. A. (1967). "An Introduction to the Statistical Theory of Classical Simple Dense Fluids." Pergamon Press, Oxford.

de Boer, J. (1949). *Rept. Progr. Phys.* **12**, 305.

Frisch, H. L., and Lebowitz, J. L. (1964). "The Equilibrium Theory of Classical Fluids." Benjamin, New York.

Henderson, D., and Davison, S. G. (1967). *In* "Physical Chemistry: An Advanced Treatise" (H. Eyring, D. Henderson, and W. Jost, eds.), Vol. II, Chapter 7. Academic Press, New York.

Pippard, A. B. (1957). "Elements of Classical Thermodynamics." Cambridge Univ. Press, London and New York.

Rice, S. A., and Gray, P. (1965). "The Statistical Mechanics of Simple Liquids." Wiley (Interscience), New York.

Rowlinson, J. S. (1965). *Rept. Progr. Phys.* **24**, 169.

Tolman, R. C. (1938). "The Principles of Statistical Mechanics," Chapter 14. Oxford Univ. Press, London and New York.

Special References

Alder, B. J., and Wainwright, T. E. (1960). *J. Chem. Phys.* **33**, 1439.

Baxter, R. J. (1967a). *Phys. Rev.* **154**, 170.

Baxter, R. J. (1967b). *J. Chem. Phys.* **47**, 4855.

Baxter, R. J. (1968a). *Ann. Phys.* **46**, 509.

Baxter, R. J. (1968b). *Australian J. Phys.* **21**, 563.

Baxter, R. J. (1968c). *J. Chem. Phys.* **49**, 2770.

Born, M., and Green, H. S. (1946). *Proc. Roy. Soc.* **A188**, 10.

Broyles, A. A., Chung, S. U., and Sahlin, H. L. (1962). *J. Chem. Phys.* **37**, 2462.

Cohen, E. G. D. (1962). *In* "Fundamental Problems in Statistical Mechanics" (E. G. D. Cohen, ed.), p. 119–122. North-Holland, Amsterdam.

de Boer, J., van Leeuwen, J. M. J., and Groeneveld, J. (1964). Physica **30**, 2265.

de Dominicis, C. (1962). *J. Math. Phys.* **3**, 983.

Eisenstein, A., and Gingrich, N. S. (1942). *Phys. Rev.* **62**, 261.

Green, M. S. (1960). *J. Chem. Phys.* **33**, 1403.

Hutchinson, P. (1967). *Mol. Phys.* **13**, 495.

Khan, A. A. (1964). *Phys. Rev.* **134**, A367.

Kirkwood, J. G. (1935). *J. Chem. Phys.* **3**, 300.

Kirkwood, J. G., Maun, E. K., and Alder, B. J. (1950). *J. Chem. Phys.* **18**, 1040.

Kittel, C. (1958). "Elementary Statistical Physics." Wiley, New York.

Klein, M., and Green, M. S. (1963a). *J. Chem. Phys.* **39**, 1367.

Klein, M., and Green, M. S. (1963b). *J. Chem. Phys.* **39**, 1388.

Krein, M. G. (1962). *Am. Math. Soc. Transl. Ser. 2* **22**, 194.

Levesque, D. (1966). *Physica* **32**, 1985.

Meeron, E. (1960). *J. Math. Phys.* **1**, 192.

Michels, A., Wijker, Hub., and Wijker, Hk. (1949). *Physica* **15**, 627.

Morita, T., and Hiroike, K. (1960). *Progr. Theoret. Phys. (Kyoto)* **23**, 1003.

Ornstein, L. S., and Zernike, F. (1914). *Proc. Acad. Sci. (Amsterdam)* **17**, 793.
Percus, J. K. (1962). *Phys. Rev. Letters* **8**, 462.
Percus, J. K., and Yevick, G. J. (1958). *Phys. Rev.* **110**, 1.
Reiss, H. (1965). *Advan. Chem. Phys.* **9**, 1.
Reiss, H., Frisch, H. L., and Lebowitz, J. L. (1959). *J. Chem. Phys.* **31**, 369.
Rushbrooke, G. S. (1960). *Physica* **26**, 259.
Stell, G. (1964). *In* "The Equilibrium Theory of Classical Fluids" (H. L. Frisch and J. L. Lebowitz, eds.), pp. 171–266. Benjamin, New York.
Temperley, H. N. V. (1964). *Proc. Phys. Soc. (London)* **84**, 399.
Thiele, E. J. (1963). *J. Chem. Phys.* **39**, 474.
Throop, G. J., and Bearman, R. J. (1965). *J. Chem. Phys.* **42**, 2408.
Throop, G. J., and Bearman, R. J. (1966). *Physica* **32**, 1298.
Van Leeuwen, J. M. J., Groeneveld, J., and de Boer, J. (1959). *Physica* **25**, 792.
Verlet, L. (1960). *Nuovo Cimento* **18**, 77.
Verlet, L. (1964). *Physica* **30**, 95.
Verlet, L. (1965). *Physica* **31**, 965.
Verlet, L., and Levesque, D. (1962). *Physica* **28**, 1124.
Watts, R. O. (1968a). *J. Chem. Phys.* **48**, 50.
Watts, R. O. (1969). *Can. J. Phys.* **47**, 2709.
Wertheim, M. S. (1963). *Phys. Rev. Letters* **10**, 321.
Wertheim, M. S. (1967). *J. Math. Phys.* **8**, 927.
Wood, W. W., Parker, F. R., and Jacobson, J. D. (1958). *Nuovo Cimento Suppl.* **9**, 133.
Wood, W. W., and Parker, F. R. (1957). *J. Chem. Phys.* **27**, 720.

Chapter 5

The Significant Structure Theory of Liquids

MU SHIK JHON AND HENRY EYRING

I. Introduction

Since the formal development of statistical mechanics forty years ago, there have been two main theoretical approaches to a study of the liquid state. One is the formal or mathematical approach starting from the cluster integrals or from the radial distribution function pioneered by Kirkwood and Mayer. This formal approach starts with an exact formulation, but it soon becomes necessary to introduce approximations such as simplified intermolecular potentials because of the mathematical difficulties.

An alternate procedure is the model approach. A model of the liquid state is postulated which to be successful must be faithful to reality as well as tractable. This is then translated into a mathematical equation, the partition function, from which the properties of the liquid are calculated. An acceptable model of the liquid state must lead to a quantitative theory of thermodynamic and transport properties. The significant structure theory (Eyring *et al.*, 1958; Eyring and Jhon, 1969) is such a model. For simple substances such as argon, the significant structure theory provides a model without adjustable parameters. For more complicated molecules, a knowledge of certain properties allows one to evaluate the remaining properties. There seems to be no liquid which cannot be usefully examined using this model.

In this chapter, the present status of the significant structure theory in comparison with other theories and the calculation of the following properties will be discussed: (a) thermodynamics of liquids, (b) transport phenomena, (c) surface phenomena, (d) dielectric phenomena, and (e) two-dimensional liquids.

II. The Significant Structure Model

A simple classical liquid like argon expands 12% upon melting. This expansion is accompanied by a decrease of the coordination number and a marked increase in fluidity. X-ray diffraction[†] indicates that nearest neighbors are at essentially the same distance as in the solid. This heterogeneous expansion must arise through the addition of holes to a distorted fluctuating lattice. In fact the liquid phase is distinguished from a solid phase by what we can call "lift off." A hole persists in the liquid state if locally the kinetic energy density (the pressure) which tends to make a region expand is balanced by the opposing potential energy density (the internal pressure) tending to bring about the collapse of the hole. Thus the liquid state can only occur when the kinetic energy, i.e., the temperature has reached a value where a substantial number of molecules are lifted out of the cradle formed by neighbors as a result of their high kinetic energy density much as a burning rocket lifts off when enough fuel has burned so that the specific impulse overcomes gravity and the system seeks a new equilibrium. This kind of mechanical stability is always

[†] Eisenstein and Gingrich (1942).

present in a liquid either above the melting point or in the supercooled liquid below the melting point. Below the melting point the liquid phase is thermodynamically less stable than the solid although it is mechanically stable. How fast a supercooled liquid moves toward thermodynamic stability by freezing is a problem in reaction rates and depends on how soon it can form a nucleus large enough that one more molecule added to the growing nucleus makes it more stable than the liquid phase. At this point persistance of the growing nucleus is assured.

Thus in melting some of the molecules are replaced by "fluidized" vacancies which are quite different from the "static locked in" vacancies of the solid state and are moved by cooperative action of all the neighbors. Holes of molecular size will be strongly favored because smaller holes will not provide easy access to entering molecules and so curtail the entropy while larger holes will be unnecessarily wasteful of energy without a compensating increase of entropy. The fluidized vacancies introduced at melting, increase the volume by increasing the number of empty sites leaving the nearest neighbor distances nearly constant. There is nevertheless enough variance left in the volume of the fluidized vacancies to destroy long range order. An instantaneous photograph would, accordingly, reveal an intimate mixture of solidlike and gaslike degrees of freedom. This fine grained heterogeneity is attested to by the fact that liquids contain no solidlike or gaslike nuclei large enough to form focii for either supercooling or bumping of liquids.

Our model of fluidized vacancies provides an explanation of the law of rectilinear diameters. According to this law, the mean density of a liquid and its vapor is linear and decreases slowly with rising temperature. The energy of vaporization of a molecule is half of all bonds joining the molecule to other molecules. However, the removal of a molecule from the liquid to the vapor, leaving a vacancy, breaks all its bonds to other molecules. It follows that the energy to form a vacancy equals the energy of vaporization.

Since these fluidized vacancies are moved about by neighboring molecules jumping into them, a vacancy changes three degrees of freedom that would otherwise have been vibrations into three translations. In effect, a vacancy thus acts like a vapor molecule, with not only the effective energy of a vapor molecule but with the entropy as well (see Fig. 1).

Consequently, we expect the motion of molecules in the vapor to be mirrored in the liquid as vacancies. Accordingly, there are just enough molecules in 1 cm^3 of vapor to fill the vacancies in 1 cm^3 of liquid so that

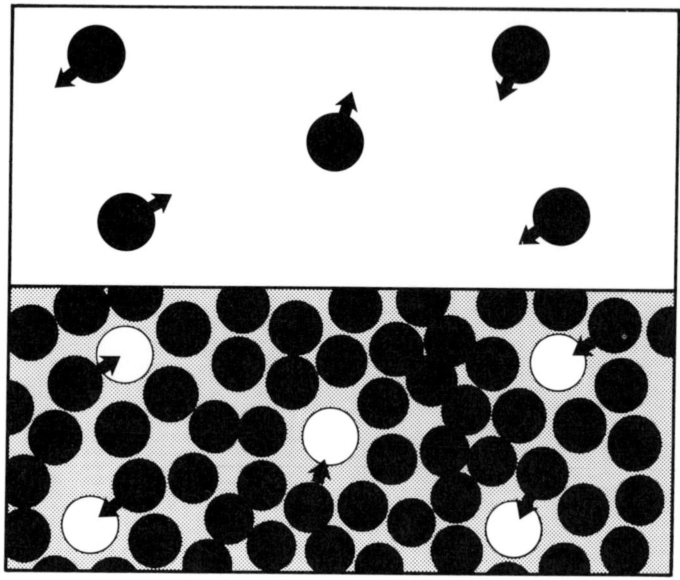

FIG. 1. Vacancies in a liquid behave like molecules in a gas. In a liquid, vacancies move among molecules. In a gas, molecules move among vacancies. (After International Science and Technology, March, 1963.)

the mean density of the two phases decreases with temperature only because of lattice expansion. According to our model the fluidized vacancies are on the average of molecular size so that the number of moles of such vacancies is $(V - V_s)/V_s$ where V and V_s are the molar volumes of liquid and solid, respectively. A vacancy completely surrounded by vacancies has no dynamic properties. Completely surrounded by molecules a vacancy converts what would otherwise be three vibrational degrees of freedom of neighbors into three translational degrees of freedom. Since the fraction of neighboring positions to a vacancy which are filled with molecules is V_s/V it follows that there are

$$(V_s/V)[(V - V_s)/V_s] = (V - V_s)/V$$

moles of gaslike molecules and the rest V_s/V moles of solidlike molecules.

In addition vacancies confer positional degeneracy on the solidlike molecules. The number of additional sites for a molecule will be the number of vacancies around solidlike molecules n_h multiplied by the probability that the molecule has the required energy ε_h to preempt the site from competing neighbors. Here n_h is proportional to the number of vacancy and is given by

$$n_h = n[(V - V_s)/V_s]. \tag{2.1}$$

And ε_h is inversely proportional to n_h and directly proportional to the energy of sublimation E_s. Thus

$$\varepsilon_h = aE_s V_s/(V - V_s). \tag{2.2}$$

The total number of the positions available to a given molecule is

$$1 + n_h e^{-\varepsilon_h/RT}. \tag{2.3}$$

In view of the above, the partition function f_N for a mole of liquid can be expressed as

$$f_N = (f_s)^{NV_s/V}(f_g)^{N(V-V_s)/V}. \tag{2.4}$$

Here N is Avogadro's number, f_s and f_g are the partition function of solid-like molecules (including the degeneracy factor) and that of gaslike molecules, respectively. For a simple liquid such as argon, we can write f_N as follows:

$$f_N = \left\{\frac{e^{E_s/RT}}{(1 - e^{-\theta/T})^3}\left[1 + n\left(\frac{V - V_s}{V_s}\right)\exp\left(-\frac{aE_s V_s}{(V - V_s)RT}\right)\right]\right\}^{NV_s/V}$$
$$\times \left[\frac{(2\pi mkT)^{3/2}}{h^3}(V - V_s)\right]^{N(V-V_s)/V}\left[\left(N\frac{(V - V_s)}{V}\right)!\right]^{-1} \tag{2.5}$$

where θ is the Einstein characteristic temperature, and E_s is the sublimation energy. In this derivation, we have used the Einstein oscillator as an adequate representation of the solidlike molecule, while the nonlocalized independent ideal gas partition function is used for the gaslike molecules.

A criticism sometimes made of this theory is that it has a number of adjustable parameters, E_s, θ, V_s, a, and n in the partition function. For simple liquid, on the contrary, the values of E_s, θ, and V_s are obtained from the solid state, either from experiment or a theoretical calculation, and the values of a and n are evaluated theoretically.

Near the melting point, where the liquid approximates a latticelike structure, the fraction of the neighboring positions Z which are empty and therefore available for occupancy is

$$Z\left(\frac{V_m - V_s}{V_m}\right) = \left(Z\frac{V_s}{V_m}\right)\frac{V_m - V_s}{V_s} = n\frac{V_m - V_s}{V_s}.$$

Hence

$$n = Z \frac{V_s}{V_m}. \tag{2.6}$$

We next calculate a (Eyring and Jhon, 1969).

The average solid like molecule has kinetic energy equal to $\frac{3}{2}kT$. If a molecule is to preempt a neighboring position in addition to its original position, it must have additional kinetic energy $\frac{1}{2}(\frac{3}{2}kT)(n-1)/Z$. This is because to preempt a position a molecule must introduce into it more kinetic energy than the $(n-1)$ other average neighbors, to the contested position, do. The factor $1/Z$ occurs because a neighbor is moving toward the vacancy that fraction of the time and when it does move toward the vacancy it introduces half of $\frac{3}{2}kT$ into the vacancy. This value is equivalent to ε_h:

$$\varepsilon_h = \frac{aE_s V_s}{V_m - V_s} = \frac{1}{2}\left(\frac{3}{2}RT\right)\frac{n-1}{Z}.$$

Here

$$E_m = \frac{3}{2}RT = \frac{(V_m - V_s)}{V_m} E_s. \tag{2.7}$$

Equation (2.7) is true because in a liquid the average kinetic energy $\frac{3}{2}kT$ must just balance the energy of melting $E_s(V_m - V_s)/V_m N$ for mechanical stability. The energy of melting is $E_s(V_m - V_s)/V_m N$ because $(V_m - V_s)/V_m$ is the fraction of the space emptied of bonds by the melting process and because the temperature does not change in melting; only the potential energy changes. Accordingly,

$$a = \frac{n-1}{Z}\frac{1}{2}\frac{(V_m - V_s)^2}{V_m V_s}. \tag{2.8}$$

III. Some Theoretical Discussion of the Significant Structure Model

As we have seen, the theory involves no adjustable parameters for simple classical liquids (see Walter and Eyring, 1941). For polyatomic liquids involving restricted rotation and a change in the solidlike structure upon melting, a more sophisticated calculation of the parameters is required, or they can be obtained from experiment.

The usefulness of the model should be evident. However, a detailed justification of the model of significant structure theory from first principles is tied up with the fact that a fluidized vacancy is moved about by the cooperative action of all the neighbors correlating their motion in such a way as to transform three degrees of freedom which would otherwise be solidlike vibrations into three translational degrees of freedom characteristic of a gas molecule although no single gas molecule having this property may exist. Such a many body problem has much of the complexity of the exact phase integral for a liquid. Our model is in fact a theory of the nature of the normal modes in the liquid state much as Debye's solid state model is a theory of the normal modes of the solid.

Usually an Einstein oscillator has been used for the solidlike degrees of freedom. Henderson (1963), however, used the Lennard-Jones and Devonshire cell model to replace the Einstein oscillator. Accordingly, the partition function becomes

$$f_N = \left(\frac{2\pi mkT}{h^2}\right)^{3/2N} \left\{ \left[\exp\left(-Z\frac{\psi(a)}{2kT} \right) \right] \right.$$
$$\left. \times \left[\left(1 + n\frac{(V-V_s)}{V_s} \right) \exp\left(-\frac{\varepsilon_h}{RT} \right) v_f \right]^{NV_s/V} \left(\frac{eV}{N} \right)^{N(V-V_s)/V} \right\} \tag{3.1}$$

where $\psi(a)$ is the Lennard-Jones (6–12) potential, a is the nearest neighbor distance, and

$$v_f = \int_{\text{cell}} \exp\left[-Z\frac{\psi(r)-\psi(a)}{2kT} \right] dr. \tag{3.2}$$

He has applied the significant structure theory to a hard sphere system in which the interaction potential is

$$\psi(r) = \begin{cases} 0, & r > a \\ \infty, & r < a \end{cases} \tag{3.3}$$

where a is the diameter of a sphere. Then $\psi(a) = 0$ and $\varepsilon_h = 0$.

Some of the calculated results for the compressibility factor and the excess entropy from Eq. (3.1) are plotted against V_s/V with $n = 12$ and $V_s = 1.5V_0$ a value which has been observed by Wood and Jacobsen (1957) and Adler and Wainwright (1957) by means of machine calculations. The theory for the systems of hard spheres show excellent agreement with the corresponding results of the machine calculations as is shown in Fig. 2.

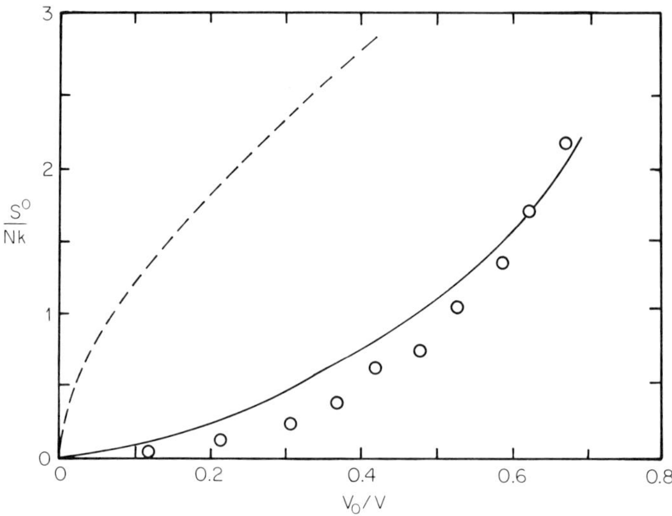

FIG. 2. Excess entropy for a system of rigid spheres. The solid line gives the results of the present calculations while the dotted line gives the results of the cell model and the points give the results of the machine calculations of Alder and Wainwright. (After Henderson, 1963.)

Other efforts have been made to improve the partition function. In one, the dense fluid under high pressure is treated. The Einstein oscillator is modified to provide for the gradual transition from a parabolic to a square well potential (Ree *et al.*, 1962) successfully. As shown later, the calculated critical pressure tends to be somewhat high. This is due to the neglect of clustering of vacancies in the partition functions. A simple modification in which the dimer term is added improves the result (Grosh *et al.*, 1967). However, one of the disadvantages of the model is that "it has not been derived from an exact partition function by any mathematically well defined approximation, but is a result of intuituion." It is instructive, therefore, to trace the correlation between this model and other models such as the cell theory whose statistical mechanical foundation bases has been discussed by Kirkwood (1950). Several efforts have been made to examine the basis of significant structure theory (Pierotti, 1965; Ree and Eyring, 1964; Jhon, 1967). One of these has been made by Pierotti (1965).

According to cell theory, the partition function for N identical monatomic molecules randomly distributed over L lattice cells is

$$f_c = \left(\frac{L!}{N!(L-N)!}\right) \frac{(2\pi mkT)^{3/2N}}{h^{3N}} (v_f)^N \exp\left(-\frac{yNZ\phi(a)}{2kT}\right). \qquad (3.4)$$

The following linear assumption for the relation between v_f and y is used:

$$v_f = yv_1 + (1 - y)v_0 = yv_1\{1 + [(1 - y)/y](v_0/v_1)\} \tag{3.5}$$

where v_1 and v_0 are the free volumes of molecules with y equal to unity and zero, respectively. Putting Eq. (3.5) into Eq. (3.4) gives

$$f_c = \left[\frac{(2\pi mkT)^{3/2}}{h^3} v_f \exp\left(-\frac{Z\phi(a)}{2kT}\right)\left\{1 + \frac{(1-y)}{y}\frac{v_0}{v_1}\right\}\right]^{yN} \times \left[\frac{(2\pi mkT)^{3/2}}{h^3} v_f\right]^{(1-y)N} \left(\frac{N}{L}\right)^{yN} \frac{L!}{N!(L-N)!}. \tag{3.6}$$

Since the total volume of the lattice V is equal to Lv_0, y is equal to V_s/V where V_s is the volume of a solidlike lattice of N cells. Thus, Eq. (3.6) can be rewritten as

$$f_c = \left[\frac{(2\pi mkT)^{3/2}}{h^3} v_f \exp\left(-\frac{Z\phi(a)}{2kT}\right)\left(1 + \frac{(V - V_s)}{V_s}\frac{v_0}{v_1}\right)\right]^{NV_s/V} \times \left[\frac{(2\pi mkT)^{3/2}}{h^3}(V - V_s + V_{ex})\right]^{N(V-V_s)/V} \times (L)^{-N}(N)^{N(N/L)} \frac{L!}{N!(L-N)!} \tag{3.7}$$

where $V_{ex} = Nv_1$ is the effective excess volume of the solid due to the nearest neighbor interactions. If we set $V_{ex} = 0$, and identify $n \exp[-aE_sV_s/(V - V_s)RT]$ with v_0/v_1, the significant structure partition function differs only in the combinatorial factor. This difference arises in the formation of the partition function. The free volume theory partition function fails to go into the partition function for an ideal gas at large volumes as the significant structure theory does.

IV. Comparison with Other Models

Next, we discuss reduced equations of state for various models and compare the calculated thermodynamic properties of simple liquids. The partition function (Ree *et al.*, 1965) according to Eq. (3.5) for a simple liquid is

$$f_N = \left[e^{E_s/RT}\left(\frac{T}{\theta}\right)^3 (1 + n_h e^{-\varepsilon_h/RT}) J(T)\right]^{NV_s/V} \times \left[\left(\frac{2\pi mkT}{h^2}\right)^{3/2} (V - V_s) J(T)\right]^{N(V-V_s)/V} \left[\left(\frac{N(V - V_s)}{V}\right)!\right]^{-1}. \quad (4.1)$$

The first set of brackets represents the solidlike partition function, and the classical approximation to the Einstein partition function is used:

$$1/(1 - e^{-\theta/T})^3 \simeq (T/\theta)^3. \quad (4.2)$$

The remaining part of Eq. (4.1) is for the gaslike degrees of freedom. Here, $J(T)$ is the partition function for the internal degrees of freedom. Assuming that the molecules in the solid state are hexagonally close packed and that the interaction potential between two molecules is given by the 6-12 Lennard-Jones potential then the sublimation energy, E_s, the Einstein characteristic temperature, θ_E, and V_s can be written as follows:

$$E_s = 6N\varepsilon[2.4090(V_s/Na^3)^{-2} - 1.0109(V_s/Na^3)^{-4}] \quad (4.3)$$

$$\theta_E = (h/k)(1/2\pi)(k_r/m)^{1/2} \quad (4.4)$$

and

$$k_r = 24\varepsilon[22.106(V_s/Na^3)^{-4} - 10.559(V_s/Na^3)^{-2}] \cdot (2^{1/3}a^2)^{-1}(Na^3/V_s)^{2/3} \quad (4.5)$$

where

$$V_s = Na^3.$$

In the above equations, a and ε are the distance and energy characteristic of the molecules and are listed by Hirschfelder *et al.* (1964). By substituting Eqs. (2.1), (2.2), and (4.2)–(4.5) into Eqs. (4.1) and applying Stirling's approximation to $[N(V - V_s)/V]!$, we obtain

$$f_N = \left[\frac{(2\pi mkT)^{3/2}}{h^3} \frac{eV}{N} J(T)\right]^N \left[\frac{e^{8.388/T^*}}{eV^*}\left(\frac{T^*}{35.01}\right)\right]^{3/2} \times \left[1 + 10.7(V^* - 1)\exp\left(-\frac{0.0436}{T^*(V^* - 1)}\right)\right]^{N/V^*} \quad (4.6)$$

where $T^* = kT/\varepsilon$ and $V^* = V/Na^3$, respectively. In the derivation of

Eq. (4.6), the values of $n = 10.7$ and $a = 0.0052$ have been used. These values were derived theoretically in Eq. (2.2). With the use of Eq. (4.6) we are able to evaluate the thermodynamic properties from the melting point to the ctitical point for simple liquids.

In Figs. 3 and 4, the reduced experimental vapor pressure ($p^* = Pa^3/\varepsilon$) and volume (V^*) for simple liquids such as Ne, Ar, N_2, and CH_4 are compared with the calculated values from the three theories. The agreement between significant structure theory and experiment is very good

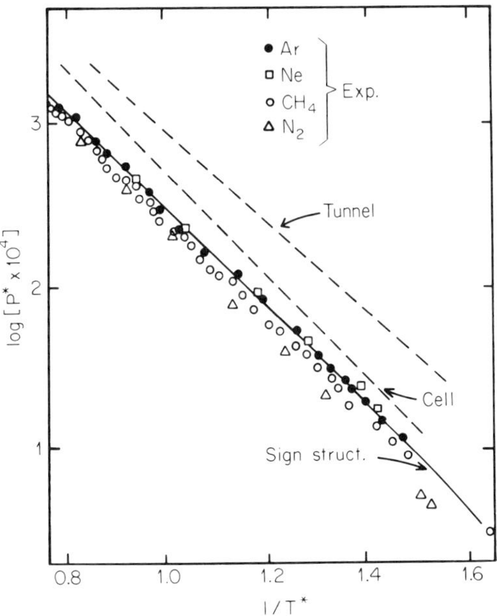

FIG. 3. Logarithm of the reduced vapor pressure P^* versus $1/T^*$ for simple liquids. The experimental values have been reduced by using ε/k and Na^3. The significant structure theory (full curve) is compared with the cell theory and the tunnel theory (broken curves). (After Ree *et al.*, 1965.)

and is far better than for the other theories. According to the significant structure approach, the reduced vapor pressure and volumes are obtained by plotting the Helmholtz free energy versus reduced volume as shown later.

In Table I, we compare the critical properties for the three theories and experimental data for simple liquids. The significant structure theory is better than the other theories. The tunnel theory (Barker, 1961) gives slightly worse results than the cell theory. This is probably due to the

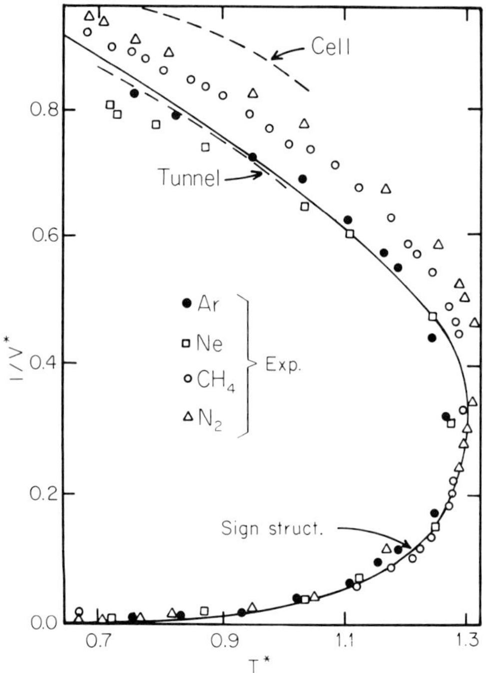

FIG. 4. Reduced density versus T^*. The experimental values have been reduced by using ε/k and Na^3. The significant structure theory (full curve) is compared with the cell theory and the tunnel theory (broken curves). (After Ree *et al.*, 1965.)

TABLE I

EXPERIMENTAL AND THEORETICAL CRITICAL CONSTANTS

	T_c^*	P_c^*	V_c^*	(PV/RT)
Mean values for Ne, Ar, N_2, CH_4	1.277	0.121	3.09	0.292
Significant structure theory	1.306	0.141	3.36	0.362
Tunnel theory	1.07	0.37	1.8	0.6
L-J and D theory	1.30	0.434	1.77	0.591

approximation of treating the motion in the tunnels as strictly one-dimensional. We also calculate the normal melting temperature and liquid melting volume. The results are shown in Table II. One sees that the L-J and D theory (Lennard-Jones and Devonshire, 1937a, 1938) gives better agreement for solid Ar that for the liquid.

TABLE II

THEORETICAL AND EXPERIMENTAL PROPERTIES AT THE MELTING POINT

	Melting temp. (T^*)	Reduced volume (V^*)	Reduced excess entropy (S^E/Nk)	Reduced excess energy (E^E/N)
Significant structure theory	0.711	1.159	−3.89	−6.19
Tunnel theory		1.184	−4.8	−5.9
Liquid argon	0.701	1.178	−3.64	−5.96
L-J and D theory		1.037	−5.51	−7.32
Solid argon	0.701	1.035	−5.53	−7.14

V. Thermodynamics of Liquids, Simple Classical Liquids, and Dense Gases

With the use of the basic partition function given in Eq. (2.5), we are in an excellent position to calculate thermodynamic properties from the melting point to the critical point for the various liquids. The Helmholtz free energy A is

$$A = -kT \ln f. \tag{5.1}$$

Since we have A as a function of volume and temperature, as shown in Fig. 5, we can calculate all other thermodynamic properties from

$$\begin{aligned} P &= -(\partial A/\partial V)_T \\ E &= -T^2(\partial(A/T)/\partial T)_V \\ S &= -(\partial A/\partial T)_V \\ C_v &= T(\partial S/\partial T)_V = -T(\partial^2 A/\partial T^2)_V \end{aligned} \tag{5.2}$$

and similar expressions for the remaining properties. The critical properties are obtained by using the usual conditions

$$(\partial P/\partial V)_T = 0 \quad \text{and} \quad (\partial^2 P/\partial V^2)_T = 0. \tag{5.3}$$

The vapor pressure is determined by finding the common tangent to the points corresponding to the liquid and vapor volume. For the liquid inert gases, the partition function is given in Eq. (2.5). The values of

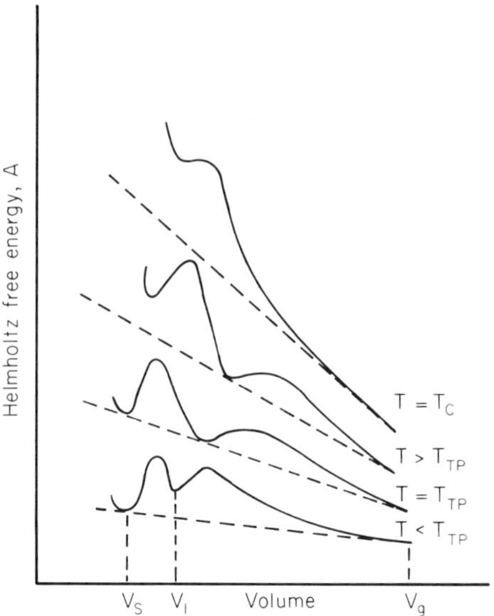

FIG. 5. Isotherms for the Helmholtz free energy A versus volume. The scales are greatly distorted. T_c is the critical temperature and T_{TP} is the triple point temperature. (After Eyring and Jhon, 1969.)

E_s, θ, and V_s are taken from the properties of the solid phase. Theoretical values of $n = 10.7$ and $a = 0.0054$ were used from Eqs. (2.6) and (2.8).

Some of the calculated results are summarized in Table III. The agreement between experiment and theory is very good except that the theory in all cases gives high critical pressures. This is due to the neglect of the clustering of the vacancies as the critical point is approached. This defect is improved by even modest attempts to take account of the clustering (Grosh *et al.*, 1967).

The significant structure model works very well for normal temperatures and pressure ranges. The extension of the theory to very high pressures and temperatures such as for dense liquids and gases has also been carried out successfully (Ree *et al.*, 1962). In the high temperature range a harmonic oscillator was used as the solidlike partition function up to the lth level after which the molecule was assumed to act like a particle in a box having a length

$$d = v_f^{1/3} = [(V_s/N)^{1/3} - (b/4N)^{1/3}] \tag{5.4}$$

Thus we have for the partition function for one degree of freedom the

TABLE III

VAPOR PRESSURES, MOLAR VOLUMES, AND ENTROPY OF VAPORIZATION

Temperature (°K)	P_{calc} (atm)	P_{obs} (atm)	V_{calc} (cc)	V_{obs} (cc)	ΔS_{calc} (eu)	ΔS_{obs} (eu)
			Argon			
83.96 (mp)	0.6874	0.6739	28.84	28.03	20.07	19.43
87.49 (bp)	1.040	1.000	29.36	28.69	18.90	18.65
97.76	2.883	2.682	31.04	30.15	15.92	—
			Krypton			
116.0 (mp)	0.7605	0.7220	34.31	34.13	20.14	—
119.93 (bp)	1.0660	1.000	34.90	—	19.15	17.99
			Xenon			
161.3 (mp)	0.8372	0.804	42.24	42.68	20.25	—
165.1 (bp)	1.0623	1.000	42.84	—	19.50	18.29

	T_c (°K)		P_c (atm)		V_c (cc)	
	Calc	Obs	Calc	Obs	Calc	Obs
Ar	149.5	150.7	52.9	48.0	83.5	75.3
Kr	209.5	210.6	62.4	54.24	101.0	—
Xe	292.4	289.8	69.6	58.2	125.0	113.8

	E_s, θ, and V_s values used		
	Ar	Kr	Xe
E_s (cal/mole)	1888.6	2740	3897.7
V_s (cc/mole)	24.98	29.6	36.5
θ (°K)	60.0	45.0	39.2

expression

$$f_{\mathrm{I}} = \sum_{i=1}^{l-1} e^{-i\theta/T} + e^{-l\theta/T} \sum_{n=1}^{\infty} \exp\left[-\frac{n^2h^2}{8d^2mkT}\right]. \tag{5.5}$$

Here v_{f}, V_{s}, b and N are the molar free volume, the molar volume, van der Waals b, and Avogadro's number, respectively. Accordingly, we replace the simple Einstein partition function in Eq. (2.5) by

$$\exp[(E_{\mathrm{s}}/RT)f_{\mathrm{I}}^{3}] = e^{E_{\mathrm{s}}/RT}\left\{\frac{1-e^{-l\theta/T}}{1-e^{-\theta/T}} + e^{-l\theta/T}\frac{(2\pi mkT)^{1/2}}{h}v_{\mathrm{f}}^{1/3}\right\}^{3}. \tag{5.6}$$

When $l\theta \gg T$, Eq. (5.6) reduces to the Einstein partition function in Eq. (2.5); and when $T \gg l\theta$, it reduces to a translational partition function.

In addition to the above effect on the vibrational partition function, the high pressure requires that the pressure dependence of V_{s} must be taken into account. Thus,

$$V_{\mathrm{sp}} = V_{\mathrm{s}}(1 - \beta\, \Delta p). \tag{5.7}$$

Here β is the solid compressibility and Δp is the pressure change. As exemplyfying the results, the second virial coefficient $B(T)$ is compared with experiment in Fig. 6. The expression may be determined with the

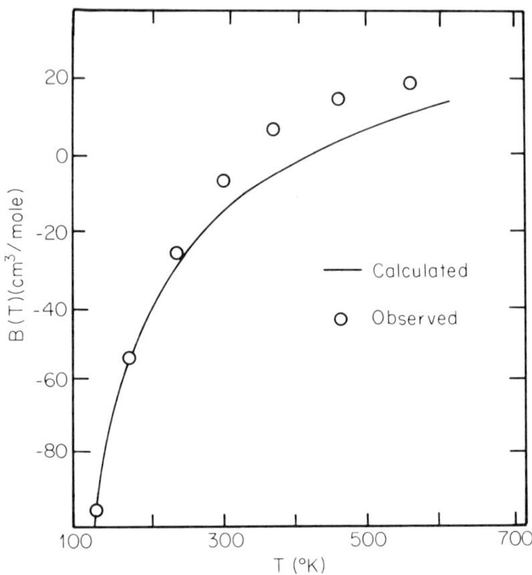

FIG. 6. The second virial coefficients of nitrogen versus temperature. (After Ree *et al.*, 1962.)

use of Eqs. (2.5) and (5.6).

$$B(T) = V_s(\gamma - \alpha_1 - \ln n). \tag{5.8}$$

Where

$$n = ZV_s/V_m$$

$$\gamma = \ln\left(\frac{(2\pi mkT)^{3/2}}{h^3}\frac{eV}{N}\right)$$

$$\alpha = \frac{E_s}{RT} + 3\ln\left\{\frac{1-e^{-l\theta/T}}{1-e^{-\theta/T}} + e^{-l\theta/T}\frac{(2\pi mkT)^{1/2}}{h}v_f^{1/3}\right\}. \tag{5.9}$$

The calculated results are also in fair agreement with experiment. The success of the significant liquid structure model throughout the entire liquid and gas region is quite satisfactory.

VI. Inorganic and Organic Liquids

The significant structure model has been applied to a number of inorganic and organic liquids (Eyring and Jhon, 1969) ranging from simple diatomic molecules to complicated ones such as pentaborane and simple hydrocarbons such as methane to more complicated molecules such as the xylenes. As exemplifying the theoretical calculations, some of the results for methane and carbon tetrachloride are given in Table IV. The following form of the partition function for liquid CH_4 and CCl_4 is used:

$$f = \left\{\frac{e^{E_s/RT}}{(1-e^{-\theta/T})^3}\left(1 + n\frac{V-V_s}{V_s}\exp\left[-\frac{aE_sV_s}{RT(V-V_s)}\right]\right)\right\}^{N(V_s/V)}$$
$$\times\left\{\frac{(2\pi mkT)^{3/2}}{h^3}\frac{eV}{N}\right\}^{N(V-V_s)/V}\left\{\prod_{i=1}^{9}\frac{1}{1-e^{-h\nu_i/kT}}$$
$$\times\frac{(8\pi^2kT)^{3/2}(\pi ABC)^{1/2}}{12h^3}\right\}^N \tag{6.1}$$

A, B, and C are the principal moments of inertia, ν_i is the ith internal molecular vibrational frequency, and the remaining notation has been defined.

Judging from the experimental heat capacity curves, it was assumed that the molecule rotates in the solid state. Since the partition function

TABLE IV

THERMODYNAMIC PROPERTIES OF LIQUID CH_4 AND CCl_4

	n	a	θ	V_s (cc/mole)	E_s (cal/mole)
CH_4	11.05	0.364×10^{-2}	71.34	31.06	2201
CCl_4	11.64	0.479×10^{-3}	53.53	89.39	4334

	CH_4		CCl_4	
	Calc	Obs	Calc	Obs
T_m (°K)	90.65	90.65	250.22	250.22
V_m (cc)	33.63	33.63	92.12	92.12
P_m (atm)	0.1154	0.1154	0.0111	0.0111
ΔS_v (eu)	23.20	23.20	—	—
T_b (°K)	111.67	111.67	349.9	349.9
V_b (cc)	36.12	37.79	102.02	103.89
P_b (atm)	0.9954	1.0000	1.0317	1.0000
ΔS_b (eu)	17.514	17.51	—	—
T_c (°K)	211.29	191.05	567.3	556.3
P_c (atm)	60.59	45.8	56.85	45
V_c (cc)	103.83	99.34	250.02	275.7
ΔS_c (eu)	—	—	21.22	20.41

for the internal vibrations and the rotations occur in both portions of the partition function, we have factored them out for convenience. The criteria for free rotation or nonrotation of a solidlike molecule are frequently ambiguous. In fact, it is more reasonable to assume hindered rotational degrees of freedom for the solidlike molecule in the liquid. The following approximate hindered rotational partition function was introduced by McLaughlin and Eyring (1966).

$$f_{HR} = f_{vib} + (f_{FR} - f_{vib}) \exp[-BV_s/RT(V - V_{s0})] \qquad (6.2)$$

where f_{FR} is a free rotational partition function, f_{vib} is the vibrational partition function, V_{s0} is the solid volume at which a free rotation starts. The parameter B was taken as a constant.

VII. Molten Metals and Fused Salts

Next, we describe the applicability of the model to the liquid with high melting temperature such as molten metals (Carlson *et al.*, 1960b) and fused salts (Carlson *et al.*, 1960a; Jhon *et al.*, 1968; Lu *et al.*, 1968). Since the conducting electrons of a metal are spread out over the positive ions, only the size of ionic vacancies appear in the liquid. Since the ions are only about a third the size of atoms, the vacancies are also smaller.

Thus, the expansion of the metals upon melting is 3 or 4%, which is about a third as much as a normal liquid. However, the entropy of melting is almost the same as that of a normal liquid, so n must have a value about three times that of a normal liquid. For the construction of the partition function of a liquid metal, an Einstein partition function is assumed for the solidlike degrees of freedom, and monoatomic and diatomic terms are included in the gaslike part of the partition function, since an appreciable concentration of the dimer species exists in the vapor.

A useful partition function for the liquid metal to evaluate all thermodynamic properties takes the following form:

$$f_N = \left\{ \frac{e^{E_s/RT}}{(1-e^{-\theta/T})^3} \left[1 + n \frac{(V-V_s)}{V_s} \exp\left(-\frac{aE_sV_s}{RT(V-V_s)} \right) \right] \right\}^{NV_s/V} \times \left\{ \left(f_1 \frac{eV}{N} \right)^2 + \left(f_2 \frac{eV}{N/2} \right) \right\}^{N(V-V_s)/2V} \tag{7.1}$$

where

$$f_1 = \frac{(2\pi m_1 kT)^{3/2}}{h^3}, \qquad f_2 = \frac{(2\pi m_2 kT)^{3/2}}{h^3} \frac{8\pi^2 IkT}{2h^2} \frac{e^{D/RT}-1}{1-e^{-h\nu/kT}}$$

and m_1, m_2, I, D, and ν are the masses of the monatomic atoms, the diatomic atoms, the moment of inertia, the dissociation energy, and vibrational frequency of the dimer, respectively. In recent investigations of liquid metals, soon to be published, the temperature dependence of the solidlike volume V_s and the dependence of E_s, θ, and V_s are taken into account with striking improvement in the calculated results. As $T \to T_c$, the atoms no longer behave as if they were ions. Near T_c, n probably approaches that of a normal liquid.

Very little has been done to develop the theory of fused salt, in spite of its importance in industry and in science. Typical investigations, other than those using the significant structure model, are those of McQuarrie (1962) using the method of cell theory (1937), and of Kirkwood and Monroe (1941), of Bogolyubov (1946), and of Kirkwood and Boggs (1942), who used a radial distribution function. The significant liquid structure model has also been applied to fused salt (Carlson *et al.*, 1960a; Lu *et al.*, 1968). The above cited papers deal with highly ionized molten alkali halides. The matter is more complicated for other molten salts. Recently, Jhon *et al.* (1968) developed the significant structure model to treat slightly ionized salts. Early investigations neglect the dimer concentrations in the vapor which, as experiment shows, are appreciable. Here, we will discuss the most recent version (Lu *et al.*, 1968) of the significant structure model applied to fused salt. For highly ionized molten alkali halides, the observed expansion percentage is approximately twice that of Ar. This is an expected result since the entropy of melting comes from the randomness introduced by the excess volume, and only half the extra space is available to positive and half to negative ions. Thus, it should require twice the percentage of expansion to obtain the same entropy increase for each kind of ion. Because of the fact that only half the excess volume ($V - V_s$) provides vacancies for each kind of ion, we expect the factor n in the positional degeneracy term to be only about half as large for salts as it is for Ar. In constructing the partition function, the solid like part of the partition function is raised to the power $2N(V_s/V)$, since a mole of alkali halide contains $2N$ particles.

A common Einstein temperature is used for both ions as an approximation. And the solid E_s has been multiplied by the factor $(V/V_s)^{1/3}$ to account for the long range coulombic interionic potential, and is divided by two to get the E_s per ion. Since appreciable amounts of the dimer exist in the alkali halide vapor phase, such terms were included in the gas like partition function. These considerations lead to the following partition function.

$$
\begin{aligned}
f = \Bigg\{ & \frac{\exp[(E_s/2RT)(V/V_s)^{1/3}]}{(1 - e^{-\theta/T})^3} \\
& \times \left[1 + n\left(\frac{V - V_s}{V_s}\right) \exp\left(\frac{aE_s(V/V_s)^{1/3}V_s}{2RT(V - V_s)}\right)\right]\Bigg\}^{2N(V_s/V)} \\
& \times \left\{\left(\frac{F_1 eV}{N}\right)^{N(V-V_s)/V} \left(1 + 2\frac{n_2}{n_1}\right)^{n_1} \left[\frac{KN}{eV}\left(\frac{n_1}{n_2} + 2\right)\right]^{n_2}\right\}
\end{aligned}
\tag{7.2}
$$

where

$$
F_2 = \left(\frac{2\pi mkT}{h^2}\right)^{3/2} \frac{8\pi^2 I_1 kT}{h^2} \frac{1}{1 - e^{-h\nu/RT}}
$$

and

$$
K = F_2/F_1 = e^{-\Delta H/RT} e^{\Delta S/R}.
$$

TABLE V

CALCULATED AND OBSERVED THERMODYNAMIC PROPERTIES OF THE MOLTEN SALTS[a]

T (°K)	V (cc)		P (atm)		S (e.u.)		C_p (cal/mole)	
	Calc	Obs	Calc	Obs	Calc	Obs	Calc	Obs
				NaCl				
1074 (mp)	37.42	37.56	0.00015	—	36.37	40.77	15.69	16.00
1290	40.38	40.62	0.00753	—	38.95	—	15.71	16.00
1783 (bp)	46.99	48.80	1.00046	1.0000	44.10	—	24.46	—
(2219)	(152.59)	—	(141.96)	—	—	—	—	—
				KCl				
1043 (mp)	48.77	48.80	0.00022	—	39.00	42.21	16.58	16.00
1200	51.96	51.91	0.00433	—	41.14	44.45	16.70	16.00
1680 (bp)	63.06	64.47	1.00149	1.0000	47.38	—	26.12	—
(2144)	(151.13)	—	(211.43)	—	—	—	—	—

[a] Parentheses indicate calculated values.

Here K is the equilibrium constant between monomer and dimer. F_2 denotes the dimer partition function per unit volume, ΔH and ΔS are the heat of reaction and the entropy change, n_1 and n_2 are the number of monomer and dimer molecules, respectively; the remaining notation is as previously defined. Some calculated results are shown in Table V. Only the values of C_p at high temperature seem too high but this indicates needed modification of the partition function.

VIII. Water and Electrolyte Solutions

One of the most exciting problems being studied is the structure and properties of water. Water is the most common liquid on earth, but it has many abnormal properties such as a maximum density at 4°C, and the peculiar decrease in viscosity with pressure up to about 1000 atm etc. In the past, many theories have been developed to explain the properties of water and to elucidate its structure. Unfortunately, none of these theories explains all its properties quantitatively. The current theories of water are divided into two categories, the first of these assumes the "uniformist" average model such as has been developed by Bernal and Fowler (1933), Pople (1951), and Wall and Hornig (1965), while the second category considers "mixture" models such as the theories of Eucken (1947), Frank and Wen (1957), and Nemethy and Scheraga (1962).

One of the more successful theories in explaining the maximum density of water is that of Nemethy and Scheraga. However, they did not write down a partition function from which all the properties could be calculated. The significant structure model also has been applied to water by Eyring and Marchi (1964). This early model did not lead to a maximum in the density.

Recently, Jhon *et al.* (1966) have developed a new model in a series of papers (Eyring and Jhon, 1969) designed to explain all the properties of water. Here, we will discuss briefly the picture of this model. The model is visualized as containing at least two solidlike clusters in equilibrium with each other and with the gaslike molecules. One of these structures is a cluster of about 46 molecules with similar ice-I-like structure and density. These are dispersed in the ice-III-like structures which is 20% more dense than ice-I but still has hydrogen bonds. When a cluster of ice-I-like molecules disappear, they change their structure cooperatively to a cluster of 46 ice-III-like molecules according to the model. Although others of the nine like forms of ice may well be present there is a yet no indication that they are present in substantial amounts.

Introducing these concepts and the usual routine statistical mechanical considerations, we can write a partition function of water:

$$f = \left\{ \frac{e^{E_s/RT}}{(1 - e^{-\theta/T})^6} \times \left\{1 + n\left(\frac{V - V_s}{V_s}\right) \exp\left[- \frac{aE_sV_s}{(V - V_s)RT} \right]\right\} (K)^{K/(1+K)q} \prod_{i=1}^{3} \frac{1}{1 - e^{-h\nu_i/kT}} \right\}^{NV_s/V} \left\{ \frac{(2\pi kmT)^{3/2}}{h^3} \frac{eV}{N} \frac{8\pi^2(8\pi^3ABC)^{1/2}(kT)^{3/2}}{2h^3} \times \prod_{i=1}^{3} \frac{1}{1 - e^{-h\nu_i/kT}} \right\}^{N(V-V_s)/V}. \tag{8.1}$$

In this derivation, we assumed that solidlike molecules in water do not rotate, and the equilibrium constant is written as

$$K = \frac{\text{ice-I-like}}{\text{ice-III-like}} = \left(\frac{f_{sI}}{f_{sIII}}\right)^q = (e^{-\Delta H/RT}\, e^{\Delta S/R}\, e^{-p\Delta V/RT})^q \tag{8.2}$$

and

$$V_s = [K/(1 + K)]V_{sI} + [1/(1 + K)]V_{sIII}. \tag{8.3}$$

Here q is the number of molecules in a cluster, f_{sI} and f_{sIII} are the partition function of ice-I-like molecules and ice-III-like molecules. The remaining notation has been defined previously.

Some of the typical calculated results are shown in Tables VI and

TABLE VI

MOLAR VOLUMES AND VAPOR PRESSURES OF WATER

T (°K)	V_{calc} (cc)	V_{obs} (cc)	Error (%)	P_{calc} (atm)	P_{obs} (atm)	Error (%)
273.15 (mp)	17.923	18.019	−0.53	0.006051	0.006030	−0.35
277.15	17.888	18.016	−0.71	0.008036	0.008007	−0.36
283.15	17.891	18.021	−0.72	0.01210	0.01218	−0.66
293.15	17.961	18.048	−0.48	0.02298	0.02307	−0.39
313.15	18.171	18.157	0.08	0.07234	0.07279	−0.62
353.15	18.673	18.538	0.73	0.4623	0.4672	−1.05
373.15 (bp)	18.959	18.799	0.85	0.9852	1.0000	−1.48
423.15	19.800	19.641	0.81	4.549	4.698	−3.17

VII. The theory successfully predicts the maximum density at 4°C, and a minimum in the heat capacity versus temperature. The agreement between theory and experiment for the second derivatives of the partition function such as C_p constitute a reliable and severe test of the model. The application of significant structure theory to electrolyte solutions has also been developed (Jhon and Eyring, 1968a; Sung and Jhon, 1970a, 1970b),

TABLE VII

SPECIFIC HEATS OF WATER

T (°K)	C_p (obs) (cal/mole)	C_p (calc) (cal/mole)	Error (%)
273.15 (mp)	18.15	17.40	− 4.13
283.15	18.06	15.84	−12.3
293.15	17.99	16.61	− 7.67
313.15	17.98	17.39	− 3.28
353.15	18.06	18.94	− 0.11
373.15 (bp)	18.14	18.33	1.05
423.15	—	19.04	—

IX. Binary Mixtures

Another interesting development involves liquid mixtures (Ma and Eyring, 1965; Miner and Eyring, 1965; Ahn *et al.*, 1965; Vilen and Misdolea, 1966). The significant structure theory works very well when used to evaluate the properties of binary mixtures. Besides the knowledge of the pure components, some additional assumptions such as (a) nonrandom mixing may be neglected, and (b) parameters in the mixture partition functions are only simple concentration dependent functions of those of the pure components.

Then, the partition function of the mixture is

$$f_{\text{mixture}} = \frac{(N_1 + N_2)!}{N_1! N_2!} f_1 \, f_2 \times \left\{1 + n \frac{(V - V_s)}{V_s} \exp\left(-\frac{aE_s V_s}{RT(V - V_s)}\right)\right\}^{(N_1+N_2)V_s/V} \times \exp\left(\frac{E_s}{RT} (N_1 + N_2) \frac{V_s}{V}\right). \tag{9.1}$$

The factorial terms arise from random mixing, N_1 and N_2 are the number of molecules of the respective species, and $N_1 + N_2 = N$, where N is Avogadro's number. In Eq. (9.1), E_s, V_s, a, and n are averages of those of pure components and are written as

$$\begin{aligned} E_s &= X_1{}^2E_{s1} + X_2{}^2E_{s2} + 2X_1X_2(E_{s1}E_{s2})^{1/2} \\ V_s &= X_1V_{s1} + X_2V_{s2}, \qquad a = X_1a_1 + X_2a_2, \qquad n = X_1n_1 + X_2n_2 \end{aligned} \tag{9.2}$$

where X_1 is the mole fraction of component i ($i = 1$ or 2). In the case of the argon and nitrogen mixture, f_1 and f_2 in Eq. (9.1) are written as

$$f_1 = \left[\frac{(2\pi m_1kT)^{3/2}}{h^3}\frac{eV}{N}\right]^{N_1(V-V_s)/V}\left[\left(\frac{1}{1-e^{-\theta_1/T}}\right)^3\right]^{N_1V_s/V} \tag{9.3}$$

and

$$\begin{aligned} f_2 = &\left[\frac{(2\pi m_2kT)^{3/2}}{h^3}\frac{eV}{N}\frac{1}{1-e^{-h\nu_2/kT}}\frac{8\pi^2I_2kT}{2h^2}\right]^{N_2(V-V_s)/V} \\ &\times\left[\left(\frac{1}{1-e^{-\theta_2/T}}\right)^3\frac{1}{1-e^{-h\nu_2/kT}}\frac{8\pi^2IkT}{2h^2}\right]^{N_2V_s/V}. \end{aligned}$$

Equation (9.1) satisfied the following thermodynamic requirements, ie.,:

$$A + pV = N_1\mu_1 + N_2\mu_2 = G. \tag{9.4}$$

where G is the Gibb's free energy of the mixture, and μ_1 and μ_2 are the chemical potential of species 1 and 2, respectively. In Table VIII, we compare the theoretical values from the experimental data for $Ar + N_2$.

TABLE VIII

Excess Properties of Binary Mixtures
T (°K) = 83.82, $X_1 = X_2 = 0.500$

Substances		H^E (kcal)	G^E (kcal)	V^E (cc)
$Ar + N_2$	Theoretical	8.44	5.05	−0.138
	Experimental	12.1	8.2	−0.18
$Ar + O_2$	Theoretical	9.97	1.40	0.134
	Experimental	14.4	8.8	0.14
$O_2 + N_2$	Theoretical	7.28	6.80	−0.30
	Experimental	10.0	9.2	−0.31

$Ar + O_2$, and $O_2 + N_2$ at 83.82°K and $X_1 = X_2 = 0.500$. The agreement between theory and experiment for the thermodynamic properties such as excess enthalpy H^E, excess Gibb's free energy G^E, and excess molar volume V^E is very good.

X. Liquid Hydrogen

The significant structure model has been applied to quantum liquids such as liquid hydrogen, deuterium, and hydrogen deuteride. These applications are a further test of the validity of the model since the partition function involves different forms of statistics. Because these liquids exist at extremely low temperatures, a Debye partition function was used for the solidlike molecules. At the temperature of interest, the gaslike molecules are slightly degenerate, and therefore the Bose–Einstein statistics is applied to hydrogen and deuterium and the Fermi–Dirac statistics to hydrogen deuteride. Henderson *et al.* (1962) obtained the following partition function:

$$\ln f = N\frac{V_s}{V}\left\{\frac{E_p}{RT} - \frac{9}{8}\left(\frac{\theta_D}{T}\right) - 9\left(\frac{T}{\theta_D}\right)^3 \int_0^{\theta_D/T} u^2 \ln(1 - e^{-u})\, du + \ln\left[1 + Z\frac{V - V_s}{V}\exp\left(-\frac{aE_pV}{RT(V - V_s)}\right)\right]\right\} + N\frac{V - V_s}{V}\left\{1 - \ln y \pm \frac{y}{2^{5/2}}\right\} + N \ln f_r \tag{10.1}$$

where the top sign (+) indicates a Bose–Einstein gas and the bottom sign (−) the Fermi–Dirac case; E_p is the potential energy of the lattice, θ_D is the Debye temperature of the solidlike lattice, $y = (N/V) \times (h^2/2\pi mkT)^{3/2}$, and f_r is the rotational partition function for the molecules. In the case of ortho-hydrogen and para-hydrogen the rotational partition functions are given by the expression

$$f_r^o = 3 \sum_{n=1,3}^{\infty} (2n + 1)e^{-n(n+1)\theta_r/T}$$
$$f_r^p = \sum_{n=0,2}^{\infty} (2n + 1)e^{-n(n+1)\theta_r/T} \tag{10.2}$$

Here $\theta_r = h^2/(8\pi^2 Ik)$ and the corresponding expressions for o-D_2, p-D_2 and HD are standard.

The rotational partition function for mixtures of ortho and para molecules is, then, given by

$$\ln f_r = y \ln f_r^{o} + (1 - y) \ln f_r^{p}. \tag{10.3}$$

Some calculated and observed properties at the melting, boiling and critical points are listed in Table IX. Good agreement between theory and experiment is observed in a straight forward application of the significant structure model.

TABLE IX

CALCULATED AND OBSERVED THERMODYNAMIC PROPERTIES OF LIQUID HYDROGEN

	p–H_2	n–H_2	H–D	p–D_2	n–D_2	
T_m (°K)	(13.84)	(13.94)	(16.60)	(18.63)	(18.73)	Calc
	13.84	13.94	16.60	18.63	18.73	Obs
P_m (atm)	0.07388	0.07589	0.1236	0.1706	0.1724	Calc
	0.06942	0.07085	0.1221	0.1678	0.1692	Obs
V_m (cm³ mole⁻¹)	26.213	26.093	24.491	23.262	23.155	Calc
	26.176	26.108	24.487	—	23.162	Obs
ΔS_m (cal mole⁻¹ deg⁻¹)	1.932	1.936	2.048	2.198	2.210	Calc
	2.028	—	—	2.526	—	Obs
T_b (°K)	20.58	20.70	22.29	23.65	23.75	Calc
	20.261	20.365	22.14	23.59	23.67	Obs
V_b (cm³ mole⁻¹)	28.829	28.692	26.525	24.955	24.830	Calc
	28.482	28.393	—	—	—	Obs
ΔS_b (cal mole⁻¹ deg⁻¹)	10.553	10.564	11.868	12.741	12.737	Calc
	10.602	—	—	12.459	—	Obs
T_c (°K)	35.9	36.2	37.6	39.4	39.7	Calc
	32.994	33.24	35.908	38.262	38.24	Obs
P_c (atm)	13.6	13.8	15.5	17.1	17.3	Calc
	12.770	12.797	14.645	16.282	16.421	Obs
V_c (cm³ mole⁻¹)	77.7	77.3	71.5	68.3	68.0	Calc
	65.5	—	62.8	60.3	—	Obs

XI. The Viscosity and Diffusion Coefficients of Liquids

According to early derivations (Eyring, 1936) from rate theory, the viscosity is expressed as follows:

$$\eta = \frac{hN}{V} e^{\Delta G^{\neq}/RT} = \left(\frac{hN}{V} e^{-\Delta S^{\neq}/R}\right) e^{\Delta H^{\neq}/RT}. \tag{11.1}$$

Here V is the molar volume, N is Avigadro's number, $\Delta G^{\neq}$, $\Delta H^{\neq}$, and $\Delta S^{\neq}$ are the free energy of activation, enthalpy of activation, and entropy of activation, respectively. A major success of this model is that it leads to an expression of a viscosity proportional to $\exp(\Delta H/RT)$. This Arrehenius form has been long observed to hold for many liquids. However, one of the defects of this old model is that the viscosity of many liquids change exponentially with temperature only at low temperatures. Further, the old model is unable to explain properties such as supercooling or the glass transition region.

Also, since the fluidity of a liquid (the reciprocal of the viscosity) is proportional to the excess volume $V - V_s$, i.e., to the numbers of vacancies introduced, as Batschinski pointed out, this result should be part of an acceptable theory of the liquid state.

In this respect, efforts have been made by Doolittle (1951, 1952), Beuche (1959), McCedo and Litovitz (1965) Cohen and Turnbull (1950, 1959) to develop realistic theories of viscosity. Doolittle found that the fluidity of many liquids can be represented satisfactory by a relation in terms of the free volume. Modifications of the free volume approach were made also by Williams, Landel, and Ferry to explain the glass transition region. Recently Cohen and Turnbull have developed a theory of the molecular transport in liquid and glass, much like the Doolittle equation.

The significant structure theory automatically leads to an appropriate theory for transport property. We consider these matters in the following. Let us consider that a fraction x_s of a shear surface is covered by solidlike molecules and the remaining fraction x_g is covered by gaslike molecules. Then, the viscosity η, which is the ratio of shear stress f to rate of strain $\dot{s}$, is given by

$$\eta = \frac{f}{\dot{s}} = \frac{(x_s f_s + x_g f_g)}{\dot{s}} = x_s \eta_s + x_g \eta_g = \frac{V_s}{V} \eta_s + \frac{V - V_s}{V} \eta_g \quad (11.2)$$

where V_s, V, η_s, and η_g are the specific volume of the solidlike structure, that of the liquid, and the viscosities of solidlike and gaslike structures, respectively. The term η_g is taken as equal to the gas viscosity obtained from the kinetic theory of gases, i.e.,

$$\eta_g = (2/3d^2)(mkT/\pi^3)^{1/2} \quad (11.3)$$

where m is the molecular weight, and d is the diameter of the molecules. In accord with a generalization (Ree *et al.*, 1962) of Eyring's early procedure, one of the η_s term (Ree *et al.*, 1964) is found to have the value

$$\eta_s = \left[(\pi m k T)^{1/2} N \frac{l_f}{2(V - V_s)\varkappa} \right] \exp \frac{a' V_s Z \phi(a)}{(V - V_s) 2kT} \frac{V}{V_s}. \qquad (11.4)$$

Here l_f is the free length between nearest neighbors, $\varkappa$ is the transmission coefficient, a' is a dimensionless constant, $\phi(a)$ is an intermolecular potential, Z the number of nearest molecules, k is Boltzmann's constant, and N is Avogadro's number. Substituting Eqs. (11.3) and (11.4) in (11.2) gives

$$\eta = \left[(\pi m k T)^{1/2} N \frac{l_f}{2(V - V_s)\varkappa} \right] \exp\left[\frac{a' V_s Z \phi(a)}{(V - V_s) 2kT} \right] + \frac{V - V_s}{V} \frac{2}{3d^2} \left(\frac{mkT}{\pi^3} \right)^{1/2}. \qquad (11.5)$$

There is excellent agreement between theory (Eyring and Jhon, 1969; Ree *et al.*, 1962, 1964b; Eyring and Ree, 1961; Jhon *et al.*, 1969) and experiment. And the viscosity equation obtained is self consistant since if $V \to V_s$, then $\eta \to \infty$ and if $V \to V_g$, then $\eta \to \eta_g$. It also satisfied the Batschinski relation. Generally, this equation is adequate to predict the viscosity in the low temperature region or supercooled region (Bahng and Jhon, 1970) if the appropriate V_s is used. The V_s to be used is the molar volume of the solidlike structure as it actually exists in the liquid due to thermal expansion and to structure changes. The effect of pressure

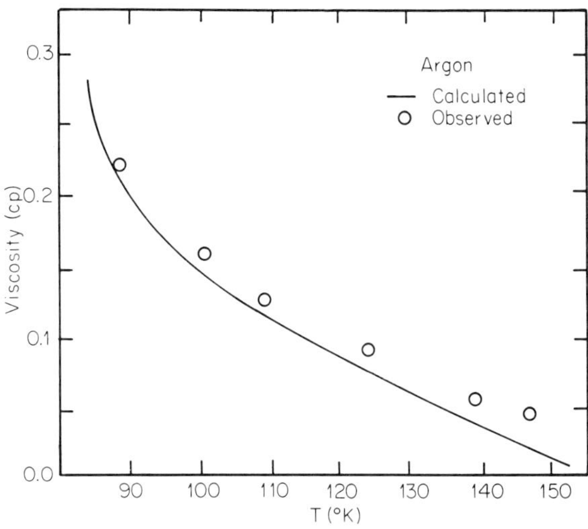

FIG. 7. Viscosity of liquid argon in centipoises versus temperature.

was taken into account in a successful calculation by Jhon *et al.* (1969). In this derivation, the explicit expression of the molar solid like volume V_s and the intermolecular force $\phi(a)$ at high pressure were introduced. The diffusion coefficient also may be calculated by means of the following equation of Eyring and Ree:

$$D = kT/\xi(V_s/V)^{1/3}\eta \tag{11.6}$$

where ξ is the effective number of neighbors of a molecule lying in the same plane. For a close-packed structure, this is six. The viscosity of argon shown in Fig. 7 gives excellent agreement between theory and experiment.

XII. Thermal Conductivity of Liquids

The significant structure theory also forms the basis for an interpretation of thermal conductivity. Thus we can write

$$\varkappa = (V_s/V)\varkappa_s + [(V - V_s)/V]\varkappa_g \tag{12.1}$$

where $\varkappa$, $\varkappa_s$, and $\varkappa_g$ are the thermal conductivity of the liquid, the solid-like and the gaslike degrees of freedom, respectively. Since the contribution of gaslike molecules to the liquid thermal conductivity is generally small, it is adequate to use the equation for the thermal conductivity of an ideal gas (Loeb, 1927):

$$\varkappa_g = [(9\gamma - 5)/4]\eta_g C_{vg} \tag{12.2}$$

where γ is the ratio of the specific heat at constant pressure to that at constant volume, and η_g and C_{vg} are the viscosity and the heat capacity at constant volume of an ideal gas, respectively.

To express $\varkappa_s$, we use phenon theory (Klemens, 1951; Callaway, 1959; Peierls, 1929)

$$\varkappa_s = \frac{C_s^2 R}{V_s(A\omega^4 + BT\omega^2)} \frac{(h\omega/kT)^2 e^{-\hbar\omega/kT}}{(e^{\hbar\omega/kT} - 1)^2} \tag{12.3}$$

where ω is the Einstein characteristic frequency, $A\omega^4$ represent the scattering by lattice imperfections; the term $BT\omega^2$ includes the normal Umklapp processes, and C_s is the velocity of sound in the solid. For the relatively high temperatures of interest, here we have $\hbar\omega/kT < 1$ so that

$$\frac{(\hbar\omega/kT)^2 e^{-\hbar\omega/kT}}{(e^{\hbar\omega/kT} - 1)^2} \simeq 1$$

Substituting Eqs. (12.2) and (12.3), we obtain

$$\varkappa = \frac{RC_s^2}{V_s(A\omega^4 + BT\omega^2)} + \frac{V - V_s}{V}\frac{(9\gamma - 5)}{4}\eta_g C_{vg}. \qquad (12.4)$$

The first term on the right-hand side of Eq. (12.4) represents the contribution to thermal conduction due to the vibration of molecules near their equilibrium position and the second term, the contribution due to the random motion of molecules. Furthermore, it is convenient to absorb C_s and R into the constants A and B. The calculated results (Lin *et al.*, 1964) are shown in Fig. 8. One sees that over small ranges of pressure, $\varkappa$ is linear with respect to pressure.

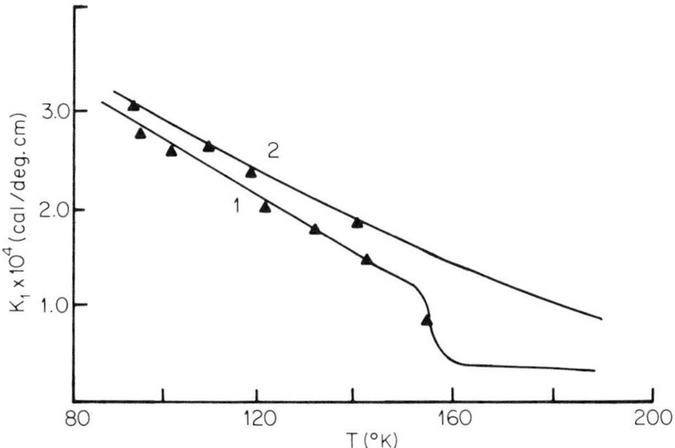

FIG. 8. Temperature dependence of K_1 of argon: (——), experimental values; (▲), calculated values; curve 1, $P = 48$ atm; curve 2, $P = 120$ atm.

XIII. Surface Tension of Liquids

There are several theories for the surface tension of liquids, ranging from fairly simple treatments in terms of an intuitive physical model, such as the significant liquid structure theory (Eyring *et al.*, 1958; Eyring and Jhon, 1969), to detailed formal theories. The latter theories have been throughly reviewed by Onno and Kondo (1960) and others.

Here, we describe the significant structure model approaches. Among them, one is an iteration procedure which calculates the sum of the contributions of successive molecular layers (Chang *et al.*, 1962), the other is a monolayer approximation which yield a simplified calculation (Ree *et al.*, 1964a; Lu *et al.*, 1967).

Since the partition function reduces to the gas partition function for $V \gg V_s$, it is quite reasonable to expect that this partition function will also represent the surface layer, if the appropriate density gradient is taken into account. For a close-packed liquid such as Ar, it is expected that each molecule should have close to six neighbors in the same layer, three neighbors in the layer below and three above. Then, only E_s, the energy of sublimation in the bulk is appreciably different for a surface.

The energy E_{si} of a molecule in the ith surface layer has the following relationship.

$$E_{si} = E_s\left(\frac{6}{12}\frac{\varrho_i}{\varrho_i} + \frac{3}{12}\frac{\varrho_{i+1}}{\varrho_i} + \frac{3}{12}\frac{\varrho_{i-1}}{\varrho_i}\right) \tag{13.1}$$

where ϱ_i, ϱ_{i+1}, and ϱ_{i-1} are the densities of the ith layer and the layers immediately below and above. The Gibbs free energy per mole $G_i = A_i + PV_i$, for each surface layer may be calculated by means of Eq. (13.1) by using an iteration method to determine the ϱ_i's (Chang *et al.*, 1962).

The surface tension γ is the excess Gibbs free energy per unit area over that of the bulk liquid. Therefore, the surface tension may be calculated from the relation

$$\gamma = \sum_i (G_{si} - G_l)(d/V_i) \tag{13.2}$$

where $G_{si} - G_l$ is the difference in free energy of the ith surface layer and the bulk liquid, V_i the molar volume of the ith layer, d becomes $0.9165(V_s/N)^{1/3}$ since cubic closest packing can be referred to as a face centered cubic lattice. Other types of packing would give a different factor. Surface tensions of several inorganic and organic liquids have been tested succesfully.

Exemplifying the model, the results of nitrogen and methane are given in Table X. There is good agreement between experiment and theory.

However, the deviations between theory and experiment are quite large in the case of polar liquids such as water and alcohol. But a model, which takes account of orientation at the surface, leads to results in good agreement with experiment. In this case, the additional orientation term in the partition function is given as $(kT/\mu X)\sinh(\mu X/kT)$ (Jhon *et al.*, 1967), where μ is the dipole moments of the molecule and X is the field strength. While the iteration procedure explained already is very useful in its application to various liquids, this approach is unable to provide a simple equation in closed form.

TABLE X

SURFACE TENSION OF LIQUID NITROGEN AND METHANE

T (°K)	% Contribution of each layer			γ_{calc} (dynes cm^{-1})	γ_{obs} (dynes cm^{-1})	Error (%)
	1st	2nd	3rd			
			Nitrogen			
63.30 (mp)	90.24	9.76	0.00	11.78	12.05	−2.24
68.41	89.18	10.82	0.00	10.63	10.89	−2.39
77.34 (bp)	85.22	13.52	1.26	8.73	8.91	−2.02
			Methane			
90.65 (mp)	91.25	8.75	0.00	17.14	18.20	−5.82
99.67	90.20	9.80	0.00	15.30	16.24	−5.79
111.67 (bp)	87.09	11.91	1.00	13.01	13.70	−5.04
123.15	84.34	14.26	1.40	10.73	11.34	−5.38

According to Raman and Ramdas' experiment (1927), the boundary between a liquid and its vapor is sharp and extends over about one molecular layer. Assuming that the surface of liquid consists of a monomolecular layer whose molecules have energies different from the bulk liquid, the following partition function is then derived.

$$f^N = f^{N'} f_B^{N_B} \tag{13.3}$$

where f' is the partition function for the surface molecules, f_B is that for the bulk liquid molecules, and $N = N' + N_B$. The surface tension γ is calculated from the Helmholtz free energy A as follows:

$$\gamma = (\partial A/\partial \Omega)_{N,V,T} = \omega^{-1}(\partial A/\partial N_c)_{N,V,T}. \tag{13.4}$$

Here, Ω is the surface area, N_c is the total number of cells available on the surface and is related to N' by the relation $N' = (V_s/V)N_c$, since a random distribution of holes is assumed. The symbol ω represents the area occupied by one molecule and is equal to $\sqrt{3}/2\ (V_s\sqrt{2}/N)^{2/3}$ for close packing. For monatomic liquids such as Ar and Ne, Eq. (13.4) leads to

$$\gamma \simeq \omega^{-1}\left(\frac{V_s}{V}\right)^2 kT \frac{E_s - E_s'}{RT} + 3 \ln \frac{1 - \exp(-\theta'/T)}{1 - \exp(-\theta/T)}. \tag{13.5}$$

Here, the prime (′) denotes the values in the surface layer, E_s' has the value $(9/12)E_s$, and θ' is approximately $(9/12)^{1/2}\theta$. Very recently, both techniques explained above have been extended to the calculation of surface tension of liquid mixtures with success (Kim *et al.*, 1968, 1969).

XIV. The Dielectric Constant of Hydrogen-Bonded Liquids

The fundamental electrostatic equations to predict the dielectric constant is

$$(\varepsilon - 1)/4\pi = P/E. \tag{14.1}$$

Here ε is the dielectric constant of the liquid, P the polarization of the dielectric, and E the field intensity. However, since dielectric phenomena incorporate the dielectric relaxation process, an acceptable model should explain both the dielectric constant and the dielectric relaxation.

This brings us to a consideration of the problem of hydrogen-bonded liquids. We were led to a domain theory of the dielectric constants of hydrogen-bonded liquid in which the following assumptions are made.

(1) Liquid water and other hydrogen-bonded liquids are made up of domains with the dipoles having average resultant moment $U \cos^2 \theta$ along the direction of maximum polarization for the domain, while the direction of maximum polarization of neighboring domains tend to be rotated through 180° with respect to the first in the same way magnets juxtapose south poles against north poles.

(2) When an electric field is applied, favorably oriented domains grow at the expense of the less favorably oriented domains until a steady state is reached.

Since the relaxation process only involves molecules at the interface between domains, the rate of relaxation measures the size of domains. The use of the foregoing assumptions and the concepts of significant structure theory lead to the following basic dielectric constant equation, which has been applied to various liquids including water (Hobbs *et al.*, 1966), various form of ice (Hobbs *et al.*, 1966), steam (Jhon and Eyring, 1968b), alcohols (Jhon *et al.*, 1967), a variety of liquid mixtures (Jhon and Eyring, 1968a), and electrolyte solution (Sung and Jhon, 1970a):

$$\frac{(\varepsilon - n^2)(2\varepsilon + n^2)}{3\varepsilon} = 4\pi \frac{N}{V}\left(\frac{n^2 + 2}{3}\right)^2 \left(\frac{V_s}{V}\frac{\mu^2 G}{kT} + \frac{V - V_s}{V}\frac{\mu^2}{3kT}\right) \tag{14.2}$$

where $G = \cos^2 \theta$ and is close to unity. Here ε, n, and μ are the dielectric constant, the index of refraction, and the dipole moment, respectively, and V_s and V are the molar volume of the solidlike structure in the liquid and the molar volume of the liquid, respectively. Some of the results are shown in Table XI and XII. The striking agreement of the

TABLE XI

DIELECTRIC CONSTANTS OF LIGHT AND HEAVY WATER ACCORDING TO SIGNIFICANT STRUCTURE THEORY

T (°K)	ε_{obs}	ε_{calc}	V (cc)	V_s (cc)	n^2
		H_2O			
273	88	88	18.0	17.85	1.78
373	56	60	18.8	17.7	1.74
473	35	38	20.8	17.7	1.64
573	20	22	25.3	17.7	1.51
		D_2O			
278	85.8	85.3	18.11	17.88	1.78
293	80.1	80.9	18.13	17.85	1.78
313	73.1	75.4	18.20	17.85	1.77
333	66.7	69.8	18.36	17.85	1.76

TABLE XII

DIELECTRIC CONSTANT OF THE WATER–METHANOL SYSTEM IN VARIOUS PROPORTIONS

Mole % (MeOH)	283°K		298°K		313°K	
	ε_{calc}	ε_{obs}	ε_{calc}	ε_{obs}	ε_{calc}	ε_{obs}
0	83.02	83.83	78.52	78.48	74.18	73.15
0.2432	—	—	62.72	65.55	—	—
0.2725	63.79	66.05	—	—	55.84	56.20
0.4575	54.09	56.20	50.40	51.67	46.96	47.40
0.8350	39.86	41.55	35.89	37.91	34.12	34.60
1.0000	34.75	35.75	32.68	32.61	30.37	29.80

theoretical and experimental dielectric constants over a wide range of temperature and density is a striking confirmation of a model which previous theories have failed to achieve. Discussion of the dielectric relaxation is omitted here but may be seen in the original paper (Hobbs *et al.*, 1966).

XV. Two-Dimensional Liquids

Very few theories of two-dimensional liquids have been developed. Examples are Devonshire's cell theory (Devonshire, 1937) and the scaled particle theory (Frisch, 1964). Recently, the significant structure theory also was applied to describe the two-dimensional liquid of hard disks (Wang *et al.*, 1965) since this system is a convenient and meaningful test of the two-dimensional liquid theory. The partition function for a two-dimensional hard disk liquid is given as

$$f_N = \left(\frac{2\pi mkT}{h^2}\right)^N \left\{a_f\left(1 + \frac{n(A - A_s)}{A_s}\right)\right\}^{NA_s/A} \left(\frac{eA}{N}\right)^{N(A-A_s)/A} \tag{15.1}$$

Here a_f is the free area being given by $a_f = 0.05552a^2$ (a being the distance of nearest neighbors). A is the molar area and is equal to $N(\sqrt{3}/2)a^2$ for close packing. A_s is the molar area for the two dimensional solid.

The solid–liquid phase transition for a system of hard disk is observed at $A_s = 1.321A_0$, (Alder and Wainwright, 1960), where $A_0 = N(\sqrt{3}/2)\sigma^2$ for the close packed area (σ is the collision diameter). Equation (15.1) with $n = 6$ has been applied to evaluate the equation of state and the excess entropy for a system of rigid disks. Some of the results are shown in Fig. 9. The agreement with the machine calculations (Alder and Wainwright, 1960) is quite good.

Another interesting application of the two-dimensional liquid is concerned with the problem of the physical adsorption of gases on solids. Recently, McAlpin and Pierotti (1964) applied the significant structure theory of liquids to a simplified system whose surface is a structureless uniform plane and considered that physical adsorption occurs in the submonolayer region and that the adsorbed phase is a two-dimensional fluid. The partition function of a two-dimensional adsorbed fluid is accordingly written as

$$f_{\text{ads}} = \left\{f_{2s}\left(1 + n_h \exp\left(-\frac{\varepsilon_a}{RT}\right)\right)\right\}^{NA_s/A} f_{2g}^{N(A-A_s)/A} \tag{15.2}$$

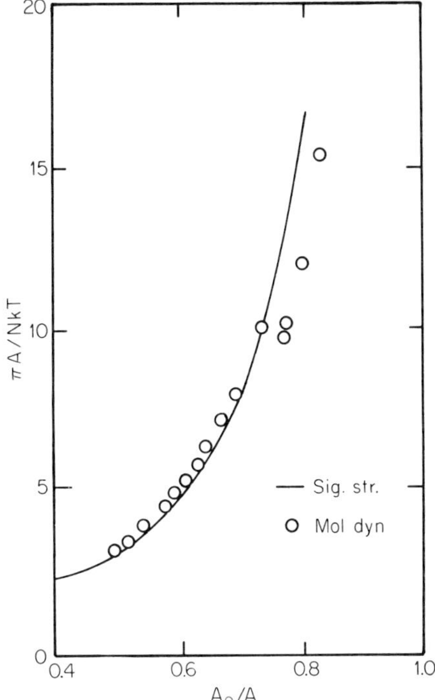

FIG. 9. Compressibility factor for a system of rigid disks (two-dimensional liquid). (After Wang *et al.*, 1965.)

where f_{ads}, f_{2s}, and f_{2g} are the partition functions for the adsorbed phase, the two-dimensional solidlike structure, (including adsorbate–adsorbent interaction terms), and the two-dimensional gaslike structure (including adsorbate–adsorbent interaction terms), respectively.

A_s is the molar area of the solid adsorbate, A is the molar area of the adsorbent phase, and N is the total number of molecules in the adsorbed phase. In Eq. (15.2), the following relations are obtained:

$${}_{2s} = \left(\frac{\exp(-\theta_E/2T)}{1-\exp(-\theta_E/T)}\right)^2 \exp\left(\frac{U_0+W}{RT}\right)\left(\frac{\exp(-h\nu_0/kT)}{1-\exp(-h\nu_0/kT)}\right) \tag{15.3}$$

$$f_{2g} = \left(\frac{2\pi mkT}{h^2}\right)\frac{eA}{N}\exp\left(\frac{U_0}{RT}\right)\left(\frac{\exp(-h\nu_0/kT)}{1-\exp(-h\nu_0/kT)}\right) \tag{15.4}$$

and

$$n_h \exp(-\varepsilon_a/RT) = 6[(A - A_s)/A]. \tag{15.4}$$

Here, W is the two-dimensional lattice energy obtained by summing the Sinanoglu and Pitzer modification (1960) of the Lennard-Jones 6-12 potential over the 18 nearest neighbors in the hexagonal lattice and integrating over the rest of the lattice; θ_E is the Einstein characteristic temperature obtained by multiplying the ratio of the perturbed and unperturbed frequencies determined from the curvature of the potential energy curves by the Einstein temperature of the bulk adsorbate; U_0 is the adsorbate adsorbent interaction energy; ν_0 is the frequency of the adsorbate molecule vibrating normal to the surface; ε_a is assumed to be negligible in the calculations.

By equating the chemical potentials for the adsorbed phase and the gas phase, the following adsorption isotherm is derived:

$$\ln P = \ln[(2\pi mkT)^{3/2}h^{-3}kT] - 2\theta \ln\{f_{2s}/f_{2g})[1 + 6(1-\theta)]\} - \theta^2 \frac{7 + 12\theta}{1 + 6(1-\theta)} - \ln f_{2g} + 1. \tag{15.5}$$

Here, the partition function for a monatomic gas has been used; P is the gas pressure; θ is the fraction of the adsorbed surface, and A_s is defined as A at $\theta = 1$ since $A = A_s\theta$. The isosteric heat of adsorption q_{st} is given by

$$q_{st} = RT^2\left(\frac{\partial \ln P}{\partial T}\right)_\theta. \tag{15.6}$$

The detailed expression for Eq. (15.6) is omitted here. Figures 10 and

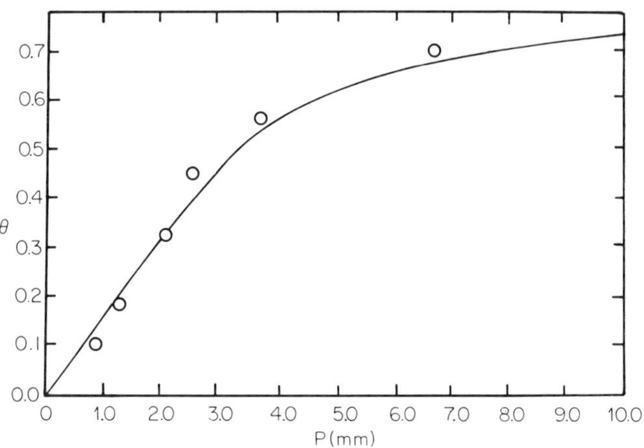

FIG. 10. Argon–graphite isotherm at 90.1°K for Ar on P-33. Solid line is theoretical. Points are experimental data. (After McAlpin and Pierotti, 1964.)

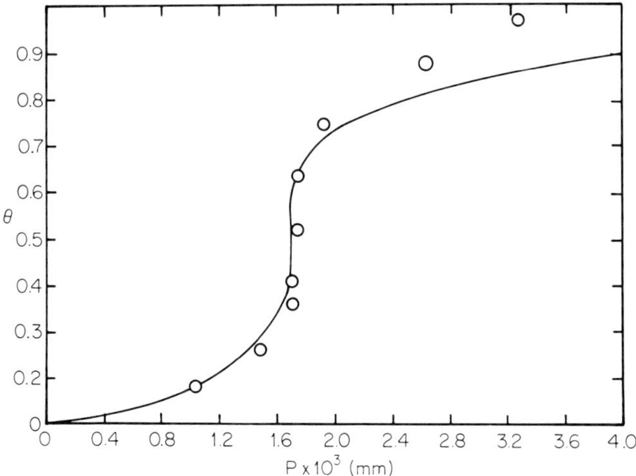

FIG. 11. Krypton–graphite isotherm at 77.8°K for Kr on P-33. Solid line is theoretical. Points are experimental data. (After McAlpin and Pierotti, 1964.)

11 compare the theoretical isotherm with experiment. The agreement for both cases is very good, but the results for Kr are not as good as for Ar.

ACKNOWLEDGMENT

The authors wish to thank the U.S. Army (Durham) Grant DA ARO-D-31-1240G1110 for support of this research.

REFERENCES

AHN, W. S., PAK, H., and CHANG, S. (1965). *J. Korean Chem. Soc.* **9**, 215.
ALDER, B. J., and WAINWRIGHT, T. E. (1957). *J. Chem. Phys.* **27**, 1208.
ALDER, B. J., and WAINWRIGHT, T. E. (1960). *J. Chem. Phys.* **33**, 1439.
BAHNG, J. S., and JHON, M. S. (1970). *J. Korean Chem. Soc.* **14**, 193.
BARKER, J. S. (1961). *Proc. Roy. Soc.* (*London*) **A259**, 442.
BERNAL, J. D., and FOWLER, R. H. (1933). *J. Chem. Phys.* **67**, 1304.
BEUCHE, F. (1959). *J. Chem. Phys.* **30**, 748.
BOGOLYUBOV, N. N. (1946). "Problemy Dinamischeskoi Teorii Statisticheskoi Fizike." Gostekhizdet, Moscow.
CALLAWAY, J. (1959). *Phys. Rev.* **113**, 1046.
CARLSON, C. M., EYRING, H., and REE, T. (1960a). *Proc. Natl. Acad. Sci.* (*U.S.*) **46**, 333.
CARLSON, C. M., EYRING, H., and REE, T. (1960b). *Proc. Nat. Acad. Sci.* (*U.S.*) **46**, 649.
CHANG, S., REE, T., EYRING, H., and MATZNER, I. (1962). *In* "Progress in International Research on Thermodynamics and Transport Properties" (J. F. Masi and D. H. Tsai, eds.), p. 88. Princeton Univ. Press, Princeton, New Jersey.

Cohen, M. H., and Turnbull, D. (1950). *J. Chem. Phys.* **29**, 1049.
Cohen, M. H., and Turnbull, D. (1959). *J. Chem. Phys.* **31**, 1164.
Devonshire, A. F. (1937). *Proc. Roy. Soc.* (*London*) **A163**, 132.
Doolittle, A. (1951). *J. Appl. Phys.* **22**, 1471.
Doolittle, A. (1952). *J. Appl. Phys.* **23**, 238.
Eisenstein, A., and Gingrich, N. (1942). *Phys. Rev.* **62**, 261.
Eucken, A. (1947). *Nachr. Ges. Wiss. Gottingen* 33.
Eyring, H. (1936). *J. Chem. Phys.* **4**, 283.
Eyring, H., and Jhon, M. S. (1969). "Significant Liquid Structures." Wiley, New York.
Eyring, H., and Ree, T. (1961). *Proc. Nat. Acad. Sci.* (*U.S.*) **47**, 526.
Eyring, H., Ree, T., and Hirai, N. (1958). *Proc. Nat. Acad. Sci.* (*U.S.*) **44**, 683.
Frank, H. S., and Wen, W. Y. (1957). *Discuss. Faraday Soc.* **24**, 133.
Frisch, H. L. (1964). *Advan. Chem. Phys.* **6**, 229–289.
Grosh, J., Jhon, M. S., Ree, T., and Eyring, H. (1967). *Proc. Nat. Acad. Sci.* (*U.S.*) **57**, 1566.
Henderson, D. (1963). *J. Chem. Phys.* **39**, 1857.
Henderson, D., Eyring, H., and Felix, D. (1962). *J. Phys. Chem.* **66**, 1128.
Hirschfelder, J. O., Curtiss, C. F., and Bird, R. B. (1964). "Molecular Theory of Gases and Liquids." Wiley, New York.
Hobbs, M. E., Jhon, M. S., and Eyring, H. (1966). *Proc. Nat. Acad. Sci.* (*U.S.*) **56**, 31.
Jhon, M. S. (1967). *J. Korean Chem. Soc.* **11**, 60.
Jhon, M. S., Clemena, G., and Van Artsdalen, E. R. (1968). *J. Phys. Chem.* **72**, 4155.
Jhon, M. S., and Eyring, H. (1968a). ACS Southeastern Regional Meeting Dec. 4–7. FSU, Tallahassee, Fla. H. Eyring *et al.* (to be published).
Jhon, M. S., and Eyring, H. (1968b). *J. Amer. Chem. Soc.* **90**, 3071.
Jhon, M. S., Grosh, J., Ree, T., and Eyring, H. (1966). *J. Chem. Phys.* **44**, 1465.
Jhon, M. S., Klotz, W. L., and Eyring, H. (1969). *J. Chem. Phys.* **51**, 3692.
Jhon, M. S., Van Artsdalen, E. R., Grosh, J., and Eyring, H. (1967). *J. Chem. Phys.* **47**, 2231.
Kim, S. W., Eyring, H., and Lee, Y. T. (1969). *J. Chem. Phys.* **51**, 3967.
Kim, S. W., Jhon, M. S., Ree, T., and Eyring, H. (1968). *Proc. Nat. Acad. Sci.* (*U.S.*) **59**, 336.
Kirkwood, J. G. (1950). *J. Chem. Phys.* **18**, 380.
Kirkwood, J., and Boggs, E. (1942). *J. Chem. Phys.* **10**, 394.
Kirkwood, J., and Monroe, E. (1941). *J. Chem. Phys.* **9**, 514.
Klemens, P. G. (1951). *Proc. Roy. Soc.* (*London*). **A208**, 108.
Lennard-Jones, J. E., and Devonshire, A. F. (1937a). *Proc. Roy. Soc.* (*London*) **163**, 53.
Lennard-Jones, J. E., and Devonshire, A. F. (1937b). *Proc. Roy. Soc.* **169A**, 317.
Lennard-Jones, J. E., and Devonshire, A. F. (1938). *Proc. Roy. Soc.* (*London*) **165**, 1.
Liang, K., Eyring, H., and Marchi, R. P. (1964). *Proc. Nat. Acad. Sci.* (*U.S.*) **52**, 1107.
Lin, S. H., Eyring, H., and Davis, W. J. (1964). *J. Phys. Chem.* **68**, 3017.
Loeb, L. B. (1927), "Kinetic Theory of Gases." McGraw Hill, New York.
Lu, W. C., Jhon, M. S., Ree, T., and Eyring, H. (1967). *J. Chem. Phys.* **46**, 1075.
Lu, W. C., Ree, T., Gerrard, G., and Eyring, H. (1968). *J. Chem. Phys.* **49**, 797.
Ma, S. M., and Eyring, H. (1965). *J. Chem. Phys.* **42**, 1920.

McAlpin, J. J., and Pierotti, R. A. (1964). *J. Chem. Phys.* **41**, 68.
McCedo, P., and Litovitz, T. A. (1965). *J. Chem. Phys.* **42**, 245.
McLaughlin, D. R., and Eyring, H. (1966). *Proc. Nat. Acad. Sci.* (*U.S.*) **55**, 1031.
McQuarrie, D. A. (1962). *J. Phys. Chem.* **66**, 1508.
Marchi, R. P., and Eyring, H. (1964). *J. Phys. Chem.* **68**, 221.
Miner, B. A., and Eyring, H. (1965). *Proc. Nat. Acad. Sci.* (*U.S.*) **53**, 1227.
Nemethy, G., and Scheraga, G. H. (1962). *J. Chem. Phys.* **36**, 3382.
Onno, S., and Kondo, S. (1960). "Handbuch der Physik" (S. Flügge, ed.), Vol. X, "Structure der Flussikeiten," p. 134. Springer Verlag, Berlin.
Peierls, R. (1929). *Ann. Phys.* **3**, 1055.
Pierotti, R. A. (1965). *J. Chem. Phys.* **43**, 1072.
Pople, J. A. (1951). *Proc. Roy. Soc.* **A205**, 163.
Raman, C. V., and Ramdas, L. A. (1927). *Phil. Mag.* **3**(7), 220.
Ree, T. S., and Eyring, H. (1964). *Proc. Nat. Acad. Sci.* (*U.S.*) **51**, 3441.
Ree, T. S., Ree, T., and Eyring, H. (1962). *Proc. Nat. Acad. Sci.* (*U.S.*) **48**, 501.
Ree, T. S., Ree, T., and Eyring, H. (1964a). *J. Chem. Phys.* **41**, 524.
Ree, T. S., Ree, T., and Eyring, H. (1964). *J. Phys. Chem.* **68**, 3262.
Ree, T. S., Ree, T., Eyring, H., and Perkins, R. (1965). *J. Phys. Chem.* **69**, 3322.
Singanoglu, O., and Pitzer, K. S. (1960). *J. Chem. Phys.* **32**, 1279.
Sung, Y. K., and Jhon, M. S. (1970a). *J. Korean Chem. Soc.* **14**, 185.
Sung, Y. K., and Jhon, M. S. (1970b). *J. Oceanolog. Soc. Korea.* **5**, 1.
Vilen, R., and Misdolea, C. (1966). *J. Chem. Phys.* **45**, 3414.
Wall, T. F., and Hornig, D. F. (1965). *J. Chem. Phys.* **43**, 2079.
Walter, J. and Eyring, H. (1941). *J. Chem. Phys.* **9**, 393.
Wang, Y. L., Ree, T., Ree, T. S., and Eyring, H. (1965). *J. Chem. Phys.* **42**, 1926.
Wood, W. W., and Jacobson, J. D. (1957). *J. Chem. Phys.* **27**, 1207.

Chapter 6

Perturbation Theories

DOUGLAS HENDERSON AND J. A. BARKER

I. Introduction

As has been pointed out in the earlier chapters of this volume, the thermodynamic properties of a classical simple fluid consisting of N molecules can be calculated from the partition function:

$$Z_N = \lambda^{-3N}(1/N!) \int \exp\{-\beta\Phi_N\} \, d\mathbf{r}_1 \cdots d\mathbf{r}_N, \tag{1.1}$$

where $\lambda = h/(2\pi mkT)^{1/2}$ (m the molecular mass and T the absolute temperature), $\beta = 1/kT$, Φ is the potential energy of the molecules, $\mathbf{r}_n$ is the position of the nth molecule, and h and k are Planck's and Boltzmann's constants, respectively. The potential energy can be written as a sum of pair, triplet, quadruplet, etc., interactions:

$$\Phi_N(\mathbf{r}_1, \ldots, \mathbf{r}_N) = \sum_{i<j} u(ij) + \sum_{i<j<k} u(ijk) + \cdots, \tag{1.2}$$

where $u(ij) = u(R_{ij})$ is the pair potential of molecules i and j, etc., and $R_{ij} = | \mathbf{r}_i - \mathbf{r}_j |$.

It is usually assumed that only pair interactions are important, i.e.,

$$\Phi_N(\mathbf{r}_1, \ldots, \mathbf{r}_N) = \sum_{i<j} u(ij). \tag{1.3}$$

This is known to be false for real systems (Barker and Pompe, 1968). However, in simple fluids the pair interactions provide most of the potential energy. Thus, Eq. (1.3) provides a point of departure, and we use this assumption throughout much of this chapter. However, we also consider the effects of multibody forces. Indeed, one of the most attractive aspects of perturbation theory is that multibody forces can be incorporated into the theory without excessive difficulty.

The theory of fluids is most fully developed for the hard-sphere potential:

$$\begin{aligned} u(R) &= \infty, & r &< \sigma, \\ &= 0, & r &> \sigma. \end{aligned} \tag{1.4}$$

In the perturbation theory of liquids the hard-sphere results are used to obtain the properties of fluids in which the pair potential has some other form.

The fundamental concepts of the perturbation theory of liquids are that the structure of a simple liquid is determined primarily by the repulsive forces between molecules, and that the main effect of the attractive forces between molecules is to provide a uniform background potential in which the molecules move.

These concepts have been used for some time. They are the basis of the equation of state of van der Waals (1873). If, following van der Waals, we accept Eq. (1.3) and assume that the molecules have a hard core, i.e.,

$$\begin{aligned} u(R) &= \infty, & R &< \sigma, \\ &= u_1(R), & R &> \sigma, \end{aligned} \tag{1.5}$$

then, using these concepts, the Helmholtz free energy A of the liquid is the same as that of a system of hard spheres of diameter σ, except that it is lowered because of the background potential field. Thus,

$$A - A_0 = \tfrac{1}{2} N \psi, \tag{1.6}$$

where A_0 is the free energy of the hard-sphere gas,

$$\psi = 4\pi\varrho \int_{\sigma}^{\infty} u_1(R) g_0(R) R^2 \, dR, \tag{1.7}$$

where $\varrho = N/V$, and V is the volume of the fluid. The factor of 1/2 in Eq. (1.6) arises because the energy ψ is shared by two molecules, and hence would be counted twice if this factor were not inserted. The function $g(r)$ is the pair distribution function, and thus the average number of molecules in a spherical shell of thickness dR surrounding a molecule at the origin is zero for $R < \sigma$ and $\varrho g(R) 4\pi R^2 \, dR$ for $R > \sigma$, and Eq. (1.6) follows immediately. We have put a subscript zero on $g(R)$ in Eq. (1.7) to emphasize that, as a result of our assumptions, the pair distribution function is the same as that of a hard-sphere gas.

Van der Waals made the further assumption that the molecules are randomly distributed [i.e., $g_0(R) = 1$]. Thus,

$$\psi = -2\varrho a, \tag{1.8}$$

where

$$a = -\,2\pi \int_{\sigma}^{\infty} u_1(R) R^2 \, dR. \tag{1.9}$$

Until very recently the properties of hard spheres were not known. Thus, van der Waals had to approximate A_0 by assuming it to be the free energy of a perfect gas with V replaced by a smaller free "volume," V_{f}, because the molecules themselves occupy a finite volume. Therefore,

$$A_0/NkT = 3(\ln \lambda) - 1 - \ln V_{\mathrm{f}} + \ln N, \tag{1.10}$$

where

$$V_{\mathrm{f}} = V - Nb, \tag{1.11}$$

and

$$b = (2\pi/3)\sigma^3. \tag{1.12}$$

The factor b is given by Eq. (1.12) because when two molecules collide the center of mass of one of the molecules is excluded from a volume of $4\pi\sigma^3/3$. This excluded volume is divided by two because it is shared by two molecules.

Combining Eqs. (1.6) and (1.8)–(1.12) and differentiating with respect to V yields the van der Waals equation of state:

$$[p + (N^2 a/V^2)](V - Nb) = NkT. \tag{1.13}$$

Equation (1.13) gives results which are in poor agreement with experimental data. This is particularly true if a is calculated from Eq. (1.9). The situation can be improved somewhat by regarding a as a parameter chosen to fit some experimental data. However, even in this case the results are unsatisfactory.

Longuet-Higgins and Widom (1964) and Gueggenheim (1965a,b) have shown that the main defect in the van der Waals theory lies in the use of Eq. (1.10) for the hard-sphere free energy. It has already been pointed out in Chapter 4 that the Percus–Yevick (PY) compressibility isotherm is very good for hard spheres (Percus and Yevick, 1958; Thiele, 1963; Wertheim, 1963). Thus, Eq. (1.13) is replaced by

$$p = p_0 - (N^2 a/V^2), \tag{1.14}$$

where

$$\frac{p_0}{\varrho kT} = \frac{1 + \eta + \eta^2}{(1 - \eta)^3} \tag{1.15}$$

and $\eta = \pi\varrho\sigma^3/6 = b/4$. If a is regarded as an adjustable parameter, this isotherm is in good agreement with the experimental results for argon—indicating that Eq. (1.8) is an good approximation. However, it is to be borne in mind that the value of a which results from Eq. (1.9) is quite different from that which is required to fit experimental data.

It is interesting to note that Eqs. (1.8) and (1.9) have been obtained rigorously for a weak, long-range potential (Kac *et al.*, 1963; Uhlenbeck *et al.*, 1963; Hemmer *et al.*, 1964; Hemmer, 1964). This result may be seen intuitively from Eq. (1.7). If $u(R)$ is extremely long-ranged, then the major contribution to the integral comes from large values of R, where g_0 is unity.

Zwanzig (1954) obtained Eqs. (1.6) and (1.7) by assuming that the intermolecular potential can be written as a sum of the hard-sphere potential and a perturbation potential:

$$u(R) = u_0(R) + u_1(R), \tag{1.16}$$

where $u_0(R)$ is the hard-sphere potential for hard spheres of diameter d. If the partition function, and thus the free energy, are expanded in powers of β, then, to first-order, Eqs. (1.7) and (1.8) follow. Zwanzig, and later Smith and Alder (1959), calculated the free energy and equation of state using the 6–12 potential and the Born–Green–Yvon results (which at the time were the best available) for A_0 and g_0. More recently Frisch *et al.* (1966) made similar calculations using the Percus–Yevick (PY) results for A_0 and g_0. The results of these calculations are in quite reasonable agreement with experiment at high temperatures. However, these results are very sensitive to the choice of d, for which no satisfactory criterion is provided.

An alternative approach has been given by Rowlinson (1964a,b) for repulsive potentials. He expanded the free energy of a system of molecules with pair potential

$$u(R) = 4\varepsilon[(\sigma/R)^n - (\sigma/r)^{n/2}] \tag{1.17}$$

in powers of n^{-1}. If this approach is taken to first order, and n set equal to 12 to give the 6–12 potential, good results are obtained for the equation of state of gases at high temperatures (reduced temperatures above 12).

McQuarrie and Katz (1966) combined the Zwanzig and Rowlinson techniques by treating the attractive term in Eq. (1.17) as a perturbation on the repulsive term and treating the attractive term by means of the n^{-1} expansion. This procedure yields a satisfactory equation of state for reduced temperatures above 3.

Thus, the situation in 1967 was that much of the evidence indicated that perturbation theories appeared to work only at high temperatures. However, the work of Longuet-Higgins and Widom and the work of Riess and others on the scaled-particle theory (Frisch, 1964; Reiss, 1965) indicated that the hard-sphere fluid was an excellent reference system for the properties of liquids, even at the lowest temperatures, although firm conclusions could not be reached because of the presence of adjustable parameters in these approaches.

It is clearly important to determine whether the apparent failure of the Zwanzig approach at low temperatures is due to the perturbation approach itself or to the inadequate treatment of the finite steepness of the repulsive potential. In the square-well potential,

$$u(R) = \begin{cases} \infty, & R < \sigma, \\ -\varepsilon, & \sigma < R < 3\sigma/2, \\ 0, & R > 3\sigma/2, \end{cases} \tag{1.18}$$

the effect of the attractive forces is not complicated by the "softness" of the repulsive part of the potential, which is infinitely steep. Thus, the convergence of the perturbation expansion may be examined directly by comparison with the quasiexperimental machine calculations (Rotenberg, 1965; Alder, 1967; Lado and Wood, 1968).

For this reason the application of perturbation theory to the square-well potential is considered in the next section. We find that the perturbation treatment of the attractive forces gives good results at low temperatures. In later sections we consider extensions of perturbation theory to systems in which the repulsive potential has finite steepness and to real systems in which multibody forces and quantum effects are present.

II. Potentials with a Hard Core

A. Formal Expressions for First- and Second-Order Terms

We assume that the intermolecular potential has a hard core of diameter σ. Thus, the intermolecular potential can be written as the sum of two terms:

$$u(R) = u_0(R) + u_1(R), \tag{2.1}$$

where $u_0(R)$ is the hard-sphere potential:

$$u_0(R) = \begin{cases} \infty, & R < \sigma, \\ 0, & R > \sigma, \end{cases} \tag{2.2}$$

and $u_1(R)$ is the perturbation potential. Our procedure is to expand the partition function, and thus the free energy, in terms of u_1. This expansion has useful convergence if $u_1(R)$ is *small* in some sense. The condition is not that $u_1(R)$ should be small (compared with kT, for example), but rather that the effect of $u_1(R)$ on the *structure* of the fluid is small. As has been mentioned, we are really only interested in the square-well potential in this section, as quasiexperimental machine calculations are not available for any other potential with a hard core. However, for generality, we allow $u_1(R)$ to be arbitrary. Indeed, it might be a repulsive perturbation.

1. *First Method*

Following our earlier treatment (Barker and Henderson, 1967a; Barker *et al.*, 1969a) we suppose the range of intermolecular distances to be divided into the intervals R_0 to $R_1, \ldots, R_i$ to $R_{i+1}, \ldots$, etc., and suppose that the number of intermolecular distances in the interval R_i to R_{i+1} is N_i. By taking the limit as the interval width tends to zero, the continuous discription is recovered. However, the discussion is simpler in terms of discrete divisions. If the perturbing potential can be regarded as having the constant value $u_1(R_i)$ in the interval R_i to R_{i+1}, then the partition function can be written

$$\begin{aligned} Z_N &= Z_0 \sum_{N_1, N_2, \ldots} p(N_1, \ldots) \exp\Big\{-\beta \sum_i N_i u_1(R_i)\Big\} \\ &= Z_0 \Big\langle \exp\Big\{-\beta \sum_i N_i u_1(R_i)\Big\}\Big\rangle, \end{aligned} \tag{2.3}$$

where Z_0 is the hard-sphere partition function, $p(N_1, \ldots)$ is the probability that the hard-sphere system has N_i intermolecular distances in the interval R_i to R_{i+1}, etc. The angular brackets mean "average over the configurations of the unperturbed system."

Expanding the exponential in Eq. (2.3) yields

$$Z_N = Z_0\Big\{1 - \beta \sum_i \langle N_i\rangle u_1(R_i) + \tfrac{1}{2}\beta^2 \sum_{ij} \langle N_i N_j\rangle u_1(R_i) u_1(R_j) + \cdots\Big\}. \tag{2.4}$$

Thus, taking the logarithm and expanding, we find

$$\begin{aligned} (A - A_0)kT &= \beta \sum_i \langle N_i\rangle u_1(R_i) \\ &\quad - \tfrac{1}{2}\beta^2 \sum_{ij} (\langle N_i N_j\rangle - \langle N_i\rangle\langle N_j\rangle) u_1(R_i) u_1(R_j) + \cdots. \end{aligned} \tag{2.5}$$

We may regard the perturbation expansion as an expansion in powers of β.

It is worth noting that for the special case of the square-well potential,

$$u_1(R) = \begin{cases} -\varepsilon, & \sigma < R < 3\sigma/2, \\ 0, & \text{elsewhere}, \end{cases} \tag{2.6}$$

a single interval with R between σ and $3\sigma/2$ could be chosen and each of the sums in Eq. (2.5) would contain only one term.

The $\langle N_i\rangle$ are, of course, related to the pair distribution function $g_0(R)$. Thus,

$$\langle N_i\rangle = 2\pi N\varrho \int_{R_i}^{R_{i+1}} R^2 g_0(R)\, dR. \tag{2.7}$$

Hence, the first-order term is the same as that given in Eqs. (1.6) and (1.7). As we shall see, the $\langle N_i N_j\rangle$ are related to the hard-sphere three- and four-body distribution functions.

The first-order term makes no contribution to the entropy, and thus, to first-order in the free energy, the structure of the fluid is unaffected and the only effect of the attractive forces is to provide a background potential in which the molecules move as hard spheres. The second-order term represents the effect of the attractive potential on the structure of the fluid in compressing the molecules into energetically favorable regions. We would expect the effect of the second- and higher-order terms to be least important at high densities, where the fluid is nearly incompressible and changes in structure difficult. Thus, we expect perturbation theory to be most useful in the most interesting region of high densities.

There is one interesting low-density region. This is the critical-point region. In this region the convergence of the perturbation expansion must be slow, and thus, although formally correct, perturbation theory is not useful unless drastic simplifying assumptions (such as the molecules having no kinetic energy and occupying the sites of a lattice) are made. Since the critical region is discussed in Chapter 10 and 11, we will not pursue these topics further.

We have evaluated the $\langle N_i \rangle$ and $\langle N_i N_j \rangle$ by the Monte Carlo (MC) method. The distances R_λ were chosen to be given by $R_\lambda = (1 + 0.07\lambda)^{1/2} d$, where d is the diameter of the hard spheres and $\lambda = 1, 2, \ldots, 60$. In the present application $d = \sigma$. However, we will see that for potentials with a soft core d will be different from σ. A number of molecules (108 in this work) were confined to a cubic unit cell by periodic boundary conditions. The molecules were started in a cubic close-packed arrangement and chains of about 10^6 configurations were generated by tentatively displacing a molecule chosen at random by the vector (x, y, z), where x, y, and z are random numbers uniformly distributed on the interval $(-\Delta, \Delta)$. If, in its new position the molecule lies within a distance d of any other molecule, the new configuration is identical with the old; otherwise, the new configuration is the one with the chosen molecule displaced. For hard spheres it is known (Metropolis *et al.*, 1953; Wood, 1968) that in the limit of long chains averaging over the configurations of such a chain is equivalent to averaging in the canonical ensemble. In our calculations the number Δ was chosen by preliminary calculations so that the displaced configuration was acceptable on roughly half the occasions. In order to minimize the effect of the initial configuration, the first 10,000 configurations were rejected in calculating averages. In calculating intermolecular distances we used the "minimum-image-distance" convention (Wood and Parker, 1957), according to which the distance between molecules i and j is the distance between i and the nearest periodic image of j. This convention introduces errors for distances greater than half the edge of the cubic cell. At all relevant densities this distance is greater than $(5.2)^{1/2}d$, which is the greatest distance for which we used the Monte Carlo results (see below).

Since the hard sphere diameter d has been left arbitrary and since we have divided R into small intervals, these calculations are completely general and can be applied to any potential. To minimize the effect of the finite width of the intervals, we averaged $u(R)$ over each interval using the PY hard-sphere distribution function as weighting function before using Eq. (2.5) for a continuous potential. For distances greater than $(5.2)^{1/2}d$ we used PY values of $\langle N_\lambda \rangle$, computed from Eq. (2.7) using the PY $g_0(R)$. In evaluating the second-order term we neglected all contributions from $R > (5.2)^{1/2}d$. This is certainly a good approximation, because the long-range part of the potential is not only small, but also fluctuates very little.

2. *Second Method*

We consider a system of molecules such that $u(R)$ depends on a parameter γ:

$$u(R) = u(\gamma; R). \tag{2.8}$$

We may then obtain a perturbation series, similar to Eq. (2.5), by expanding the free energy in powers of γ:

$$A = A_0 + \gamma(\partial A/\partial\gamma)_{\gamma=0} + \tfrac{1}{2}\gamma^2(\partial^2 A/\partial\gamma^2)_{\gamma=0} + \cdots, \tag{2.9}$$

and setting $\gamma = 1$. This is essentially the method used by Zwanzig (1954), who assumed the particular form

$$u(R) = u_0(R) + \gamma u_1(R). \tag{2.10}$$

Equation (2.10) is appropriate for potentials with a hard core. However, we include potentials of the form Eq. (2.8) for generality.

Zwanzig obtained results for $\partial A/\partial\gamma$ and $\partial^2 A/\partial\gamma^2$. Unfortunately, his result for the second-order term, although formally correct, is not useful for numerical compution because it is obtained for the canonical ensemble and is valid only for a finite system. To obtain results which are useful, one must take the *thermodynamic limit* ($N \to \infty$, N/V fixed), and, unfortunately, this involves the unknown asymptotic behavior of the four-body distribution function when two of the molecules involved are remote from the other two.

For this reason we derive the necessary result in the grand canonical ensemble, where there are no problems with the asymptotic behavior of the distribution functions. The results for the canonical ensemble can then be derived by suitable transformations.

In the grand canonical ensemble the probability that there are molecules in the element $d\mathbf{r}_1 \cdots d\mathbf{r}_h$ is

$$\varrho_h(1 \cdots h) = \frac{1}{\Xi} \sum_{N=h}^{\infty} \frac{\exp(N\beta\mu)}{(N-h)!} \lambda^{-N} \int \exp(-\beta\Phi_N)\, d\mathbf{r}_{h+1} \cdots d\mathbf{r}_N, \quad (2.11)$$

where Ξ is the grand partition function:

$$\Xi = \sum_{N=0}^{\infty} \frac{\exp(N\beta\mu)}{N!} \lambda^{-N} \int \exp(-\beta\Phi_N)\, d\mathbf{r}_1 \cdots d\mathbf{r}_N, \quad (2.12)$$

which is related to the pressure of the fluid by

$$pV = kT \ln \Xi. \quad (2.13)$$

Differentiating, we find

$$[\partial(\ln \Xi)/\partial\gamma]_\mu = -\tfrac{1}{2}\beta \int \varrho_2(12) u_1(12)\, d\mathbf{r}_1\, d\mathbf{r}_2, \quad (2.14)$$

where $u_1(R) = \partial u(R)/\partial\gamma$. Differentiating once more,

$$\begin{aligned}
[\partial^2(\ln \Xi)/\partial\gamma^2]_\mu = &- \tfrac{1}{2}\beta \int \varrho_2(12)[\partial u_1(12)/\partial\gamma]\, d\mathbf{r}_1\, d\mathbf{r}_2 \\
&+ \tfrac{1}{2}\beta^2 \int \varrho_2(12) u_1^2(12)\, d\mathbf{r}_1\, d\mathbf{r}_2 \\
&+ \beta^2 \int \varrho_3(123) u_1(12) u_1(23)\, d\mathbf{r}_1\, d\mathbf{r}_2\, d\mathbf{r}_3 \\
&+ \tfrac{1}{4}\beta^2 \int [\varrho_4(1234) - \varrho_2(12)\varrho_2(34)] \\
&\times u_1(12) u_1(34)\, d\mathbf{r}_1\, d\mathbf{r}_2\, d\mathbf{r}_3\, d\mathbf{r}_4. \quad (2.15)
\end{aligned}$$

These results are sufficient to provide an expansion of the pressure to second order in γ at constant chemical potential.

The derivatives of the free energy can now be evaluated by essentially thermodynamic arguments, treating the parameter γ as a thermodynamic variable. We have

$$A = \bar{N}\mu - pV = \bar{N}\mu - kT \ln \Xi \quad (2.16)$$

and

$$\bar{N} = kT[\partial(\ln \Xi)/\partial\mu]_\gamma. \quad (2.17)$$

Thus,

$$\begin{aligned}[\partial A/\partial\gamma)_\varrho &= \bar{N}(\partial\mu/\partial\gamma)_\varrho - kT[\partial(\ln \Xi)/\partial\gamma]_\varrho \\ &= \bar{N}(\partial\mu/\partial\gamma)_\varrho - kT[\partial(\ln \Xi)/\partial\gamma]_\mu - kT[\partial(\ln \Xi)/\partial\mu]_\gamma(\partial\mu/\partial\gamma)_\varrho \\ &= -\, kT[\partial(\ln \Xi)/\partial\gamma]_\mu. \end{aligned} \tag{2.18}$$

Substituting Eq. (2.14) into Eq. (2.18) and introducing the conventional normalized pair distribution function $g(12) = \varrho_2(12)/\varrho^2$ gives

$$(\partial A/\partial\gamma)_{\gamma=0} = \tfrac{1}{2}N\varrho \int g_0(12)u_1(12)\, d\mathbf{r}_2. \tag{2.19}$$

We could have obtained this result directly in the canonical ensemble.

For the second derivative we obtain, using Eq. (2.18),

$$\left(\frac{\partial^2 A}{\partial\gamma^2}\right)_\varrho = -\, kT\left[\frac{\partial^2(\ln \Xi)}{\partial\gamma^2}\right]_\mu - kT\left[\frac{\partial^2(\ln \Xi)}{\partial\gamma\,\partial\mu}\right](\partial\mu/\partial\gamma)_\varrho. \tag{2.20}$$

Differentiating Eq. (2.17) yields

$$d\bar{N} = kT\left[\frac{\partial^2(\ln \Xi)}{\partial\gamma^2}\right] d\mu + kT\left[\frac{\partial^2(\ln \Xi)}{\partial\gamma\,\partial\mu}\right] d\gamma. \tag{2.21}$$

Hence,

$$\left(\frac{\partial\mu}{\partial\gamma}\right)_\varrho = -\,\frac{\partial^2(\ln \Xi)}{\partial\gamma\,\partial\mu} \Big/ \frac{\partial^2(\ln \Xi)}{\partial\mu^2}. \tag{2.22}$$

Substituting Eq. (2.22) into Eq. (2.20) yields

$$\left(\frac{\partial^2 A}{\partial\gamma^2}\right)_\varrho = -\, kT\left[\frac{\partial^2(\ln \Xi)}{\partial\gamma^2}\right]_\mu + \frac{kT[\partial^2(\ln \Xi)/\partial\gamma\,\partial\mu]^2}{\partial^2(\ln \Xi)/\partial\mu^2}. \tag{2.23}$$

Now,

$$kT\,\frac{\partial^2(\ln \Xi)}{\partial\gamma\,\partial\mu} = -\,\frac{\partial}{\partial\mu}\left[\frac{1}{2}\, N\varrho \int g(12)u_1(12)\, d\mathbf{r}_2\right] \tag{2.24}$$

and

$$kT\, \partial^2(\ln \Xi)/\partial\mu^2 = \partial\bar{N}/\partial\mu. \tag{2.25}$$

Also,

$$\partial/\partial\mu = (\partial\varrho/\partial\mu)\, \partial/\partial\varrho = \varrho(\partial\varrho/\partial p)\, \partial/\partial\varrho. \tag{2.26}$$

Hence,

$$\left(\frac{\partial^2 A}{\partial \gamma^2}\right)_\varrho = -kT\left[\frac{\partial^2(\ln \Xi)}{\partial \gamma^2}\right]_\mu$$

$$+ N\left(\frac{\partial \varrho}{\partial p}\right)_\gamma \left[\frac{\partial}{\partial \varrho}\left(\frac{1}{2}\varrho^2 \int g(12)u_1(12)\, d\mathbf{r}_2\right)\right]^2. \qquad (2.27)$$

Substituting Eq. (2.5) into Eq. (2.27) and introducing the conventional normalized distribution functions $g(1\cdots h) = \varrho_h(1\cdots h)/\varrho^h$ yields

$$\left(\frac{\partial^2 A}{\partial \gamma^2}\right)_{\gamma=0} = \frac{1}{2} N\varrho \int g_0(12)\frac{\partial u_1(12)}{\partial \gamma}\, d\mathbf{r}_2 - \frac{1}{2}\beta N\varrho \int g_0(12)[u_1(12)]^2\, d\mathbf{r}_2$$

$$- \beta N\varrho^2 \int g_0(123)u_1(12)u_1(23)\, d\mathbf{r}_2\, d\mathbf{r}_3$$

$$- \frac{1}{4}\beta N\varrho^3 \int [g_0(1234) - g_0(12)g_0(34)]u_1(12)u_1(34)\, d\mathbf{r}_2\, d\mathbf{r}_3\, d\mathbf{r}$$

$$+ N\left(\frac{\partial \varrho}{\partial p}\right)_\gamma \left[\frac{\partial}{\partial \varrho}\left(\frac{1}{2}\varrho^2 \int g_0(12)u_1(12)\, d\mathbf{r}_2\right)\right]^2. \qquad (2.28)$$

This is the desired expression.

Since the perturbation function $u_1(12)$ is arbitrary, Eq. (2.19) in conjunction with Eq. (2.28) gives an expression for the first derivative of the radial distribution function with respect to γ:

$$\left[\frac{\partial g(12)}{\partial \gamma}\right]_{\gamma=0} = -\beta g_0(12)u_1(12) - 2\beta\varrho \int g_0(123)u_1(23)\, d\mathbf{r}_3$$

$$- \frac{1}{2}\beta\varrho^2 \int [g_0(1234) - g_0(12)g_0(34)]u_1(34)\, d\mathbf{r}_3\, d\mathbf{r}_4$$

$$+ \frac{2}{\varrho}\left(\frac{\partial \varrho}{\partial p}\right)\left\{\frac{\partial}{\partial \varrho}\left[\frac{1}{2}\varrho^2 g_0(12)\right]\right.$$

$$\left.\times \frac{\partial}{\partial \varrho}\left[\frac{1}{2}\varrho^2 \int g_0(34)u_1(34)\, d\mathbf{r}_4\right]\right\}. \qquad (2.29)$$

Thus, our results are sufficient to give the perturbed free energy to second order and the perturbed radial distribution function to first order in γ.

Zwanzig (1954) gave an expression for the second derivative of $\ln Z_N$ which is identical to Eq. (2.15). This is correct if the ϱ_h are those for a *finite* system. However, for a finite system $\varrho_4(1234)$ approaches a limit given by

$$\varrho_4(1234) \to \varrho_2(12)\varrho_2(34)[1 + (q/N) + O(1/N^2)] \qquad (2.30)$$

when the pairs (12) and (34) are widely separated. In the thermodynamic limit the term of order $1/N$ makes a finite contribution, which is in fact just the last term of Eq. (2.28).

Earlier Barker (1957) considered this point, but assumed that q was simply a number, which was then determined by an indirect argument. This is incorrect because q depends on the relative coordinates of molecules 1 and 2 and of 3 and 4. The effect of this error is to omit the terms involving $\partial g_0/\partial \varrho$ in Eq. (2.28).

We obtained Eqs. (2.28) and (2.29) independently. However, since obtaining these formulas it has been pointed out to us that they may be constructed from results given in a paper of Buff and Schindler (1958). However, we have given our derivation here, as we believe it to be more straightforward.

Equations (2.28) and (2.29) are suitable for numerical calculations because in the grand canonical ensemble $g(1234) - g(12)g(34)$ approaches zero *exactly* when the pairs (12) and (34) are widely separated.

Equations (2.28) and (2.29) could be put into alternative form by using the well-known result (Baxter, 1964a,b; Schofield, 1966)

$$\left(\frac{\partial \varrho}{\partial p}\right) \frac{\partial}{\partial \varrho} [\varrho^2 g(12)] = 2\beta\varrho g(12) + \beta\varrho^2 \int [g(123) - g(12)] \, d\mathbf{r}_3. \tag{2.31}$$

Before (2.28) and (2.29) can be used some approximation must be introduced for the three- and four-body distribution functions. The most widely used approximation is the superposition approximation (Kirkwood, 1935):

$$g_0(1234) = g_0(12)g_0(13)g_0(14)g_0(23)g_0(24)g_0(34) \tag{2.32}$$

$$g_0(123) = g_0(12)g_0(13)g_0(23). \tag{2.33}$$

If these expressions are substituted into Eq. (2.28) and all *reducible* cluster integrals are omitted, we obtain

$$\begin{aligned}(\partial A/\partial\gamma)_{\gamma=0} = {} & \tfrac{1}{2}N\varrho \int g_0(12)[\partial u_1(12)/\partial\gamma] \, d\mathbf{r}_2 \\ & - \tfrac{1}{2}\beta N\varrho \int g_0(12)[u_1(12)]^2 \, d\mathbf{r}_2 \\ & - \beta N\varrho^2 \int g_0(12)g_0(23)u_1(12)u_1(23)h_0(13) \, d\mathbf{r}_2 \, d\mathbf{r}_3 \\ & - \tfrac{1}{4}\beta N\varrho^3 \int g_0(12)g_0(34)u_1(12)u_1(34)[2h_0(13)h_0(24) \\ & + 4h_0(13)h_0(14)h_0(24) + h_0(13)h_0(14)h_0(23)h_0(24)] \\ & \times d\mathbf{r}_2 \, d\mathbf{r}_3 \, d\mathbf{r}_4,\end{aligned} \tag{2.34}$$

where $h_0(ij) = g_0(ij) - 1$. The procedure leading to Eq. (2.34) is much more logical than the alternative possibility of using Eqs. (2.32) and (2.33) in the second and third terms of Eq. (2.28) while retaining the last term in (2.28) exactly, because one of the functions of the last term in (2.28) is to cancel the reducible cluster integrals arising from the earlier terms, and this cancellation must be maintained if a reasonable approximation is to be found.

Similarly, we obtain, by substituting Eqs. (2.32) and (2.33) into Eq. (2.29) and neglecting reducible cluster integrals,

$$\begin{aligned}[\partial g(12)/\partial\gamma]_{\gamma=0} = &- \beta g_0(12)u_1(12) - 2\beta\varrho g_0(12) \int g_0(23)u_1(23)h_0(13)\, d\mathbf{r}_3 \\ &- \tfrac{1}{2}\beta\varrho^2 g_0(12) \int g_0(34)u_1(34)[2h_0(13)h_0(24) \\ &+ 4h_0(13)h_0(14)h_0(24) \\ &+ h_0(13)h_0(14)h_0(23)h_0(24)]\, d\mathbf{r}_3\, d\mathbf{r}_4 .\end{aligned} \tag{2.35}$$

In making calculations based on Eqs. (2.34) and (2.35) we used the PY hard-sphere results for $g_0(12)$ and evaluated the cluster integrals using the same techniques used in the calculation of virial coefficients (Barker and Monaghan, 1962; Barker *et al.*, 1966; Henderson and Oden, 1966).

B. Numerical Results for the Square-Well Potential

Using our MC values of $\langle N_i \rangle$ and $\langle N_i N_j \rangle$ and using the square-well potential with $d = \sigma$, we have computed the terms

$$A_1/kT = \sum_i \langle N_i \rangle u_1{}^*(R_i) \tag{2.36}$$

and

$$A_2/kT = -\tfrac{1}{2} \sum_{ij} [\langle N_i N_j \rangle - \langle N_i \rangle \langle N_j \rangle] u_1{}^*(R_i) u_1{}^*(R_j) \tag{2.37}$$

in the series expansion of the free energy:

$$A = \sum_{n=0}^{\infty} (\beta\varepsilon)^n A_n , \tag{2.38}$$

where

$$u_1{}^*(R) = \begin{cases} -1, & \sigma < R < 3\sigma/2 \\ 0, & \text{elsewhere.} \end{cases} \tag{2.39}$$

These values of A_1 and A_2 are plotted in Figs. 1 and 2, respectively. The quantity $V_0 = N\sigma^3/\sqrt{2}$. It is interesting to note that A_1 is roughly

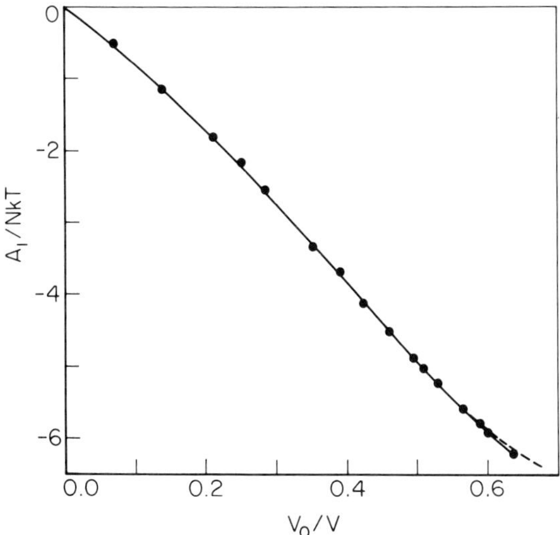

FIG. 1. First-order coefficient for the free energy using the square-well potential. The points give the MC results and the solid curve gives the smoothed fit of these results. The broken curve gives the results obtained from Eq. (2.19) using the PY $g_0(R)$.

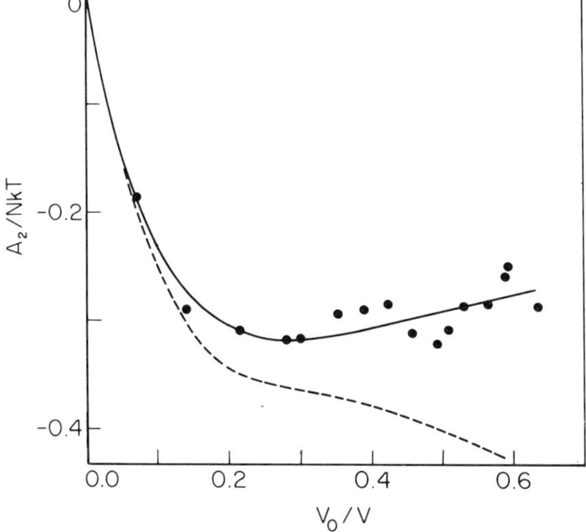

FIG. 2. Second-order coefficient for the free energy using the square-well potential. The points give the MC results and the solid curve gives the smoothed fit of these results. The broken curve gives the results obtained from Eq. (2.34) using the PY $g_0(R)$.

a linear function of the density. Recall that one of the fundamental assumptions of the van der Waals and Longuet-Higgins and Widom theories was that A_1 was proportional to the density.

The scatter of the points in Fig. 2 gives a reasonable estimate of the accuracy of our calculations, and indicates, as one might expect, that it is much harder to calculate the difference $\langle N_i N_j \rangle - \langle N_i \rangle \langle N_j \rangle$ than the $\langle N_i \rangle$. In earlier calculations (Barker and Henderson, 1968), applicable only to the square-well potential, we calculated $\langle N_1 \rangle$ and $\langle N_1{}^2 \rangle$, where N_1 is the number of intermolecular distances in the interval $\sigma < R < 3\sigma/2$, using 32 and 108 molecules. The two sets of calculations are in excellent agreement. Also, these previous calculations showed no significant difference between the 32- and 108-molecule systems.

We have smoothed these MC results by making a fit of the function

$$A_n/NkT = C_n\{1 - \exp[-\alpha_n \varrho/(\beta_n - \varrho)] - (\alpha_n/\beta_n)\varrho\} + P_n \varrho + Q_n \varrho^2 \quad (2.40)$$

to these results. We chose $\beta_n = \sqrt{2}$ (i.e., the close-packed density) and forced P_n to give the correct contribution of order ϱ, which can be calculated from the second virial coefficient. The remaining coefficients were then chosen by the least-squares criterion. The resulting values of the parameters are given in Table I.

TABLE I

PARAMETERS USED IN THE SMOOTHED FIT OF THE A_n FOR THE SQUARE-WELL POTENTIAL

n	α_n	β_n	C_n	P_n	Q_n
1	4.5	$\sqrt{2}$	3.173136	−4.974192	5.134186
2	9.75	$\sqrt{2}$	−0.384466	−2.487096	−0.047652

The smoothed fits obtained from Eq. (2.40) for A_1 and A_2 are compared with the MC results in Figs. 1 and 2, respectively. The smoothed fit of the A_2 MC results differs somewhat from that obtained earlier (Barker and Henderson, 1968) because the present calculations cover a wider density range.

If terms of order $(\beta\varepsilon)^3$ and higher are assumed to be negligible and if the Ree and Hoover (1964) Padé expression for the hard-sphere isotherm is used to obtain A_0, Eq. (2.38) may be used to calculate the thermodynamic properties.

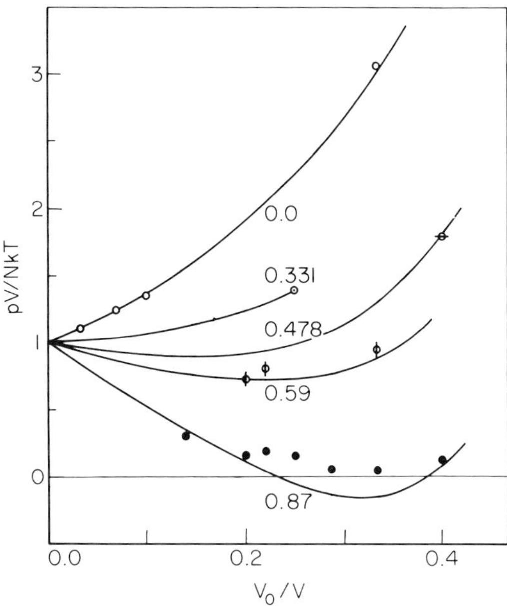

FIG. 3. Equation of state for the square-well potential. The points are the molecular-dynamics values and the curves are isotherms calculated using second-order perturbation theory. The points given by ○ were calculated by Alder and Wainwright (1960) at $\varepsilon/kT = 0$, while the points given by ⊙, ⊖, ⦶, and ● were calculated by Alder (1967) at $\varepsilon/kT = 0.331$, 0.478, ~0.59, and ~0.87, respectively.

In Figs. 3 and 4 the isotherms computed from Eq. (2.40) are compared with the machine calculations of Alder (1967) and of Rotenberg (1965), respectively. The agreement is very good. The slight errors at the lowest temperature are due to the neglect of higher-order terms. Later we shall discuss how the effect of these terms may be estimated. In Figs. 5 and 6, respectively, the densities and vapor pressure of the coexisting liquid and gaseous phases are plotted. The agreement with the few available machine calculations is very good. Finally, in Table II the critical constants calculated from second-order perturbation theory are compared with the results of Alder (1967). The agreement is very good. The small overestimation of the critical temperature results from the neglect of higher-order terms.

In Figs. 1 and 2 we give the results of our calculations (Smith *et al.*, 1970) of A_1 and A_2 from Eqs. (2.19) and (2.34). Except at high densities, the agreement with the MC results is good.

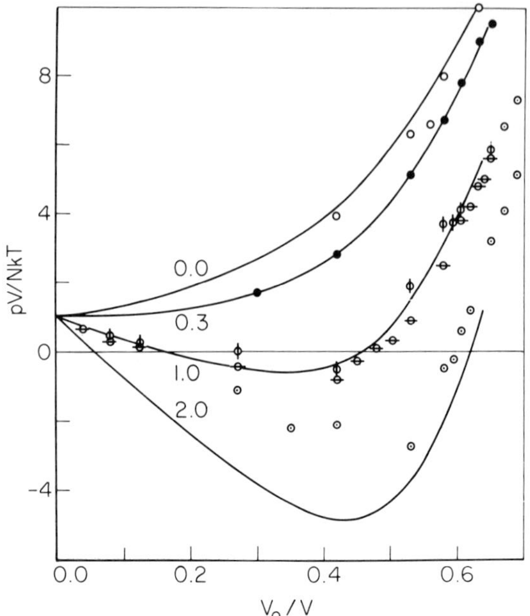

FIG. 4. Equation of state for the square-well potential. The points are Monte Carlo values of Rotenberg (1965), and the curves are isotherms calculated using second-order perturbation theory. The points given by ○, ●, ⏀, and ⊙ were calculated using 256 molecules at $\varepsilon/kT = 0$, 0.3, 1, and 2, respectively, while the points given by ⊖ were calculated using 864 molecules at $\varepsilon/kT = 1$.

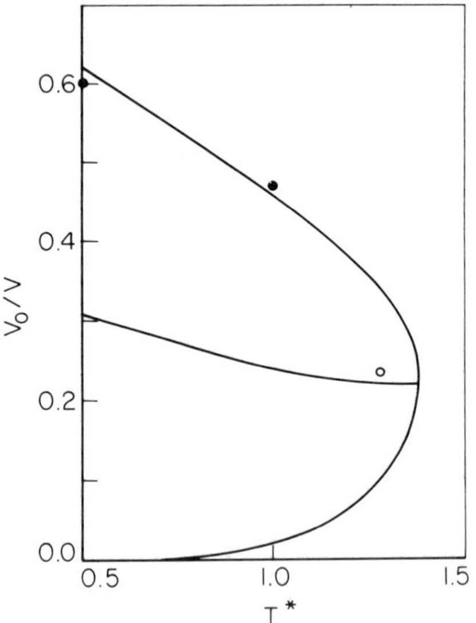

FIG. 5. Densities of coexisting liquid and gaseous phases for the square-well potential. The curve is calculated from second-order perturbation theory, while the points given by ● and ○ were taken from the machine calculations of Rotenberg (1965) and Alder (1967), respectively.

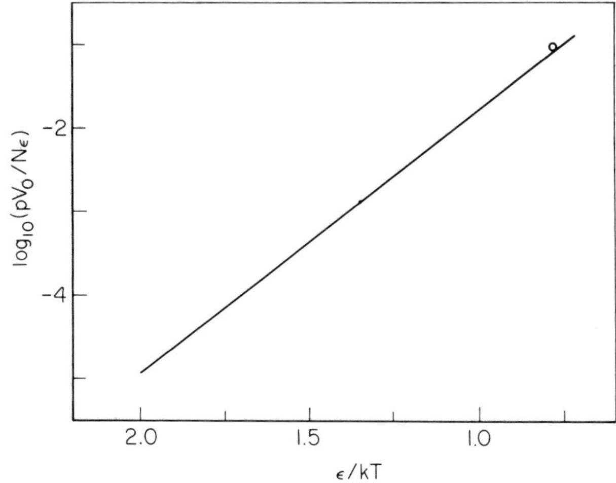

Fig. 6. Saturated vapor pressures for the square-well potential. The curve is calculated from second-order perturbation theory, while the point is taken from the molecular-dynamics calculations of Alder (1967).

TABLE II

Critical Constants for the Square-Well Potential

Constant	Alder (1967)	Present calc.
kT_c/ε	1.28	1.39
V_c/V_0	4.25	4.69
p_cV_0/NkT_c	0.072	0.0837

III. Potentials with a Soft Core

A. Perturbation Expansion

We have seen in previous sections that perturbation theory gives excellent results for the equation of state of a square-well fluid even at very low temperatures if the attractive potential is treated as a perturbation in a hard-sphere system. This suggests that the failure of earlier perturbation theories at low temperatures when applied to the 6–12 potential was due either to the lack of a satisfactory treatment of the softness of the repulsive potential, with consequent extreme sensitivity to the choice of the hard-sphere diameter (Zwanzig, 1954; Smith and Alder, 1959; Frisch *et al.*, 1966), or to the use of the large R^{-6} term as a perturbation (McQuarrie and Katz, 1966). In this section we outline a perturbation theory which is, as nearly as possible, free of these defects.

We consider an arbitrary potential function $u(R)$ and define a modified potential function $v(d, \sigma, \alpha, \gamma; R)$ by the relations

$$v(d, \sigma, \alpha, \gamma; R) = \begin{cases} u\left(d + \dfrac{R-d}{\alpha}\right), & d + \dfrac{R-d}{\alpha} < \sigma, \\ 0, & \sigma < d + \dfrac{R-d}{\alpha} < d + \dfrac{\sigma - d}{\alpha}, \\ \gamma u(R), & \sigma < R. \end{cases} \tag{3.1}$$

We assume that $u(R)$ rises to effectively infinite positive values for small values of R. For $\alpha = \gamma = 0$, $v(R)$ becomes the hard-sphere potential with diameter d, and for $\alpha = \gamma = 1$ the original potential is recovered. The inverse steepness parameter α varies the steepness of the modified potential in the repulsive region and the parameter γ varies the depth of the potential.

Our procedure is to expand the free energy in powers of α and γ about the point $\alpha = \gamma = 0$, which corresponds to the hard-sphere potential. Thus,

$$A(\alpha, \gamma) = A(0, 0) + \alpha(\partial A/\partial\alpha)_{\alpha=\gamma=0} + \gamma(\partial A/\partial\gamma)_{\alpha=\gamma=0}. \tag{3.2}$$

For $\alpha = \gamma = 1$ we obtain an expression for the free energy corresponding to the original potential $u(R)$. This procedure has useful convergence if $u(R)$ is *steep* for $R < \sigma$ and *small* for $R > \sigma$.

The differentiation with respect to α is straightforward, but rather lengthy. We have given complete details elsewhere (Barker and Henderson, 1967b). The result is

$$(\partial A/\partial\alpha)_{\alpha=\gamma=0} = -\, 2\pi NkT\varrho d^2 g_0(d)\left[d + \int_0^\sigma f(z)\, dz\right], \tag{3.3}$$

where

$$f(z) = \exp[-\beta u(z)] - 1. \tag{3.4}$$

The differentiation with respect to γ is identical to that performed in Section II. The result is

$$(\partial A/\partial\gamma)_{\alpha=\gamma=0} = 2\pi N\varrho \int_\sigma^\infty g_0(R) u(R) R^2\, dR. \tag{3.5}$$

The parameters d and σ are still at our disposal. We use for σ the value of R for which $u(R)$ is zero. An exception to this general rule is the case of purely repulsive potentials (Watts and Henderson, 1969; Henderson and Barker, 1970), where it is sometimes convenient to use other choices. However, we will not pursue this question further. We choose for d the value given by

$$d = -\int_0^\sigma f(z)\, dz. \tag{3.6}$$

Thus, d depends on the temperature, but not on the density, and with this choice the term of order α is zero at all temperatures and densities.

For molecules with a soft core $u(z) = \infty$ for $z < \sigma$, and thus $f(z) = -1$ for $z < \sigma$. Hence, from Eq. (3.6) we obtain $d = \sigma$. Thus, the above results are a natural generalization of our previous results for molecules with a hard core.

Thus, to first order, the free energy is

$$A = A_0 + 2\pi N\varrho \int_\sigma^\infty g_0(R)u(R)R^2\, dR, \tag{3.7}$$

where A_0 and $g_0(R)$ are, respectively, the free energy and the pair distribution function for hard spheres of diameter d. The term of order $\alpha\gamma$ is zero by virtue of Eq. (3.6). The γ^2 term can be calculated by the methods used in Section II. Any realistic potential rises very steeply for $R < \sigma$. This is particularly true at the low temperatures characteristic of liquids. Thus, we expect that for liquids the α^2 term will be negligibly small, and do not consider this term. Levesque (1969) has recently presented strong evidence supporting this conjecture. However, it should be kept in mind that at high temperatures the α^2 term may be needed.

For a steeply repulsive potential $\sigma \to \infty$ and the free energy is that of a system of hard spheres of diameter d. If Eq. (1.17) is assumed for the pair potential and if the integral in Eq. (3.6) is evaluated to order n^{-1}, then Rowlinson's n^{-1} expansion is obtained. This expansion is only useful at high temperatures. However, it is interesting to see its close connection to the present results.

B. Numerical Results for the 6–12 Potential

Calculations are most complete for the 6–12 potential,

$$u(R) = 4\varepsilon[(\sigma/R)^{12} - (\sigma/R)^6]. \tag{3.8}$$

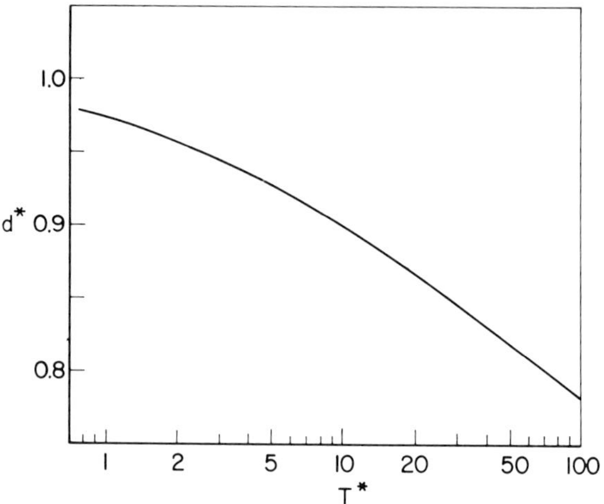

FIG. 7. Plot of $d^* = d/\sigma$ as a function of temperature for the 6–12 potential.

In Fig. 7 we show the temperature ($T^* = kT/\varepsilon$) dependence of $d^* = d/\sigma$, and in Fig. 8 we compare the values of the second virial coefficient calculated from perturbation theory with the exact results. The agreement is good, and indicates that this treatment deals in a satisfactory way with the "softness" of the repulsive potential.

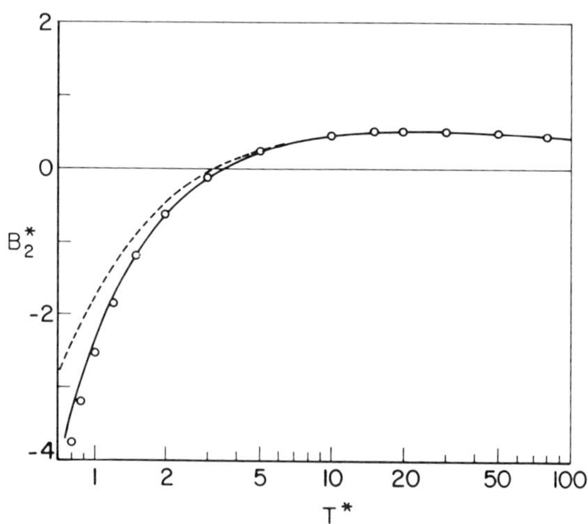

FIG. 8. Second virial coefficient for the 6–12 potential. The points are exact values (Barker *et al.*, 1966). The solid and broken curves, respectively, give the perturbation-theory results with and without the term of order γ^2.

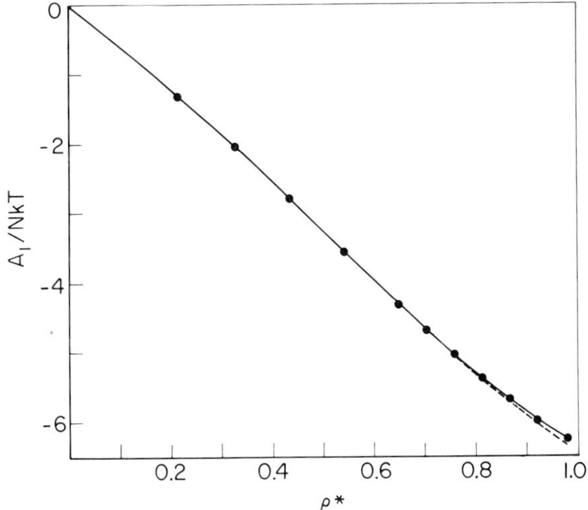

FIG. 9. First-order (γ) coefficient for the free energy using the 6–12 potential. The points give the MC results and the solid curve gives the smoothed fit of these results. The broken curve gives the results obtained from Eq. (2.19) using the PY $g_0(R)$.

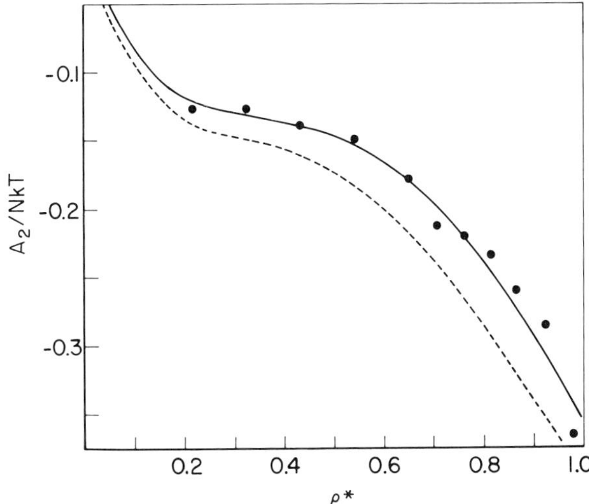

FIG. 10. Second-order (γ^2) coefficient for the free energy using the 6–12 potential. The points give the MC results and the solid curve gives the smoothed fit of these results. The broken curve gives the results obtained from Eq. (2.34) using the PY $g_0(R)$.

In Figs. 9 and 10 the values of the γ and γ^2 terms calculated from our MC data are shown for $T^* = 1$. Here $\varrho^* = \varrho\sigma^3$. It is to be noted that the γ term is, to a rough approximation, proportional to the density. We also plot our smoothed fits of these values obtained from Eq. (2.40) and the values of the γ and γ^2 calculated from Eqs. (2.19) and (2.34) using the PY $g_0(R)$ (Smith *et al.*, 1970). The results obtained from these approximate formulas are better than were the corresponding results for the square-well potential.

We calculated the thermodynamic properties for the 6–12 fluid from perturbation theory using the Ree and Hoover (1964) Padé approximant to obtain the free energy of the hard spheres of diameter d and the smoothed fits of the MC γ and γ^2 terms. The α term is included through the value of d. The pressure is then obtained by analytic differentiation with respect to density, and the energy and entropy are obtained by numerical temperature differentiation.

We now proceed to compare these results with molecular-dynamics calculations (Verlet, 1967), Monte Carlo calculations (Wood and Parker, 1957; Wood, 1968; McDonald and Singer, 1967a,b), and experimental results for liquid argon (Guggenheim, 1945; Michels *et al.*, 1949; 1958; Levelt, 1960; Rowlinson, 1959; van Itterbeek and Verbeke, 1960; van Itterbeek, *et al.*, 1963) and solid argon (Dobbs and Jones, 1957; Guggenheim and McGlashan, 1960). In reducing the experimental results for liquid argon we used the parameters $\varepsilon/k = 119.8°\text{K}$ and $\sigma = 3.405$ Å derived by Michels *et al.* (1949) from second virial coefficients. In making this comparison we do not imply that the 6–12 potential with these parameters gives a close representation of the true pair potential for argon; evidence is now strong that it does not (Rowlinson, 1965). However, the good agreement between the machine calculations made with this potential and the experimental data for liquid argon shows that it does represent a good *effective* pair potential for argon at high densities, apparently absorbing some many-body interactions into a pseudo pair potential. It is on this basis that we regard comparison of our results with the properties of argon as meaningful.

The calculated thermodynamic properties are compared with machine-calculation results and experimental data in Figs. 11–17. The reduced pressure is $p^* = p\sigma^3/\varepsilon$ and the internal entropy S_i and energy U_i are defined by

$$S_\text{i} = S - \tfrac{5}{2}Nk + 3Nk\ln\lambda + \ln\varrho, \tag{3.9}$$

$$U_\text{i} = U - \tfrac{3}{2}NkT. \tag{3.10}$$

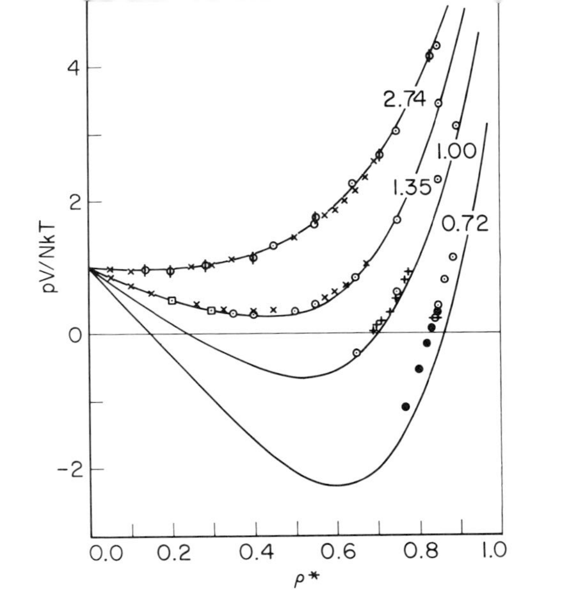

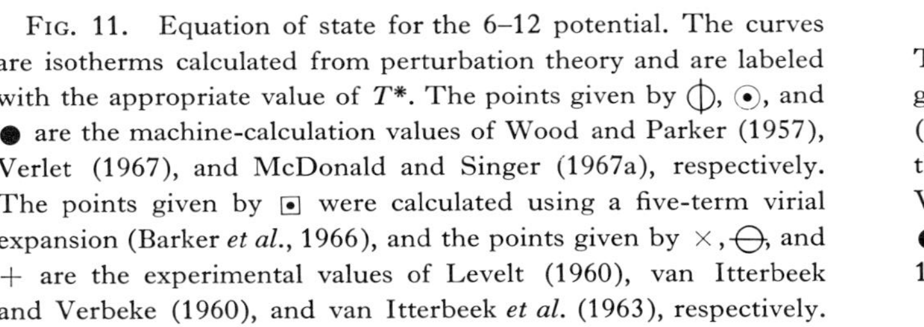

FIG. 11. Equation of state for the 6–12 potential. The curves are isotherms calculated from perturbation theory and are labeled with the appropriate value of T^*. The points given by ⌽, ⊙, and ● are the machine-calculation values of Wood and Parker (1957), Verlet (1967), and McDonald and Singer (1967a), respectively. The points given by ⊡ were calculated using a five-term virial expansion (Barker *et al.*, 1966), and the points given by ×, ⊖, and + are the experimental values of Levelt (1960), van Itterbeek and Verbeke (1960), and van Itterbeek *et al.* (1963), respectively.

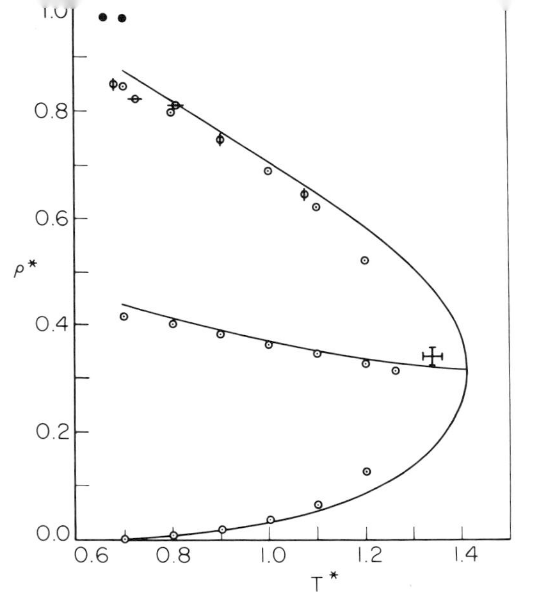

FIG. 12. Densities of coexisting phases for the 6–12 potential. The curve gives the perturbation-theory results, and the points given by ⌽ and ⊖ are the machine-calculation values of Verlet (1967) and McDonald and Singer (1967a). The large cross gives the machine-calculation estimate, together with error bounds, of Verlet (1967) for the critical point. The points given by ⊙ and ● are the experimental values for liquid argon (Guggenheim, 1945) and solid argon (Dobbs and Jones, 1957), respectively.

The agreement is very good at high densities. At low densities, where perturbation theory converges more slowly, slightly larger errors exist. When differences between the machine calculations and the experimental data exist it is pleasing that our results follow the machine calculations.

Only in the neighborhood of the critical point do appreciable errors exist. Higher-order terms must be included in order to flatten the coexistence curve plotted in Fig. 12 in the critical region. However, as may be seen in Table III, even with these higher-order terms neglected, perturbation theory locates the critical point with good accuracy.

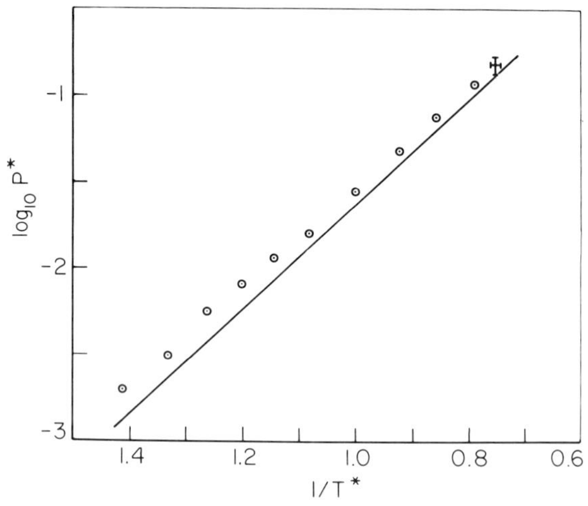

FIG. 13. Reduced vapor pressures for the 6–12 potential. The curve gives the perturbation-theory results, and the large cross gives the machine-calculation estimate, together with error bounds, of Verlet (1967) for the critical point. The points are the experimental values for argon (Rowlinson, 1959).

To provide insight into the relative merits of perturbation theory and the PY theory, we have compared the predictions of these theories for pV/NkT and $\beta(\partial p/\partial \varrho)_T$ of liquid argon in equilibrium with its vapor with the experimental values in Figs. 18 and 19, respectively. Perturbation theory is a considerable improvement over the PY theory. The small errors in the perturbation results at the lowest temperatures are due to the neglect of the γ^3 term and possibly higher-order terms. We will consider methods of estimating these terms in Section V.

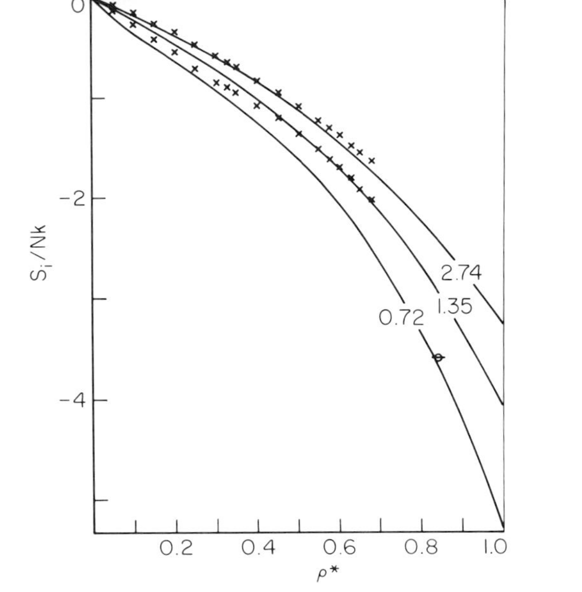

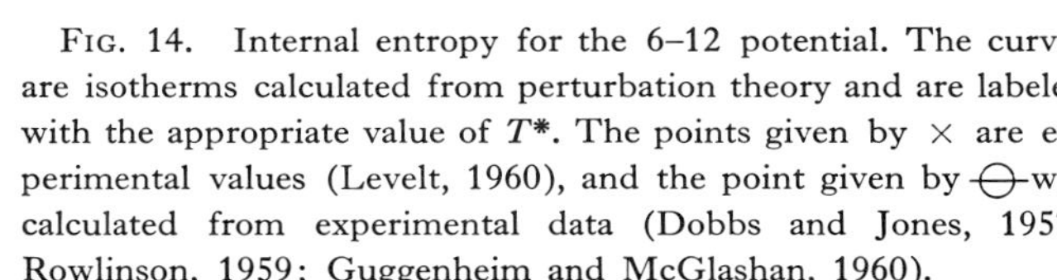

FIG. 14. Internal entropy for the 6–12 potential. The curves are isotherms calculated from perturbation theory and are labeled with the appropriate value of T^*. The points given by × are experimental values (Levelt, 1960), and the point given by ⊖ was calculated from experimental data (Dobbs and Jones, 1957; Rowlinson, 1959; Guggenheim and McGlashan, 1960).

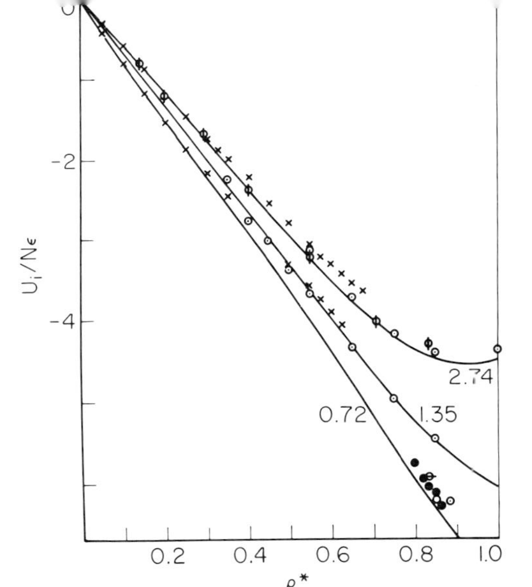

FIG. 15. Internal energy for the 6–12 potential. The curves are isotherms calculated from perturbation theory and are labeled with the appropriate value of T^*. The points given by ⏀, ⊙, and ● are the machine-calculation values of Wood and Parker (1957), Verlet (1967), and McDonald and Singer (1967a), respectively. The points given by × are the experimental values of Levelt (1960), and the point given by ⊖ was calculated from experimental data (Dobbs and Jones, 1957; Rowlinson, 1959; Guggenheim and McGlashan, 1960).

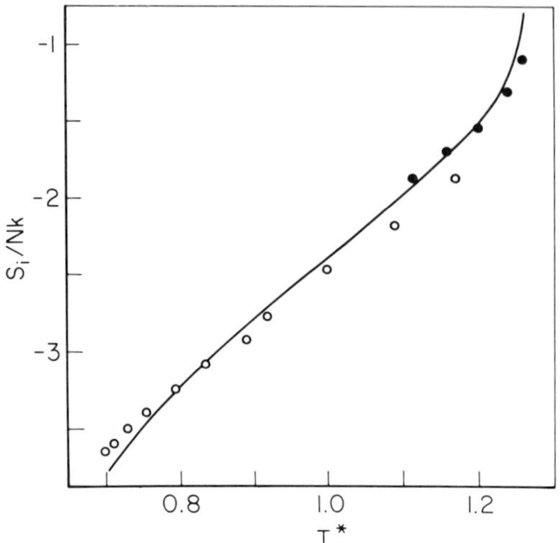

FIG. 16. Internal entropy at saturated vapor pressures for the 6–12 potential. The curve gives the results of perturbation theory when the *experimental* densities of liquid argon are used. The points given by ● are experimental values (Michels *et al.*, 1958), and the points given by ○ were calculated from experimental data (Dobbs and Jones, 1957; Rowlinson, 1959; Guggenheim and McGlashan, 1960).

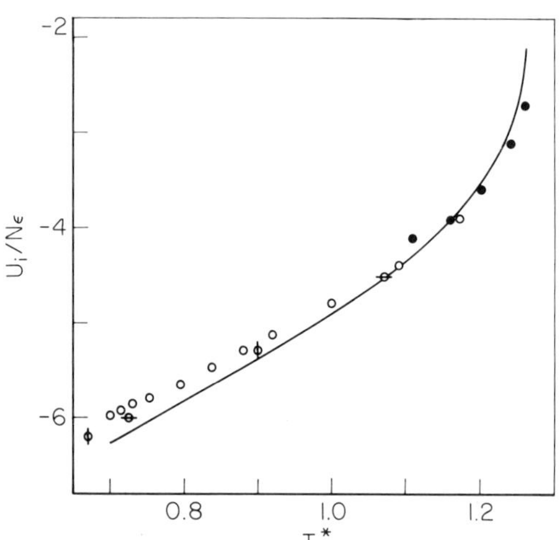

FIG. 17. Internal energy at saturated vapor pressures for the 6–12 potential. The curve gives the results of perturbation theory when the *experimental* densities of liquid argon are used. The points given by ⏀ and ⊖ are machine-calculation values of Verlet (1967) and McDonald and Singer (1967a). The points given by ● are experimental values (Michels *et al.*, 1958), and the points given by ○ were calculated from experimental data (Dobbs and Jones, 1957; Rowlinson, 1959; Guggenheim and McGlashan, 1960).

TABLE III

CRITICAL CONSTANTS FOR THE 6–12 POTENTIAL

Constant	Expt.[a]	Verlet (1967)	Present calc.
T^*	1.26	1.32–1.36	1.41
$\varrho_c{}^*$	0.316	0.32–0.36	0.32
$p_c{}^*$	0.117	0.13–0.17	0.17
$p_c V_c/NkT_c$	0.293	0.30–0.36	0.38

[a] Rowlinson (1959).

IV. Application to Real Liquids

Thus far we have considered only model liquids in which only pair interactions are present and for which classical statistical mechanics is valid. It turned out that one of these liquids, the 6–12 liquid, gives results which are in good agreement with experimental results for argon. However, this is accidental, and no substitute for a rigorous study of real liquids.

A. Triplet Interactions

To make calculations which are directly comparable with experimental results, accurate pair and multibody potentials must be used. Barker and Pompe (1968), by using all available information on pair interactions, have derived the following pair potential for argon

$$\begin{aligned} u(R) = \varepsilon\{&e^{12.5(1-x)}[0.2349 - 4.7735(x-1) - 10.2194(x-1)^2 \\ &- 5.2905(x-1)^3] - [1.0698/(x^6+0.01)] \\ &- [0.1642/(x^8+0.01)] - [0.0132/(x^{10}+0.01)]\}, \end{aligned} \tag{4.1}$$

where $x = R/R_m$ and $\varepsilon/k = 147.7°\mathrm{K}$, and $R_m = 3.756$ Å. Further, they conclude that the only important nonadditive interaction is the triple-dipole dispersion interaction:

$$u(123) = \nu(1 + 3\cos\theta_1 \cos\theta_2 \cos\theta_3)/(R_{12}R_{13}R_{23})^3, \tag{4.2}$$

where θ_1, θ_2, and θ_3, and R_{12}, R_{13}, and R_{23} are the corresponding angles and sides of the triangle formed by molecules 1, 2, and 3, and the coefficient $\nu = 73.2 \times 10^{-84}$ erg-cm^9 is calculated from known oscillator strengths (Leonard, 1968).

Perturbation theory is readily extended (Barker *et al.*, 1968, 1969a,b) to cover three-body interactions by treating ν as a third perturbation

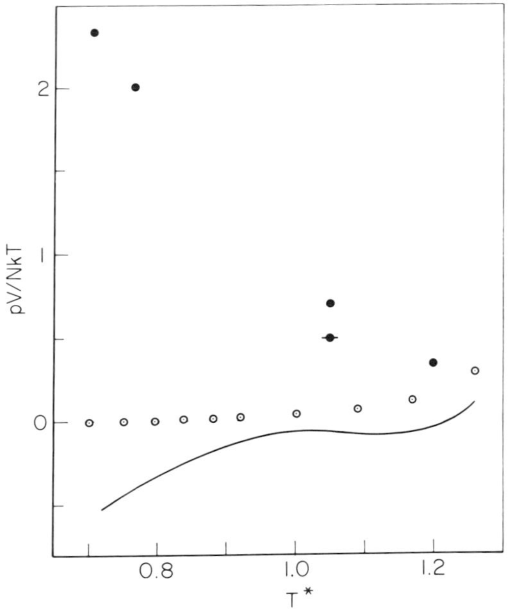

FIG. 18. Plot of pV/NkT for liquid argon in equilibrium with its vapor. The curve gives the perturbation-theory results. The points given by ⊙ are experimental values (Rowlinson, 1959), and the points given by ● and -●- are PY and PY2 values, respectively, calculated by Levesque (1966). The perturbation, PY, and PY2 results are all calculated at the *experimental* density.

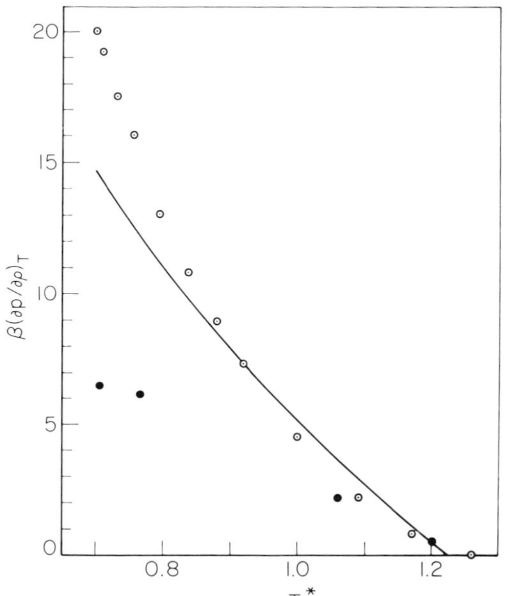

FIG. 19. Plot of $\beta(\partial p/\partial \varrho)_T$ for liquid argon in equilibrium with its vapor. The curve gives the perturbation-theory results. The points given by ⊙ are experimental values (Rowlinson, 1959), and the points given by ● are PY values calculated by Levesque (1966). The perturbation and PY results are all calculated at the *experimental* density.

parameter. Thus, to order ν, the free energy is

$$\begin{aligned} A &= A_{2b} + A_\nu \\ &= A_{2b} + \tfrac{1}{6}N\varrho^2 \int g_0(123)u(123)\, d\mathbf{r}_2\, d\mathbf{r}_3, \end{aligned} \tag{4.3}$$

where A_{2b} is the contribution to the free energy resulting from pair interactions and $g_0(123)$ is the triplet distribution function for hard spheres of diameter d.

The integral over $g_0(123)$ has been evaluated (Barker *et al.*, 1968, 1969a,b) by two methods: first, by MC averaging in a system of 108 hard spheres with periodic boundary conditions, and second, by using the superposition approximation

$$g_0(123) = g_0(12)g_0(13)g_0(23) \tag{4.4}$$

with PY $g_0(R)$. As may be seen in Fig. 20, the two methods agree closely. The higher-order terms in ν are much smaller. The $\alpha\nu$ term is zero by virtue of Eq. (3.16). The MC values for the $\nu\gamma$ terms are shown in Fig. 20. The $\nu\gamma$ term is much smaller than the ν term (except possibly at very low densities). The ν^2 term is negligible. Thus, only the ν term was used in our calculations. The ν term can be reproduced to 1 part in 10^5 by the Padé-type expression:

$$d^9 A\nu/N\nu = \varrho^2 d^6 \frac{2.70797 + 1.68918\varrho d^3 - 0.31570\varrho^2 d^6}{1 - 0.59056\varrho d^3 + 0.20059\varrho^2 d^6}. \tag{4.5}$$

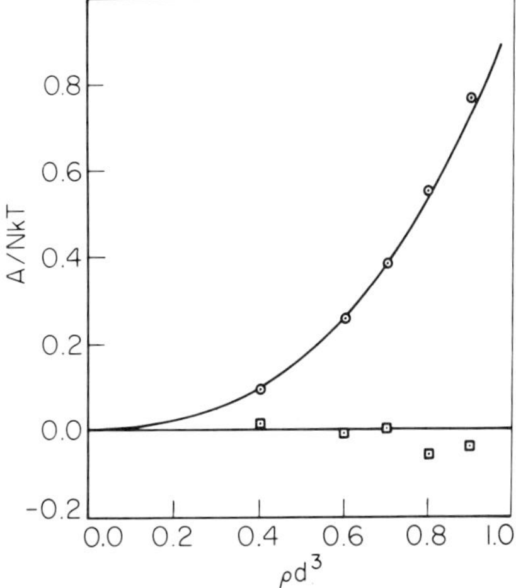

FIG. 20. Three-body contributions for argon at 86.63°K. The solid curve gives A_ν/NkT calculated from the superposition approximation using PY $g_0(R)$, while the points given by ⊙ and ⊡ give the MC estimates of A_ν/NkT and $A_{\nu\gamma}/NkT$, respectively.

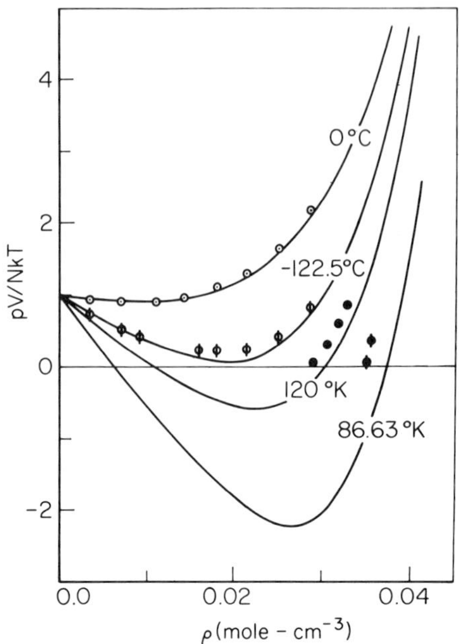

FIG. 21. Equation of state of argon. The curves are isotherms calculated from perturbation theory using the Barker–Pompe potential and are labeled with the appropriate value of T. The points given by ⊙, ⏀, ●, and ⧫ are experimental points for $T = 0°C$ (Michels *et al.*, 1949), $T = -122.5°K$ (Michels *et al.*, 1958), $T = 120°K$ (van Itterbeek *et al.*, 1963), and $T = 86.63°K$ (van Itterbeek and Verbeke, 1960).

In Fig. 21 the equation of state calculated from Eqs. (4.3) and (4.5) using the Barker–Pompe pair potential is compared with experimental results (Michels *et al.*, 1949, 1958; van Itterbeek *et al.*, 1963; van Itterbeek and Verbeke, 1960) for argon. The calculated and experimental critical constants are compared in Table IV. The agreement is good. The remaining errors are due to the neglect of the γ^3 and higher-order terms

TABLE IV

CRITICAL CONSTANTS OF ARGON

Constant	Expt.[a]	Present calc.
T_c (°K)	150.87	167
p_c (atm)	48.43	69
ϱ_c (mole-cm^{-3})	0.0134	0.013

[a] Levelt (1960).

and to the neglect of quantum effects. Possibly, the Barker–Pompe potential may need minor refinement. However, the above effects would have to be considered first before any definite conclusions can be reached.

B. Quantum Effects

If quantum effects are small, they can be included by treating them as a perturbation (Kim *et al.*, 1969). Thus, treating Planck's constant h as a perturbation parameter, we obtain, to order h^2,

$$A = A_{\text{classical}} + (h^2/24\pi mkT)N\varrho \int_d^\infty g_0(R)\, \nabla^2 u(R) R^2\, dR. \qquad (4.6)$$

In reduced units this becomes

$$A^* = A^*_{\text{classical}} + (\Lambda^{*2}/24\pi T^*)\varrho^* \int_{d^*}^\infty g_0(x)\, \nabla^2 u(x) x^2\, dx, \qquad (4.7)$$

where $A^* = A/N\varepsilon$ and $\Lambda^* = h/\sigma(m\varepsilon)^{1/2}$. Thus, the quantum correction is proportional to Λ^{*2}/T^*, and so Eq. (4.7) will be valid for small Λ^* and high temperatures. Equation (4.7) is applicable to most liquids. However, for hydrogen and helium Λ^* is large, and Eq. (4.7) is applicable to these substances only at high temperatures, i.e., the gaseous state.

Numerical calculations based on Eq. (4.7) are very limited, and in addition, were based on approximate expressions for the γ^2 term. Thus, at the present time we cannot discuss the results of Eq. (4.7) in detail. However, one important result is that the quantum correction to the pressure is positive, and thus, inclusion of quantum effects in a calculation of the properties of argon would improve the agreement with experiment.

V. Summary

Perturbation theory provides a simple and successful theory of liquids, which can deal not only with model liquids, but also with real liquids in which three-body forces and quantum effects are present.

Future developments in perturbation theory will undoubtedly be concerned with the effects of the γ^3 and higher-order terms. It is doubtful that the exact methods which we have considered will be useful in calculating these terms, because of the computational difficulties involved. Most likely, one must be content with estimates of these terms.

Presumably, when these estimates are included excellent results for the thermodynamic properties will be obtained. One exception to this will be the critical region. The critical point should be located with good accuracy, but the values of the critical exponents will not be correct. This is not too serious, since this difficulty is shared by all the distribution-function and lattice theories of liquids. Of course, if the exact higher-order perturbation terms could be calculated, then the critical exponents would be obtained correctly. However, this is possible only for the simplest systems—such as the lattice gas. Indeed, most advances in the theory of the critical phenomena have resulted from the study of the critical properties of the lattice gas.

At present the only estimates of these terms which have been made are those of Kozak and Rice (1968), who made a Padé extrapolation of the behavior of the first five virial coefficients for the 6–12 potential. This method is satisfactory at high temperatures or low densities, but fails completely for liquids, where the virial series almost certainly diverges (Barker *et al.*, 1966).

A method of estimating these higher-order terms must take into account that the convergence of the perturbation expansion is best at high densities, where the hard-sphere reference fluid is nearly incompressible and changes in structure are difficult. Thus, a promising approximation is

$$(\partial^3 A/\partial \gamma^3)_{\gamma=0} \simeq \tfrac{1}{2}\beta^2 N \varrho (\partial \varrho/\partial p)_0 \int g_0(12)[u(12)]^3 \, d\mathbf{r}_2, \tag{5.1}$$

where $(\partial \varrho/\partial p)_0$ is the compressibility of the reference hard-sphere system. Equation (5.1) gives the correct low-density behavior and assures that the contribution of the γ^3 term is small at high densities. Similar approximations could be used for the higher-order terms.

Another approach (Chen *et al.*, 1969), which is somewhat more complex, would be to use some integral equation, for example, the PY equation, to compute the derivatives of $g(R)$ with respect to α and γ. As we have seen in Section II,A,2, these results can be used to obtain the derivatives of A with respect to α and γ. Although this procedure is very similar to that used in Section II,A,2, it differs from this earlier procedure because of the approximations inherent in the PY equation. If this procedure were used to obtain all the derivatives of A with respect to α and γ, then the results would be the same as that obtained by an orthodox solution of the PY equation. However, if this procedure were used to obtain only the higher-order derivatives, with the low-order derivatives given exactly, then this would not be true.

Hopefully, both these methods would yield essentially the same results. At present we do not have numerical results for either of these approaches. A discussion of the consequences of these approaches must be the subject of future publications.

* *Note added in proof:* Most of the discrepancies between experimental and quasi-experimental results and perturbation theory reported in this chapter are due to inadequate equilibration in the Monte Carlo Calculations. Since the work reported in this chapter was completed, we have discovered that rejecting only the first 10,000 configurations lead to slight errors in the second-order term, A2, at the highest densities. We have recomputed A2 at high densities, generating up to 2 million preliminary configurations to ensure that no effect of the initial configuration remained. The recomputed A2 are somewhat less negative than are the results reported here. This results in somewhat larger pressures whose agreement with experimental and quasi-experimental results is almost exact. The results of these calculations will be reported in articles in *Accounts of Chemical Research* (November, 1971) and in *Annual Review of Physical Chemistry* (Volume 23, 1972.)

Acknowledgments

The work described in this chapter was done while the authors were at the CSIRO Chemical Research Laboratories, Melbourne, Australia and the University of Waterloo, Waterloo, Ontario, Canada. The authors are grateful for the support and the facilities provided by these institutions.

The financial support of the National Research Council of Canada, the Department of Energy, Mines and Resources of Canada, and the Office of Saline Water, Department of the Interior is gratefully acknowledged. One of us (DH) is grateful to the Alfred P. Sloan Foundation for a fellowship awarded him.

Special References

Alder, B. J. (1967). Unpublished results.

Alder, B. J., and Wainwright, T. E. (1960). *J. Chem. Phys.* **33**, 1439.

Barker, J. A. (1957). *Proc. Roy. Soc.* (*London*) **A241**, 547.

Barker, J. A. and Monaghan, J. J. (1962). *J. Chem. Phys.* **36**, 2564.

Barker, J. A., and Henderson, D. (1967a). *J. Chem. Phys.* **47**, 2856.

Barker, J. A., and Henderson, D. (1967b). *J. Chem. Phys.* **47**, 4714.

Barker, J. A., and Henderson, D. (1968). *In* "Proceedings of the Fourth Symposium on Thermophysical Properties" (J. R. Moszynski, ed.), p. 30. ASME, New York.

Barker, J. A., and Pompe, A. (1968). *Australian J. Chem.* **21**, 1683.

Barker, J. A., Leonard, P. J., and Pompe, A. (1966). *J. Chem. Phys.* **44**, 4206.

Barker, J. A., Henderson, D., and Smith, W. R. (1968). *Phys. Rev. Letters* **21**, 135.

Barker, J. A., Henderson, D., and Smith, W. R. (1969a). *J. Phys. Soc. Japan Suppl.* **26**, 284.

Barker, J. A., Henderson, D., and Smith, W. R. (1969b). *Mol. Phys.* **17**, 579.

Baxter, R. J. (1964a). *Phys. Fluids* **7**, 38.

Baxter, R. J. (1964b). *J. Chem. Phys.* **41**, 553.

Buff, F. P., and Schindler, F. M. (1958). *J. Chem. Phys.* **29**, 1075.

Chen, M., Henderson, D., and Barker, J. A. (1969). *Can. J. Phys.* **47**, 2009.

Dobbs, E. R., and Jones, G. O. (1957). *Rept. Progr. Phys.* **20**, 516.

Frisch, H. L. (1964). *Advan. Chem. Phys.* **6**, 229.

Frisch, H. L., Katz, J. L., Praestgaard, E., and Lebowitz, J. L. (1966). *J. Phys. Chem.* **70**, 2016.

Guggenheim, E. A. (1945). *J. Chem. Phys.* **13**, 253.
Guggenheim, E. A. (1965a). *Mol. Phys.* **9**, 43.
Guggenheim, E. A. (1965b). *Mol. Phys.* **9**, 199.
Guggenheim, E. A., and McGlashan, M. L. (1960). *Proc. Roy. Soc.* (*London*) **A255**, 456.
Hemmer, P. C. (1964). *J. Math. Phys.* **5**, 75.
Hemmer, P. C., Kac, M., and Uhlenbeck, G. E. (1964). *J. Math. Phys.* **5**, 60.
Henderson, D., and Oden, L. (1966). *Mol. Phys.* **10**, 405.
Henderson, D., and Barker, J. A. (1970). *Phys. Rev.* **A1**, 1266.
Kac, M., Uhlenbeck, G. E., and Hemmer, P. C. (1963). *J. Math. Phys.* **4**, 216.
Kim, S., Henderson, D., and Barker, J. A. (1969). *Can. J. Phys.* **47**, 99.
Kirkwood, J. G. (1935). *J. Chem. Phys.* **3**, 300.
Kozak, J. J., and Rice, S. A. (1968). *J. Chem. Phys.* **48**, 1226.
Lado, F., and Wood, W. W. (1968). *J. Chem. Phys.* **49**, 4244.
Leonard, P. J. (1968). Thesis, Univ. Melbourne, Australia.
Levelt, J. M. H. (1960). *Physica* **26**, 361.
Levesque, D. (1966). *Physica* **32**, 1985.
Levesque, D. (1969). *J. Phys. Soc. Japan Suppl.* **26**, 270.
Longuet–Higgins, H. C., and Widom, B. (1964). *Mol. Phys.* **8**, 549.
McDonald, I. R., and Singer, K. (1967a). *Discussions Faraday Soc.* **43**, 40.
McDonald, I. R., and Singer, K. (1967b). *J. Chem. Phys.* **47**, 4766.
McQuarrie, D. A., and Katz, J. L. (1966). *J. Chem. Phys.* **44**, 2393.
Metropolis, N., Rosenbluth, A. W., Rosenbluth, M. A., Teller, A. H., and Teller, E. (1953). *J. Chem. Phys.* **21**, 1087.
Michels, A., Wijker, H., and Wijker, H. (1949). *Physica* **15**, 627.
Michels, A., Levelt, J. M., Wolkers, G. J. (1958). *Physica* **24**, 769.
Percus, J. K., and Yevick, G. J. (1958). *Phys. Rev.* **110**, 1.
Ree, F. H., and Hoover, W. G. (1964). *J. Chem. Phys.* **40**, 939.
Reiss, H. (1965). *Advan. Chem. Phys.* **9**, 1.
Rotenberg, A. (1965). *J. Chem. Phys.* **43**, 1198.
Rowlinson, J. S. (1959). "Liquids and Liquid Mixtures." Butterworths, London and Washington, D.C.
Rowlinson, J. S. (1964a). *Mol. Phys.* **7**, 349.
Rowlinson, J. S. (1964b). *Mol. Phys.* **8**, 107.
Rowlinson, J. S. (1965). *Discussions Faraday Soc.* **40**, 19.
Schofield, P. (1966). *Proc. Phys. Soc.* **88**, 149.
Smith, E. B., and Alder, B. J. (1959). *J. Chem. Phys.* **30**, 1190.
Smith, W. R., Henderson, D., and Barker, J. A. (1970). *J. Chem. Phys.* **53**, 508.
Thiele, E. (1963). *J. Chem. Phys.* **39**, 474.
Uhlenbeck, G. E., Hemmer, P. C., and Kac, M. (1963). *J. Math. Phys.* **4**, 229.
van der Waals, J. D. (1873). Thesis, Leiden.
van Itterbeek, A., and Verbeke, O. (1960). *Physica* **26**, 931.
van Itterbeek, A., Verbeke, O., and Staes, K. (1963). *Physica* **29**, 742.
Verlet, L. (1967). *Phys. Rev.* **159**, 98.
Watts, R. O., and Henderson, D. (1969). *J. Chem. Phys.* **50**, 1651.
Wertheim, M. S. (1963). *Phys. Rev. Letters* **10**, 321.
Wood, W. W. (1968). *In* "Physics of Simple Liquids" (H. N. V. Temperley, J. S. Rowlinson, and G. S. Rushbrooke, eds.), p. 115. Wiley (Interscience), New York.
Wood, W. W., and Parker, F. R. (1957). *J. Chem. Phys.* **27**, 720.
Zwanzig, R. W. (1954). *J. Chem. Phys.* **22**, 1420.

Author Index

Numbers in italics refer to the pages on which the complete references are listed.

V

W

Y

Z

Subject Index

V

W

X